GREENGLOW

GREENGLOW

& THE SEARCH FOR GRAVITY CONTROL

RONALD EVANS

Copyright © 2015 Ronald Evans

The moral right of the author has been asserted.

Apart from any fair dealing for the purposes of research or private study, or criticism or review, as permitted under the Copyright, Designs and Patents Act 1988, this publication may only be reproduced, stored or transmitted, in any form or by any means, with the prior permission in writing of the publishers, or in the case of reprographic reproduction in accordance with the terms of licences issued by the Copyright Licensing Agency. Enquiries concerning reproduction outside those terms should be sent to the publishers.

Matador
9 Priory Business Park
Kibworth Beauchamp
Leicestershire LE8 0RX, UK
Tel: (+44) 116 279 2299
Fax: (+44) 116 279 2277
Email: books@troubador.co.uk
Web: www.troubador.co.uk/matador

ISBN 978-1784620-233

British Library Cataloguing in Publication Data.
A catalogue record for this book is available from the British Library.

Printed and bound in the UK by TJ International, Padstow, Cornwall
Typeset in Aldine by Troubador Publishing Ltd

Matador is an imprint of Troubador Publishing Ltd

To my parents, my family
and
all those people who were involved with
Project Greenglow.

EXTRACTS FROM THE LABORATORY DIARY OF PROFESSOR MICHAEL FARADAY THE ROYAL INSTITUTION OF GREAT BRITAIN, LONDON

Diary Entry Number 10018
19 March 1849

Gravity. Surely this force must be capable of an experimental relation to Electricity, Magnetism and the other forces, so as to bind it up with them in reciprocal action and equivalent effect. Consider for a moment how to set about touching this matter by facts and trial.

Diary Entry Number 10019
19 March 1849

What in Gravity answers to the dual or antithetical nature of the forms of force in Electricity and Magnetism? Perhaps the to and fro, that is, the ceeding to the force or approach of Gravitating bodies, and the effectual reversion of the force or separation of the bodies, quiescence being the neutral condition. Try the question experimentally on these grounds.

Diary Entry Number 10061
25 August 1849

I have been arranging certain experiments in reference to the notion that Gravity itself may be practically and directly related by experiments to the other powers of matter, and this morning proceeded to make them. It was almost with a feeling of awe that I went to work, for if the hope should prove well founded, how great and mighty and sublime in its hitherto unchangeable character is the force I am trying to deal with, and how large may be the new domain of knowledge that may be opened up to the mind of man.

Diary Entry Number 15785
10 February 1859

Surely the force of gravitation and its probable relation to other forms of force may be attacked by experiment. Let us try to think of some possibilities.

Diary Entry Number 15808
10 February 1859

Perhaps almost all the varying phenomena of atmospheric heat, electricity, etc. may be referable to effects of gravitation – and in that respect the latter may prove to be one of the most changeable powers instead of one of the most unchanged.

Diary Entry Number 15915
11 April 1859

It would be strange if a body should heat as gravitation increases by nearness of distance. We conceive of heat as a positive force and of gravitation as a positive force, and then instead of being the inverse of each other, they would seem to grow up together. Or else heat must be negative to gravity or the converse of gravity and gravity must be in the same negative or converse relation to heat. This is against the expectation of any thing from the heat experiment. Nevertheless make it [sic], for who knows. If gravitation depend [sic] upon forces *external* to the particles, such results might happen. Try.

FOREWORD

This book must surely be unique. It follows the history over the past 2000 years of the search to understand gravity. For the more modern times it traces some of the experimental attempts to control gravity. Ideas for further experimentation are also proposed. However, this is not a book written by an academic, but by an engineer who worked in the aerospace industry. I have some affinity with the author having begun my own professional career in the aircraft industry before moving into the academic world, and I have a close interest in post-Newtonian gravitational theory, too. Although some chapters of the book contain mathematical results, the book also contains many pictures and diagrams, allowing one to skip the maths if desired. This is a book to be read by all those people interested in the discovery of how to control gravity.

The obvious difference between gravitation and electrodynamics is that mass is the sole source of gravity fields, but positive and negative charges both create electric fields. This enables shielding from electric fields but not shielding from gravity fields. Another difference is that, for gravity, like masses attract, whilst for electrical charges it is the opposite. So, similar effects in gravitation to those occurring in electromagnetism are not possible. But if negative mass does exist in some form, as is speculated in this book, then gravity shielding, negative gravity and gravity control all become viable. The resulting unexpected interactions between positive and negative matter are well described in this book, along with methods of gravitational propulsion and the link with inertia via the quantum vacuum. Today this is science fiction but tomorrow it may become science fact, depending on the results of future experiments. The Synopsis gives a very clear indication of the subject areas covered and the Introduction supplies the details of the various approaches and experimental work that has been carried out over many years.

My own interest in gravitation began after several discussions in the mid-1970s with Professor Eric Laithwaite, who suggested that gyroscopes might hold the key to gravity control. During a lecture at the Royal Institution in 1974,

Professor Laithwaite wafted around a heavy flywheel at the end of a long axle, in a manner called precession, and claimed that the ease with which he could do it meant that the flywheel must have lost weight. He followed up his controversial claim with several papers in the *Electrical Review*. I was convinced that the flywheel's apparent loss of weight could be completely explained using Newtonian dynamics. In letters published in the *Electrical Review*, I pointed this out and suggested that Professor Laithwaite change his mind. At a meeting at the Institute of Mechanical Engineers I tried to explain the errors in Professor Laithwaite's view of gyroscopic behaviour with a practical demonstration. But, ever the showman, I was upstaged by the Professor. And there the matter rested for the time being.

I had my first contact with Ron Evans, around 1990, when he sent me a large amount of very interesting information on gravity studies, including the report of the British Aerospace sponsored University Round Table on Gravitational Research held in March 1990. It was this information that really started my own investigation of gravitomagnetism. In 1994 BBC2 televised the *Heretic* series, one of which centred on Professor Laithwaite and his controversial claim about gyroscopic behaviour. Ron Evans suggested to Tony Edwards, the producer, that I appear on the TV programme. This I was happy to do as it gave me the opportunity to show how Newtonian dynamics fully explains the curious behaviour of the gyroscope.

My interest in gravitation has continued to grow, eventually leading to the publication of my book, *Gravity: Galileo to Einstein and Back*, in which I developed a gravitational theory loosely based on the Lorentz Force equation in electrodynamics. To me, force and inertia are defined terms. Force is to dynamics as money is to commerce and mass is simply a count of the number of basic particles. From which it follows that inertial mass is always positive but a negative gravitational charge could exist. Gravity is now expressed as the relative acceleration between two bodies being a function of their separation, as in Newton's theory, with an added term, involving the relative velocity, which generates gravitomagnetism.

It is because of my activities in the realm of post-Newtonian gravity that I find Ron Evans' book very invigorating as it shows that there is more than one way forward in the search to understand gravity. Until a successful test is made then all ways are equally viable. I believe that this book will appeal to all those people fascinated by the possibility of gravity control. As Ron Evans shows, this is an exciting scientific adventure story with the goal in sight but still to be reached.

Dr H. Ron Harrison
Senior Lecturer (Retired)
Department of Mechanical Engineering & Aeronautics
City University
London

SYNOPSIS

One of Michael Faraday's goals was the unification of electromagnetism with gravity and, thereby, the control of gravity, but his experimental research in this direction was not successful. Later, Albert Einstein continued Faraday's quest with a theoretical approach but the goal eluded him, too. Nevertheless, most scientists remain convinced that electromagnetism and gravity must be intimately connected. However, since Faraday's time the discovery of two new force fields within the atom has made the task of unification even more difficult. In fact, the path to unification seems to have split in two, with particle physicists taking one route and gravitational physicists taking the other. The particle physicists, working at the microscopic scale with a gigantic device for smashing protons together and creating a debris of particles, hope to discover the secret of mass, the source of gravitation. The gravitational physicists, working at the macroscopic, or large, scale are searching for gravitational effects, in particular gravitational waves. A few experimental scientists have followed Faraday's lead in trying to understand how to manipulate gravity in a manner similar to that demonstrated for electromagnetism. Although this is a less ambitious goal than unification, even this goal has been denied, so far, with gravity remaining aloof and unchangeable.

So, 300 years after Isaac Newton discovered the universal influence of gravity and 100 years since Albert Einstein described gravity as being due to the curvature of space-time, we are still not able to control gravity. Anti-gravity remains a science fiction dream.

In the meantime, having learnt how to control the forces of electromagnetism and fluid dynamics (which encompasses aerodynamics), our everyday world has been totally changed. When the force-field patterns of these two branches of science are made visible they appear to be similar, with magnetism being seen as a vortex phenomenon. We say that electromagnetism and fluid dynamics are analogues of each other.

The beauty of spotting analogues is that it allows read-across between those

SYNOPSIS

subjects with similar patterns. Although analogues are seldom exact, so that care must be taken in using them, they do provide an excellent guidebook. Once scientists realised that gravity, electricity, magnetism and fluid flows were all analogues of each other, rapid strides were made in developing electromagnetism and fluid dynamics, but not gravity.

Einstein chose not to use the analogue approach in developing his theory of gravity, basing his idea on the equivalence that exists between gravity and acceleration. However, for weak gravitational fields, like the Earth's, and for low mass speeds, compared with the speed of light, we find that Einstein's theory of gravitation is an analogue of electromagnetism and fluid dynamics. The gravitational vortex phenomenon is associated with the very recently discovered force field, called gravitomagnetism, which arises when masses move. The gravitomagnetic field is related to twist, or torsion. Like gravity, gravitomagnetism also has an equivalence property, in this case connected with angular momentum. Whereas a body's linear momentum is its mass times its velocity, a body's angular momentum is its momentum times the distance from an axis about which the body rotates.

Angular momentum plays a special part in quantum mechanics, the subject which forms the core of the theory of atomic structure, radiation and radioactivity. Space, once thought to be an empty vacuum, is now assumed to be a turbulent region jam-packed with bursts of short-lived energetic particles and antiparticles, which possess quantum amounts of angular momentum. Many scientists believe that a breakthrough in understanding how to control gravity will only be made when the theory of gravity is joined with the theory of quantum mechanics.

This book describes an exciting, but unfinished, scientific adventure story. It tells of the epic struggle made by scientists to wrest the secrets from nature of how to control the powers of electricity, magnetism and fluid and aerodynamics. But nature has, so far, defeated our attempts to prise loose the secret of gravity control. Our guidebook contains a catalogue of the known force-field patterns of nature and this has been used by some scientists to search for the missing patterns in gravitation. They are sure to be there, but have remained hidden from our view, so far. One day the secret of gravity control will be discovered either by careful experimentation or, as is often the case, by stumbling across it by chance. It may be a research group that makes the discovery or, as I feel more likely, an individual scientist working alone. One thing is certain: the discovery will eventually be made and what a prize it will be. In the longer term, it is expected that gravity control will underpin a new method of propulsion and this book concludes with a look at some possible forms.

INTRODUCTION

My mother, the late Joan Evans, was a mathematics teacher while my father, the late Allan Evans, was an aeronautical engineer. This probably explains why the early part of my career saw me teaching undergraduate mathematics while the later part of my career saw me working for a large aerospace corporation. From 1978 until 2005, I was employed by British Aerospace and, after the merger with Marconi, by BAE Systems at their military aircraft site at Warton Aerodrome in Lancashire, near the famous seaside resort of Blackpool, in the North West of England.

I began work at Warton as a senior aerodynamicist in the Performance & Propulsion Group. My section leader was Jeffrey Newton – not Isaac but surely an omen for the future! In 1980 I was tasked with running a radar stealth research programme. Capitalising on German development of radar-absorbing materials (RAM) during the Second World War, the UK became a world leader in reducing the radar signatures of its military aircraft during the 1950s and 1960s. A recently declassified (March 2007) secret document illustrates the point. The Dawson Report, entitled *Radar Camouflage Techniques for Aircraft and Finned Missiles*, was issued in October 1960 by the Royal Aircraft Establishment at Farnborough. In 1962 there was a large exchange of information on the subject of radar stealth between the UK and US governments. In 1980 the US revealed details of its Lockheed Martin F-117 stealth fighter, now known as the Nighthawk, to the UK military establishment. This event was attended by Don Horsfield, the Chief Aerodynamicist at Warton, and eventually led to my appointment. As a postgraduate student at Leeds University I had specialised in fluid and aerodynamics, but as an undergraduate student, at the University of London, I had specialised in mathematics, which included the theory of fluid dynamics and electromagnetism. To brush up my knowledge of electromagnetism my late brother Roland, a microwave engineer with BT, gave me several of his textbooks on the subject.

After several years in the Aerodynamics Office I moved to the Advanced Technology Group to work on an experimental millimetre wave radar programme. During 1986, I started a small study of gravitational physics. It began as a general review of the subject, with the particular aim of assessing whether there were any impending breakthroughs in gravity-sensing technology which might nullify our radar stealth measures. For example, what was the likelihood that gravity radars, or gradars, might soon be developed which could detect the mass signature of an aircraft? However, as the study progressed it gradually became one of gravitation in its own right.

I had inherited many of my mother's mathematical textbooks and one of these was entitled *Newtonian Attraction*, written by A. S. Ramsey. The mathematics in this book looked very similar to that in some of my electromagnetism and fluid dynamics books but, curiously, the subject was not mentioned in my degree course. Gravitational theory had become a specialised and somewhat exclusive subject, under the title of Einstein's general theory of relativity. As an undergraduate, I had covered the subject of Einstein's special theory of relativity, which dealt with the changes in physical properties brought about merely by viewing events from observation points moving at different speeds, with the added condition that the speed of light remained constant. For example, a stationary electric charge when viewed from a moving observation point is seen to be surrounded by a magnetic field. On reading about special relativity in my brother's copy of *Electromagnetic Fields and Waves* (2nd Edition, 1970), by Paul Lorrain and Dale Corson, I was fascinated by their suggestion that there ought to be a gravitational equivalent of magnetism associated with moving mass. My interest was particularly stimulated by the idea that this was linked to a vortex-type model of gravity in three-dimensional space. Since vortex theory plays a large part in the theory of flight, this was something that I was familiar with. I wrote to Professor Lorrain at the University of Montreal, expressing my interest, and received an encouraging letter in return.

In late December 1974, during the 145th Christmas Lecture at the Royal Institution in London, Eric Laithwaite, Professor of Heavy Electrical Engineering at Imperial College London, placed a gyroscopic device on a balance and showed that it lost weight when it was activated. It was clear for all to see that the device generated vertical pulsations, but Laithwaite claimed that the up-and-down forces were unequal, resulting in the device experiencing an upward inertial thrust. Others dismissed this controversial claim and suggested that the apparent anti-gravity effect arose because the weighing system was too crude and couldn't react quickly enough to the large vertical oscillations of the device. But Eric was not alone in his claim. There were already a number of patented force-precessed gyroscopic devices which their inventors claimed produced inertial thrust. When Eric tried to show that conventional gyroscope theory might be used to explain the existence of inertial thrust he met a barrage of criticism. In another demonstration a young boy was given a non-spinning flywheel which he couldn't lift. However, when the flywheel was spinning and the boy turned he raised the flywheel almost without any effort. The flywheel must have lost weight, claimed the Professor. "No," said the academics, "it's precession and it's predicted by standard gyro theory." Eric also held the view that there was an unknown inertial force field associated with moving mass, rotating mass in particular, but this aspect was ignored in the ensuing furore. Aware of the strong feelings generated by the lecture, the Royal Institution decided, contrary to normal practice, not to publish it.

Eric Laithwaite was not the first person to experiment with gyroscopes during a public lecture at the Royal Institution. The Reverend Baden Powell, the Savilian Professor of Geometry at Oxford University and the brother of the founder of the Boy Scout Movement, first demonstrated the weird properties of the gyroscope in the Lecture Theatre in March 1854. Michael Faraday, who was in attendance, afterwards made his own gyroscope, carrying out a series of experiments to familiarise himself with its properties. I learnt, later, that Professor Laithwaite had used Faraday's gyroscope in his Christmas Lecture.

The whole story of the controversy surrounding Eric Laithwaite's claims about gyroscopes was admirably covered in the 1994 BBC TV *Heretic* programme, written and produced by Tony Edwards.

I wrote to Eric in 1986, saying that I wanted to discuss with him an extension of the Newtonian static model of gravity which included effects for moving mass. It involved a predicted, but hitherto undetected, force field and a missing 'curl' term. Also, I wanted to hear his side of the argument about inertial thrust. The 'curl' is a mathematical expression to do with rotation, or spin, and Eric had his own view that spin was associated with his predicted inertial field. We met in his office at Imperial College just before Christmas 1986 and so began a series of interesting meetings. Eric based his inertial force idea on a gravitational analogy with electromagnetism. This was the same idea that I had been exploring and which had led me to puzzle over the missing curl term associated with gravitational dynamics. I was surprised to learn, during my first meeting with Eric, that Oliver Heaviside, an English mathematician, had suggested the same idea nearly 100 years before, in 1894.

Prior to my meeting with Eric, a Scottish inventor, Sandy Kidd, had demonstrated a gyroscopic device to him which seemed to confirm the existence of inertial thrust. My brother, Clifford, who lives in St Andrews, told me that Sandy had been interviewed on the TV news in Scotland. I contacted Ron Thompson, the interviewer from Grampian TV, and made arrangements to see Sandy Kidd. During the autumn of 1987, I met Sandy and his wife Janet at their home in Dundee, together with the late Ron Thompson. In Sandy's book *Beyond 2001*, describing the adventures with his inertial thrust machine, I'm mentioned as the taciturn engineer from British Aerospace. Several of my colleagues at Warton were also intrigued by Sandy's device which had received a lot of UK media attention. Sandy agreed that we could test his device and a series of trials were carried out at Warton by engineers from the Wind Tunnel Department. The device was suspended from a load cell by an elastic cord, used to dampen the vertical thrusts. Our interest was in detecting any weight change of the device, not in measuring the magnitude of the vertical thrusts (accelerations). Apart from one unexplained and unrepeatable blip, no weight change was detected, indicating that the overall inertial thrust was zero.

Although Eric Laithwaite's 1974 Christmas Lecture was not published by the Royal Institution, the BBC published most of it in a book entitled *Engineer Through the Looking Glass*. When I got to know the Professor better he kindly sent me an autographed copy of his book. During the early 1990s, Eric began to change his mind about gyroscopic inertial thrust and said that he'd probably been wrong to support the idea that it could be used to generate a unidirectional force. Nevertheless, he still maintained that a spinning flywheel harboured an unknown inertial force field, which was related to angular momentum. He assumed that this inertial field would only exist very close to the surface of the spinning disk. On a later visit, he took me to his laboratory at Imperial College and showed me the preparations for an experiment using two flywheels which could be brought into extremely close proximity, separated by a very thin metal sheet to prevent any wind effect. By passing the spinning flywheel rapidly across the face of the other, he thought that there might be an interaction between the two flywheels, due to the change in inertial field. Eric told me that he had shown

INTRODUCTION

his experiment to an interested senior member of the Royal Family, rather in the way that Michael Faraday showed Prince Albert some of his experiments at the Royal Institution. As far as I'm aware, Eric Laithwaite never carried out this experiment. More than twenty years later, in April 2011, NASA confirmed that spinning bodies do generate an inertial angular momentum-type field and that there should be an interaction between two spinning bodies, although only detectable if at least one body is huge (planet-sized). But Eric's proposed experiment was not looking for an effect between two fixed closely spaced rotating flywheels. He wanted to see whether there was any effect due to a change in the angular momentum field, by moving one flywheel in an arc across the face of another.

The late Dr Robert Forward, at one time the senior scientist at the Hughes Aircraft Company Research Laboratory in Malibu, California, USA, was a leading advocate of gravitational research. He, too, had pointed out in a paper in 1961 that the analogy between gravity and electromagnetism meant that anti-gravity was theoretically possible. With a flow of super-dense fluid piped around a doughnut-shaped solenoid, or toroid, he explained how anti-gravity might be generated. In August 1989 I took the opportunity to meet Bob and his wife Martha at their holiday cottage on the coast near Dounreay, in the far north of Scotland. Later, he sent me a copy of his book, *Future Magic*, speculating about technologies of the future, and wrote on the title page, "To Ronald Evans – with hopes that it will inspire him to invent a true space drive!" By his leadership, Bob Forward encouraged many scientists and aerospace engineers to pursue an active interest in the subject of gravitation.

In March 1990 I organised a two-day university-industry round-table meeting attended by academics from five UK universities and Dr Anders Hansson, a space consultant. The purpose of the meeting was to learn about the latest work on gravitational research in the UK. Dr Hansson's far-sighted contribution was about the possible link between gravity and quantum mechanics, a subject I knew little about. As part of a US/UK exchange system for aerospace research documents, a copy of my report on the round-table meeting was sent to NASA by the Warton Chief Librarian, so their scientists were fully aware of our interest in gravitation.

The Director of Strategic Projects at Warton (formerly the Technical Director), the late Brian Young, was interested in the gravitational study and gave it his backing. He arranged for a small speculative study to be carried out by Advanced Projects engineers, to explore the benefits of using an anti-gravity engine. The engine chosen was based on the toroidal mass current idea proposed by Dr Forward, even though it was realised that it was not a practicable design. Parametric performance data for the fictitious anti-gravity engine was generated, thereby giving us a model to play with. It was rather an audacious study, which caused some eyebrows to be raised. But it did provide a useful focus as we discussed our ideas about how anti-gravity propulsion might be used. After several meetings we finally decided on a stealthy vertical take-off and landing (VTOL) combat aircraft as an example of an anti-gravity propelled machine.

Brian Young exposed our early thoughts about gravity to an academic audience in his inaugural lecture, in November 1991, as the Visiting Professor of Aerospace at the University of Salford. His lecture was entitled 'Anti-gravity. The End of Aerodynamics?', a rhetorical question to which he maintained the answer was a definite "No". The publicity led to further media attention and Professor Young was interviewed by Alun Lewis on BBC Radio 4's *Science Now* programme in May 1992. Nick Cook, then the Aviation Editor of the prestigious trade magazine *Jane's Defence Weekly*, contacted Brian and an article on British Aerospace's future anti-gravity concepts was published. I was introduced to Nick and discovered that he had a long-term interest in the possibility of gravitational propulsion. This culminated in Nick's book, *The Hunt for Zero Point*, published in 2001, which includes some details of our programme of gravitational research.

Following Faraday's thoughts that heat and gravity might be linked, in one of our early experiments we investigated whether the thermal conductivity of an iron bar held in a fixed temperature gradient might be altered by introducing a parallel magnetic field. The thought behind the experiment was that spin was linked with magnetism and that gravity was possibly linked with an unknown vortical field. The experiment was done during September 1996 at the University of Central Lancashire, under the direction of Professor Phil Bissell. Careful analysis of the results showed that there was no effect. If we had persevered and tried with the magnetic field orthogonal to the temperature gradient, then we would have seen an effect. In a lucky manner we would have rediscovered the Leduc effect, first recorded in 1887. However, the effect has nothing to do with gravity.

INTRODUCTION

Over the next decade, with the support of several senior colleagues at Warton, the gravitational study gradually grew into an industrially sponsored university research project called Greenglow. The name Greenglow followed from Brian Young's earlier tongue-in-cheek comment that anti-gravity propulsion might give rise to a green glow around an aerospace vehicle. The idea of an atmospheric glow also came from our work on creating plasma sheaths around aerodynamic bodies.

After the British Aerospace merger with Marconi to form BAE Systems, Project Greenglow was expanded to become a headquarters-backed university research programme. The subjects covered a wide range, from the theoretical investigation of extracting energy from the Earth-Moon's gravitational field using tethered satellite pairs; to examining the possibility that superconductivity and gravity were linked, experimenting with microwave thrust and exploring the effect of cavity design on the Casimir force at zero-point temperature. But, always, at the back of my mind was the thought of gravity control and the idea of gravitational propulsion.

Professor Robin Tucker, of Lancaster University, was the Academic Adviser for Project Greenglow. Walter Johnston (BAE Systems, Warton) had the job, while completing his PhD at Lancaster University, of liaising between academia and industry. Professor Colin McInnes, Dr Spencer Ziegler and Professor Matthew Cartmell, of Glasgow University, provided a Goals and Metrics Report for Project Greenglow which recommended that the future direction for research in advanced propulsion should focus on investigating links with the zero-point field (ZPF) of quantum mechanics. Professor John E. Allen, of Kingston University, formerly the Chief of Future Projects at British Aerospace Kingston, was the Technical Consultant for Greenglow. John has a long track record of interest in the possibility of anti-gravity, which is hardly surprising given that the Harrier VTOL aircraft was designed and built at Kingston. Additional backing for the Greenglow venture came from Dr Andrew May and Dr Gari Owen, both at the UK Ministry of Defence (MoD).

Marc Millis, of NASA Lewis, was instrumental in setting up the NASA BPP (Breakthrough-in-Propulsion Physics) Program in July 1996. In September 2000, after first visiting the European Space Agency (ESA) centre in Holland, Marc Millis brought a NASA BPP team with him to BAE Systems Warton to meet the academics supporting Project Greenglow and to exchange details of each other's research programmes. Because of its speculative nature, Project Greenglow remained a small BAE Systems research programme – noticeably so when compared with some of the company's other research activities. Indeed, Greenglow was officially only a part-time activity for me, although I spent a lot of my own time thinking about it. Nevertheless, the meeting with the NASA BPP team gave the Greenglow team a feeling that we were part of a much bigger research network investigating ideas for gravitational propulsion.

A three-day international Field Propulsion conference, was arranged to take place at Sussex University at the end of January 2001, with the proposed focus being the search for new ideas for space vehicle propulsion, with particular emphasis on force-field control. Although it appeared interesting I was rather dubious about attending, especially as a presenter. On the one hand I thought that the event might be a UFO convention in disguise and on the other hand I doubted that I would be believed if I said that BAE Systems had not made any breakthroughs in Project Greenglow and that we had no anti-gravity black programme. However, the conference organiser contacted senior BAE Systems management at the London HQ and I was instructed to attend. I learnt, later, that the conference was sponsored by the British National Space Centre (BNSC), through the offices of Mike Geer, then the Technology Manager.

The conference was co-chaired by Graham Ennis and Dr Anders Hansson. To mention just a few of the presenters, there was: Dr Alan Holt of NASA; Dr Hal Puthoff of Advanced Studies at Houston; Prof. John Allen of Kingston University; Stavros Dimitriou of the Technical Institute in Athens; Prof. Jean-Pierre Vigier of the Laboratoire de Gravitation et Cosmologie Relativistes; Dr Anders Hansson of the 3rd Commission of the International Academy of Astronautics (IAA), Paris; Tony Cuthbert, a UK inventor; Dr Claudio Maccone of ALENIA; Richard Obousy of Qinetiq; Dr Jean-Paul Petit, the Director of the CNKS Laboratoire at Marseille; and many others, including me. And to mention some of the attendees, there was: Nick Pope, formerly the MoD desk officer responsible for assessing UFO sightings in the UK; several aviation journalists, including Nick Cook of *Jane's Defence Weekly*, Malcolm English of *Air International* and Alexandre Szames who writes for *Air & Cosmos*; and several science reporters, including Ian Sample who writes for *New Scientist* and Jonathan Leake of *The Sunday Times*. Tony Edwards a TV producer of science programmes was there. And there were a number of authors of popular science books present, too.

INTRODUCTION

I enjoyed the event, which was well run. However, it did have the trappings of a media extravaganza. On the first day, Graham Ennis was interviewed by John Humphrys on the BBC Radio 4 *Today* programme. On the last day, a full-page article about the conference appeared in *The Guardian* newspaper, written by James Meek. A comprehensive write-up of the topics presented at the conference appeared in the March/April 2001 issue of *UFO Magazine*, prepared by the Editor, the late Graham Birdsall. Mark Pilkington also reported on the conference in his article entitled, 'Fields of Dreams', in the February 2001 issue of the *Fortean Times*. There were no bombshells! But, like many conferences, it was a good place to meet people and discuss ideas.

In June 2002 I received a letter from Dr Robert Baker, a US scientist based in Los Angeles, California. He wrote that he was involved with the United States Army in research work investigating the possible generation of high-frequency gravity waves (HFGW) using high temperature superconductors. On this subject he had been in contact with Professor Mike Cruise and Dr Clive Speake at Birmingham University and from them had learnt about Project Greenglow, the BAE Systems' gravitational research programme.

Dr Baker explained that he was arranging a four-day international conference on the study of high-frequency gravity waves, to be held from 6 to 9 May 2003, at the MITRE Corporation Offices in Virginia. The conference had the full support of the US Department of Defense (DoD). The three major applications of HFGW to be explored in the meeting were communications, propulsion and imaging.

The co-chairmen for the conference were Dr Paul Murad of the DoD and Dr Robert Baker, with Dr Ning Li from Alabama University invited as an Honorary Chairman. About fifty scientists and engineers attended the conference, from nine countries, including Russia and China. Boeing's interest was represented by Jamie Childress and Gary Stephenson. BAE Systems' interest was covered by Walter Johnston.

Prior to the presentations, Dr Baker paid tribute to Dr Robert Lull Forward, who died on 21 September 2002. Bob Forward, he said, was a pioneer in many areas of advanced physics, including gravitational wave detectors, gravity gradiometry, solar sails, space tethers, advanced space vehicle propulsion and the quantum vacuum, including interest in the possible exploitation of zero-point energy.

Full conference proceedings were issued, which I found heavy reading, but no breakthroughs were reported. Although he wasn't an attendee, Graham Ennis published a more colourful overview of the conference in the May/June 2003 issue of *UFO Magazine*.

Project Greenglow officially ended in 2005, with my retirement from BAE Systems. However, I was still hooked by the search for gravity control and continued the Greenglow study in private. This book is the culmination of my effort. One important subject not covered by Greenglow was particle physics. Some scientists believe that the secret of gravity control lies deep within the atom. So, for completeness, I have included some details of particle physics research which overlaps into the area of mass and gravity. But I am not a particle physicist, so I have only provided an overview. As it turned out, the simple vortex-type model of gravity that I was interested in does have close links with Einstein's own model of gravity. Indeed, some academics feel that the vortex model of gravity is just a branch of Einstein's more comprehensive model. But there are differences. Consequently, my view of gravitational theory may not accord wholly with the views of academic experts on the subject. Indeed, some of my speculative ideas may seem absurd to the experts in the fields of gravitation and quantum mechanics. However, presumably, all sensible ideas have already been investigated and they have not led to a breakthrough. This is a time where imagination, speculation and experiment must take the lead. I hope that my book will encourage some of you to apply your imagination, your theoretical knowledge and your experimental skills in the effort needed to make the breakthrough in understanding the nature of gravity and making gravitational propulsion a reality. It is an exciting venture but requires caution experimentally, as progress and developments are likely to be fraught with danger. But it's on its way; it's the next great leap forward in aerospace technology.

CONTENTS

1. Gravity, acceleration and curvature — 1
2. Gravity and time: from Ancient Greece to the Renaissance — 11
3. Analogues, poles and dipoles — 29
4. Diverging and converging flows — 42
5. The vortex phenomenon — 50
6. Electricity and magnetism — 56
7. Electromagnetism — 70
8. Gravity and the missing curls — 79
9. Frames of reference — 99
10. Sensing rotation — 107
11. Searching for gravitomagnetism — 118
12. Waves, particles and atom interferometry — 128
13. Twinned fields and some experimental ideas — 139
14. The luminiferous ether — 162
15. The quantum of energy — 167
16. Atomic particles, spin and the atomic clock — 176
17. Zero-point energy — 199
18. The quantum vacuum — 207
19. Elements of quantum gravity — 219
20. Speculative gravitomagnetism — 231
21. Forms of gravitational propulsion — 249
22. The search for gravity control — 281

Acknowledgements — 290
Self Publishing — 294
Postscript — 295
Reference Books — 296
Names Index — 299
Subject Index — 305

1
GRAVITY, ACCELERATION AND CURVATURE

Many people are fascinated by magnets which wield their influence across space to affect iron objects without the need for any physical contact. We talk about a magnetic force field emanating from the pole of a magnet. Similarly, electric force fields can be used to influence charged objects at a distance, causing them to react. We now know that these two natural force fields operate together and we have learnt how to build machines to control and exploit the combined forces of electromagnetism. And what a difference it has made to our lives!

Like magnetism and electricity, gravity is also a force field which extends across space to influence masses at a distance. Great scientific minds have thought very deeply about the phenomenon of gravity. Although some progress has been made in understanding its properties, we remain unable to control the force of gravity in the way that we can control electromagnetism. We suspect that gravity interacts with electromagnetism, unifying the force fields together, but we do not know what the linking mechanism is.

Before the idea that all masses possessed a gravitational force field became accepted, the Ancient Greek view of gravity was held to be true, namely that it was natural for all bodies to fall towards the centre of the Earth, because the Earth was the centre of the Universe. At ground level the force associated with preventing bodies from falling was their heaviness, or weight. If released, according to Aristotle in around 350 BC, the heavier the body the faster it fell. It wasn't until the late 16th century AD that Aristotle's view of gravity was challenged by several scientists. The Italian mathematician Galileo Galilei showed experimentally that masses of different weight, when released together, fell together under gravity; weight was not a discriminating factor. What is more, he showed that all falling bodies actually fell towards the ground with the same acceleration.

A cannon ball fired horizontally from the top of a hill (Fig. 1.1) also accelerates vertically downwards due to gravity with the result that it follows a parabolic curve until it hits the ground. The English mathematician Isaac

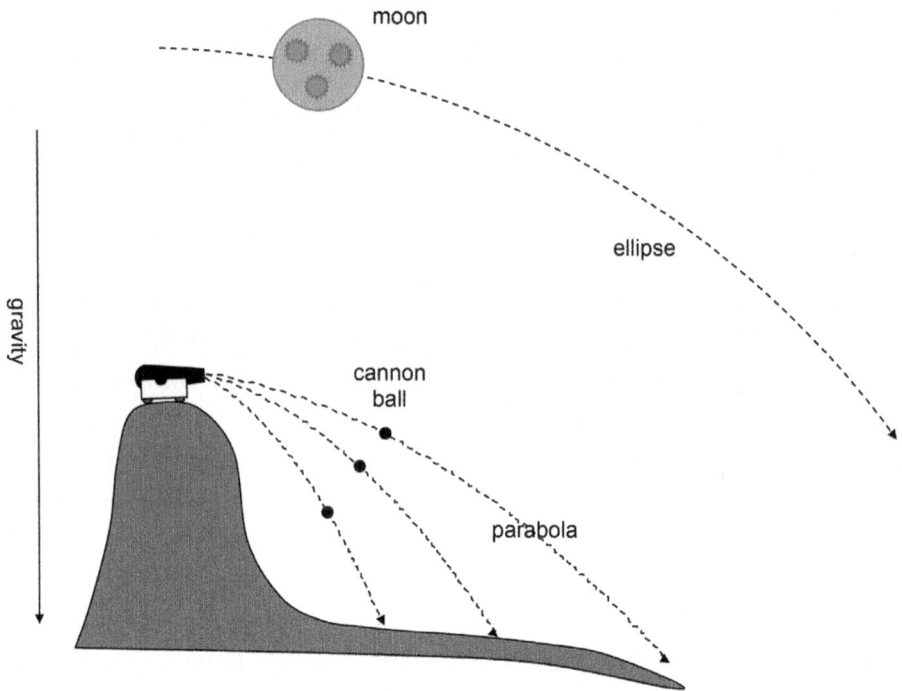

Fig. 1.1   Cannon ball fired from the top of a hill.

Newton wondered whether if the initial velocity of the cannon ball was great enough, then, as the ball fell, the curvature of the Earth would compensate for the drop in height and the ball would continue to fall round and round the Earth in a circular orbit. By extension, perhaps that's what the Moon did.

In 1679 Newton obtained data of the Moon's orbit which supported this idea. What was true for the Earth-Moon pairing was probably true for the Sun-planet pairing, too. Newton concluded that gravity was a universal force, not just an earthly force, and that all masses had their own gravitational influence, or force fields, and that all masses were attracted towards each other. Indeed, the capture and movement of the planets around the Sun was due to the interaction of their gravitational fields. Finally, in 1687 he was persuaded to publish details of his theory of gravitation. Initially, many people, including some scientists, were sceptical about the idea that masses possessed a gravitational influence, or field, which permeated across space. The Saxon German scientist Gottfried Leibniz thought that the idea that one mass could influence another at a distance was based on mystical nonsense. Newton even had to contend with cartoons poking fun at his crazy idea. One reason for resisting Newton's idea of universal gravitational influence was that the force of gravity is quite weak and not noticeable for small bodies. People passing one another in the street don't feel as though they are being pulled towards each other.

Newton developed his idea of planetary motion from his theory of gravity. But, the motion of the planets across the sky could already be predicted fairly accurately using a method of epicycles, where each planet moved on a circle whose centre rolled around the circumference of a larger circle, which had the Earth at its centre. The method, developed by the Ancient Greeks, was based on many years of collected data and on the observation that planetary motions were cyclic. The method is attributed to Hipparchus of Rhodes, circa 2nd century BC, but was improved by Ptolemy of Alexandria in the 1st century AD and refined in the *Almagest* by Arab scholars in Baghdad during the 9th century AD. So, this was another reason for resisting Newton's idea. But then scientists began to realise that Newton's model was actually simpler than the Ancient Greek model and it fitted in with a Sun-centred solar system and, also, it explained why the speeds of the planets were what they were, something that the earlier model did not do. So Newton's idea of universal gravitational influence coupled with his model of planetary motion gradually gained acceptance.

But Newton made no attempt to explain why gravity exists and said that it was a riddle. Several hundred years later, in 1915, the German mathematician Albert Einstein proposed a theory to explain why gravity occurs but, even so, this has not led to the development of machines able to control and manipulate gravity.

When a body accelerates in one direction it experiences a force in the opposite direction. We call this resisting force inertia. When the acceleration is parallel to the Earth's gravitational field we notice something remarkable. If the body is totally free to move in the Earth's gravitational field it accelerates downwards towards the centre of the Earth, nullifying the gravitational force experienced by the body in the process. In free fall the resultant force on a body is zero, so the body is in the same state as if it was in a force-free environment where it has no weight. But, if the body is in a lift cage which accelerates downward at a rate less than the acceleration of free fall, the body merely gets lighter, as though the Earth's gravitational field has been reduced. On the other hand, if the lift accelerates upwards, then the body gets heavier, as though the Earth's gravitational pull downwards has been increased. Somehow or other, the effect of acceleration on a body is equivalent to that caused by gravity.

Suppose we were able to accelerate a lift cage in a region of space totally absent of any gravitational fields. Due to inertia, the attendant in the lift cage would feel as though he had weight. Suppose that the lift cage had open sides and that as it passed by, a particle was fired across the cage (Fig. 1.2). In free space the path of the particle would be a straight line. However, the path of the particle observed in the accelerating lift cage would be bent, following a parabolic curve.

If the lift cage sits on the Earth's surface then the attendant clearly has weight. And if a particle is fired across the cage, then gravity acts on the mass

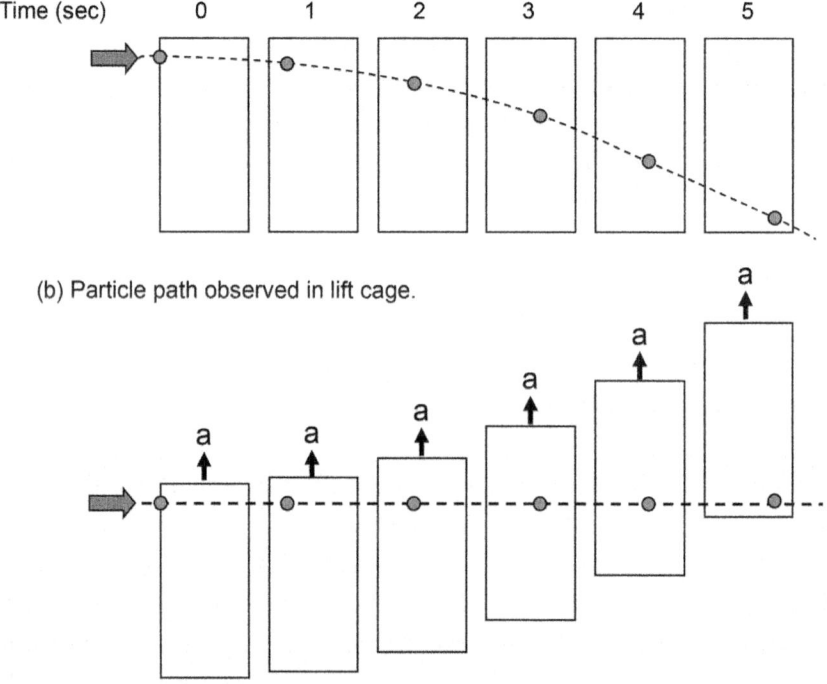

Fig. 1.2   *Particle fired across lift cage accelerating in free space.*

GRAVITY, ACCELERATON AND CURVATURE

causing it to accelerate downwards, following a parabolic curve. So, conditions in the accelerated lift cage and the lift cage fixed in a gravitational field appear to be equivalent.

Finally, let's consider the case of a lift cage falling in a gravitational field and a particle being fired across the cage as it passes by. Outside the cage the projectile's path is parabolic, but as observed inside the lift cage accelerating downwards the particle appears to be force-free and its flight path is a straight line.

Even a particle of light, which has a tiny effective mass, is influenced by gravity and acceleration. A ray of light passing near a massive body would be bent towards it by the body's strong gravitational field. Several hundred years ago scientists, using Newton's theory of gravity, predicted that a ray of starlight grazing the Sun's surface would be deflected by a tiny angle $\delta = 0.875$ arcseconds.

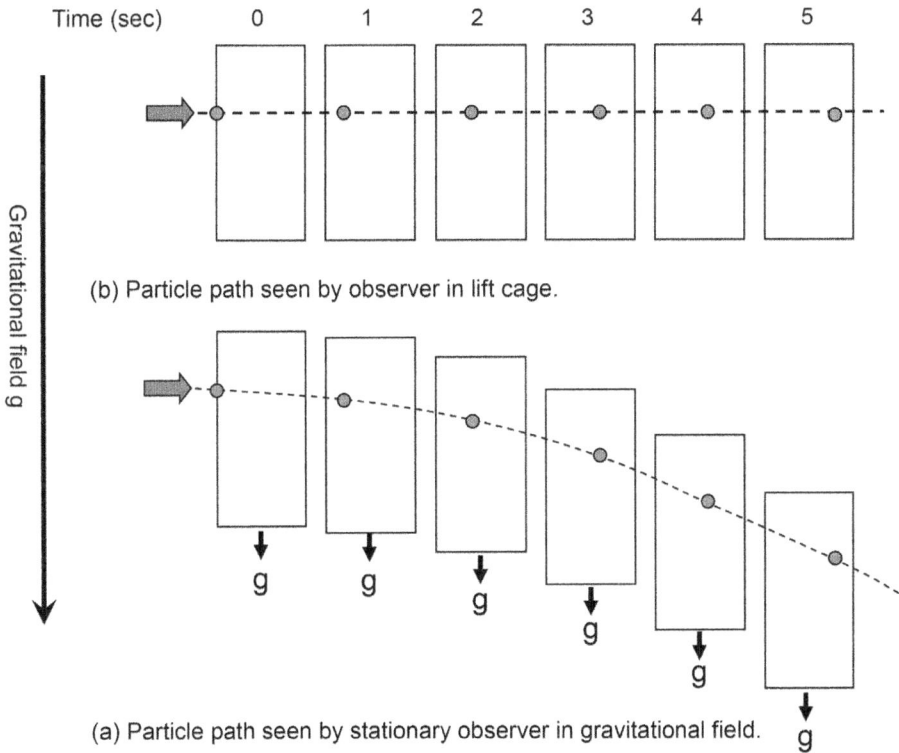

Fig. 1.3   Particle in gravitational field fired across a freely falling lift cage.

When cornering in a car we tend to get thrown outwards. This effect, called centrifugal force, is in response to radial acceleration inwards, called centripetal acceleration. The motorbike rider on the wall of death can make use of centrifugal force, together with friction, to defy Earth's gravity. As Newton realised, more than three centuries ago, the Moon falling around the Earth develops a centrifugal force which exactly counters the pull of Earth's gravity. Today, when we see orbiting astronauts floating in space we know that this is due to the centrifugal force they experience being exactly countered by the Earth's gravitational force of attraction, with the result that they are weightless.

We could develop a model which used curved paths to simulate the effect of gravity. If a body was constrained to move along a path, any curvature of the path would cause the body to undergo radial acceleration which, due to equivalence, could be interpreted as acceleration caused by gravity. To picture this idea, imagine that you are strapped into your seat on a roller coaster (Fig. 1.4) moving at constant speed in space where there is no gravity. While you move along any straight part of the track you are floating and have no weight. However, when you move along a curved part of the track you feel as though you are being pulled downwards by a gravitational force if you are on the inside

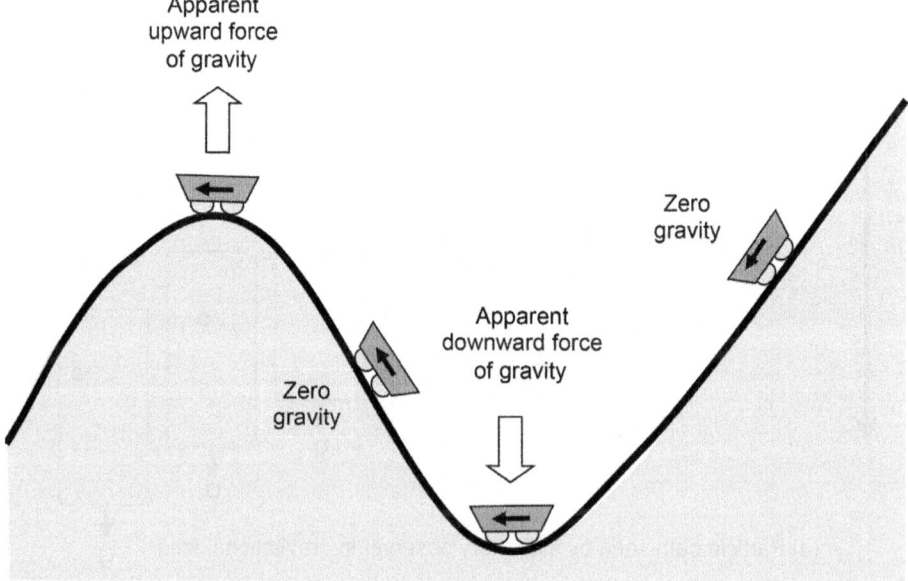

Fig. 1.4   *The roller coaster in free space.*

of the curve, but feel as though you are being pulled upwards by a gravitational force if you are on the outside of the curve.

So, in this way, the effect of gravity on a body can be modelled by using curved paths of motion in three-dimensional (3-D) space. There is just one problem. If the body doesn't move, we can't simulate gravity. Einstein got round this problem by introducing another length coordinate, or dimension, along which a body always moves. Einstein assumed that the speed of light in a vacuum (usually denoted by the letter c) is the same, anywhere in the vacuum of the Universe. He then multiplied the speed of light c by time t to get a length ct. Adding this new dimension to the three dimensions of space gave Einstein his four dimensions (4-D) of space-time.

Perhaps the easiest way to comprehend 4-D space-time is to consider the 3-D form of space-time first (Fig. 1.5). Here we have a flat surface, where any point has x and y coordinates. Time is introduced via a third length ct, perpendicular to the x-y plane. Thus, any point in 3-D space-time has coordinates (x,y,ct). We see that a particular place P in the 2-D plane has a vertical path in 3-D space-time. Driving across the flat 2-D surface from point A to point B directly with constant speed gives a straight, but slanted, path in 3-D space-time. Changing speed and direction on the way will result in a wiggly path. If the 2-D plane contains a fixed mass and we release a test mass in its vicinity, then, due to gravitational attraction, the free mass will accelerate across the surface and the 3-D space-time path will be curved. If the fixed mass is the Sun and the free mass is a planet, then the orbital path of the planet in 3-D space-time forms a helix around the Sun's time-line. But there is no 3-D space-time curvature.

Note that time-lines are open-ended, stretching from the distant past into the distant future. The possibility that space-time may be made to rotate, so that the ends of timelines curl round and join up to form closed loops, is of particular interest to those scientists exploring the idea of time travel.

The 4-D coordinates of space-time determine the place and time of an event – say a lightning flash on Earth or a supernova explosion somewhere in the Universe – whereas the 3-D coordinates of space only determine a place. Time marches onwards, so a place or body always changes its position in space-time coordinates, even though its space coordinates might not change. In Einstein's theory of general relativity, published in 1915, the effect of gravity is imposed through the curvature of the space-time path taken by a body. This was a new model of gravity which superseded Newton's original model.

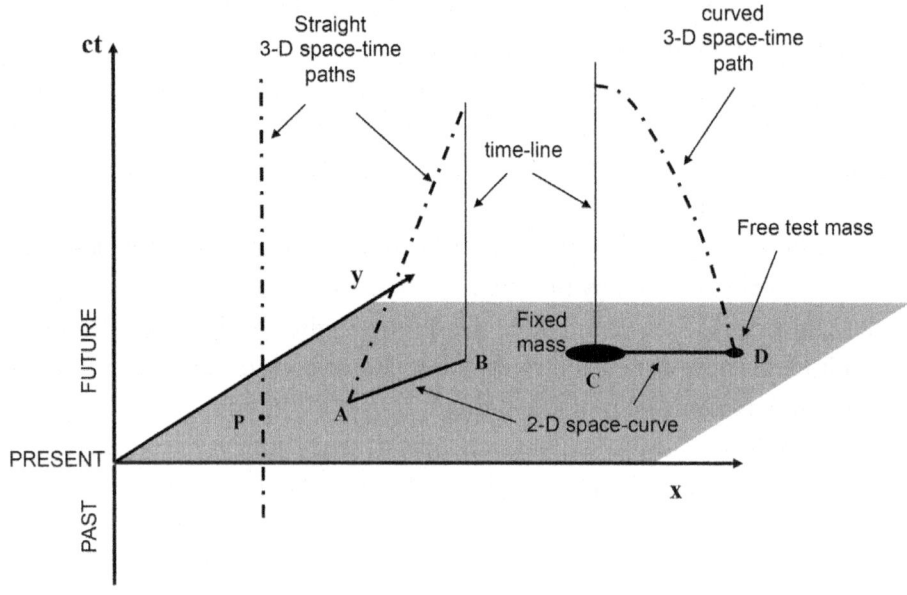

Fig. 1.5   3-D space-time.

Advocates of Einstein's theory claim that curvature doesn't just simulate the effect of gravity, it actually explains what gravity is. In their view, the Universe really is four-dimensional in nature and any matter present distorts, or warps, the 4-D space-time metric around it, resulting in path curvature for any free neighbouring mass, the resulting acceleration being properly interpreted as that caused by gravity. Unfortunately, the amount of space-time path curvature is not just dependent on the gravitational field of a mass present, but also on the energy stored in the gravitational field itself, so the situation becomes very complex.

From his theory of general relativity, Einstein predicted that due to the curvature of space-time around the Sun a ray of starlight grazing the Sun's surface would be deflected by an angle 2δ, rather than an angle δ, as predicted by Newtonian theory. The event which sparked Einstein's rapid rise to fame was the experimental confirmation of this prediction by the British astronomer Arthur Eddington, during an eclipse of the Sun in 1919. As the Moon blocked out the Sun's light, a photograph of the eclipsed Sun revealed stars at the Sun's periphery which would have been behind the Sun if their light had not been bent around it. And the measured angle of deflection was about 2δ.

Visualising Einstein's idea, that the curved path taken by a mass passing through space-time causes it to feel a force of gravity, is fairly simple. But to describe his 4-D space-time geometry, Einstein used tensor algebra, which is not particularly easy to follow. It becomes difficult to extract a picture of the underlying physical interactions, or dynamics, arising when masses move relative to one another. As a result, the development of Einstein's fundamental idea has been left, largely, to those mathematical physicists fully conversant with tensor theory and its esoteric notation. Making the theoretical model of gravitational dynamics simpler would have the advantage of opening up the subject to more scientists and engineers from other areas of science, thereby increasing the chance of someone making a breakthrough in understanding, leading to practical exploitations of gravitation, like those achieved in the phenomenon of electromagnetism.

Like Newton's original theory of gravity, Einstein's theory of gravity is only provisional, although it has passed all tests to date. Even so, there may be an element missing from it. In 1923 the French mathematician Élie Cartan suggested to Einstein that his gravitational theory could be modified to include spin angular momentum sources. Imagine a path where there is no space curvature along which a mass spins, or twists, as though caught in a vortex. Now, the mass experiences radial acceleration, due to spin, or torsion, not curvature. This form of radial acceleration is missing from Einstein's theory. Also, Einstein's theory of gravity deals with gravitational fields extending across the Universe. It deals with gravitational fields on a gigantic scale. But gravitation plays a part at the atomic level, too, on the really small scale, in the theory of quantum mechanics. The link between Einstein's gravitational theory and quantum mechanics, a subject kick-started by Einstein, is only now being addressed.

To move a body by gravitational means we must either control the space-time curvature around the body or control the body's inertia. So, we have two regions of interest, either external or internal, in which to search for a mechanism capable of generating a force on a body. However, Einstein's theoretical model of gravity is very complex, making the search for the physical principle, on which such a mechanism might function, rather difficult to uncover. What we really need is a more simplified model to begin the search.

We can extend Newton's theory of gravity more simply by using a model based on the model for electromagnetism derived by the Scottish mathematical physicist James Clerk Maxwell back in 1865. When an electric charge moves it creates a magnetic field around itself (Fig. 6.3). Reading across from Maxwell's electromagnetic model, if we replace an electric charge with a mass and replace the charge's electric field with the mass's gravitational field, then we expect a mass to create a field akin to magnetism as it moves along (Fig. 8.1). Because of the similarity with magnetism we call this new field gravitomagnetism. As it happens, the equations resulting from the read-across model are almost the same as those obtained from Einstein's model of gravity in the case of weak, low energy, gravity fields and low mass speeds, but not quite. Under such conditions the space-time curvature flattens out and gravitational and gravitomagnetic fields become distinct force fields. Theoretically, gravitomagnetism can cause a moving mass to veer off course, in the same way that the path of a moving electric charge is affected by a magnetic field. But, in practical terms, this requires sizeable gravitomagnetic fields. It is gravitomagnetism that causes the extra angle of deflection, δ, of a light ray grazing the Sun's surface.

With the Maxwell-type model of gravitation the predicted force-field patterns are similar to those observed in electromagnetism, which are also similar to the streamline and vortex patterns observed in fluid dynamics. These repeated patterns in nature are called analogues. This is the approach that we will use as a guide, on our journey of scientific discovery, as we search for the elusive means of gravity control – a prerequisite for gravitational propulsion. But we are running ahead too fast. We need to go back and look at the beginning of gravitational research, during the European Renaissance.

2

GRAVITY AND TIME: FROM ANCIENT GREECE TO THE RENAISSANCE

The Ancient Greeks were certainly interested in the influence that gravity had on bodies. However, during that time the pursuit of natural philosophy was like an intellectual parlour game for the aristocracy. The Greek philosophers were the celebrities of the time. They expounded their ideas about natural phenomena and mathematics in public lectures to show off their cleverness, while trying to attract good fee-paying students to their own schools, or academies, in the process. And some of their ideas were, indeed, brilliant.

Although only a few artefacts have survived (retrieved from wrecks on the seabed around Greece), we know that Ancient Greek craftsmen were capable of building very sophisticated geared instruments, most probably to do with astronomy and navigation. However, the setting up of laboratories with basic equipment to test leading-edge philosophical ideas, where feasible, was not of interest. As a rule, carrying out laborious work was a job for underlings and slaves and, generally, beneath the dignity of Ancient Greek philosophers. In any event, "Who will pay for fundamental blue-sky research?" was a question as pertinent then as it is today.

Aristotle, a student of Plato's Academy, the tutor of Alexander the Great and the founder of the Lyceum school in the city-state of Athens in 355 BC, was one of the greatest Ancient Greek philosophers. It was Aristotle's view of gravity that all bodies naturally fell to Earth because the Earth was the centre of the Universe. He was also of the opinion that heavier bodies fell more quickly than lighter ones. Due to air resistance, it is true that a stone falls to the Earth more quickly than a feather, so Aristotle's view seemed to accord with common sense.

The Ancient Greeks were quite familiar with the measurement of time. The idea of a twenty-four-hour day had been introduced more than 1000 years earlier by the Ancient Babylonians, who had also split an hour up into sixty minutes. The Ancient Egyptians adopted the twenty-four-hour day and developed sundials to provide a rough measurement of time during daylight hours, based on the position of an easterly moving shadow cast by an upright obelisk. This was the first link between space and time which has continued ever since. Indoors, or during the night, other methods were used to measure time. These included sandtimers, the rate at which candles burnt and the rate at which water dripped from a container. The Ancient Greeks developed sohisticated water clocks, called clepsydra, which were used to measure time in Europe up until the late Middle Ages. With the more accurate means of measuring time came the idea of splitting a minute into sixty seconds.

No doubt the Ancient Greeks understood the concept of velocity, as the rate of change of position with time, but the concept of acceleration, as the rate of change of velocity, does not seem to have been considered by them. Consequently, the thought that a dropped body might accelerate was beyond their comprehension. Although they had water clocks with which they might have timed the fall of a body over equal distances and discovered that it was accelerating, they didn't. It's not an easy experiment to do, since bodies fall very quickly under gravity and, in any case, as already noted, the Ancient Greek academics mostly avoided experimentation. There seems to be a parallel here with some old aristocratic families' disdain of trade. Trade is paramount to creating new wealth, enabling some families to climb up the social ladder, disturbing the established order in the process. Similarly, experimental evidence creates valuable data which allows scientists to probe and test existing ideas. Occasionally this leads to a major shift in understanding and the necessity to replace an old theory with a new one. This process is often disturbing and may be met with resistance by established academics as it challenges their authority. But progress means change and established aristocratic families and established academics probably see change as a threat to their position in the hierarchy.

Defence was an important and costly aspect of life in the turbulent times of the Ancient Greek city-states which had to be funded by the rulers – a fact which remains true in our modern turbulent world. Archimedes, who lived in the Ancient Greek colony of Syracuse, was a notable exception to the rule that Greek philosophers did not get involved in the arduous work of invention and experimentation. As well as being a brilliant mathematical physicist he was also a hands-on engineer who designed and built advanced military weapons.

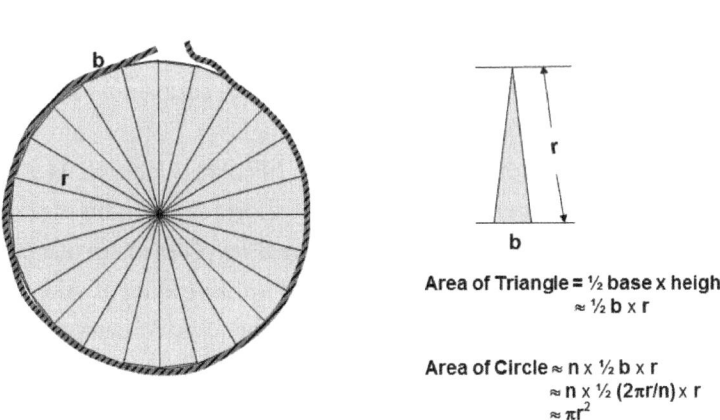

Fig. 2.1 *The area of a circle.*

Archimedes introduced his famous principle of buoyancy for bodies immersed in a medium in a gravitational field; he described the theory of the balance used to weigh bodies in Earth's gravitational field and he introduced the idea that bodies have a centre of gravity. He was also conversant with the early form of infinitesimal calculus invented by the Ancient Greeks. The Ancient Greeks were familiar with the magic number π associated with circles and knew that the circumference of a circle was 2π times its radius. As an example of the early form of calculus, by dividing a circle into infinitesimal equal triangles (Fig. 2.1), and then summing them, Archimedes calculated the area of a circle of radius r as πr^2.

Using a similar process, Archimedes also calculated the volumes of spheres and cylinders. This mathematical technique was reinvented 2000 years later by Newton and Leibniz. As an engineer, Archimedes devised and supervised the building of a number of practical mechanical devices, often involving levers and pulleys, for the defence of Syracuse, a city-state on the coast of Sicily, against attacks by the Romans. Archimedes was killed by a Roman soldier during the sacking of Syracuse in 212 BC.

From references to Hero of Alexandria, circa 1st century BC, we also know that the Ancient Greeks were developing rudimentary steam powered and pneumatic systems but, as their civilisation was overrun, these forward-looking engineering concepts remained dormant for more than 1000 years.

With the capture of Constantinople by the Moslem Turks in 1453, many Christians fled westward into mainland Europe. The academics amongst them brought their scholarly possessions with them, including technical instruments and many valuable texts linked with Ancient Greek scientific thoughts. As the refugees found new jobs, details of the Ancient Greek works, coupled with advances made since by Arab scholars, began to spread throughout Europe and so began the renaissance of scientific thought. Around 1514, Nicolaus Copernicus, a Polish cleric with an amateur interest in astronomy, issued a short report on his idea that the Sun was the centre of the Universe and that the planets revolved in circular paths around it. It seems that Copernicus' idea was stimulated by his reading of a document about Aristarchus of Samos, an Ancient Greek philosopher circa 3rd century BC, who had first suggested the idea of a Sun-centred, or solar, system, although it had been ridiculed and dismissed by his Greek contemporaries. In Copernicus' view, a solar system was much simpler than an Earth-centred system and, since nature was usually simple in character, it could well be right. However, it was Georg von Lauchen, a German professor of mathematics, who publicised the solar system hypothesis and arranged for the publication of Copernicus' famous book, *The Revolution of the Heavenly Spheres*, in 1543, the year after Copernicus died.

Dr John Dee, a graduate of Cambridge University, was a leading scientist during the reign of Queen Elizabeth I of England. Dee was an expert on ocean navigational methods and was employed by several maritime trading companies to teach their ships' officers the latest techniques for navigation. This was very important at the time, since many sea-going European countries were vying with each other to explore the Atlantic coastline of the New World of America, in order to establish trading posts with the natives and, in time, set up colonies. But just being able to find the places and return to them later was quite difficult as navigational techniques were so poor. Dee also travelled widely throughout Europe and met and worked with a number of other leading scientists. During his travels, Dee learnt of Copernicus' hypothesis of the solar system and, while abroad, purchased two copies of Copernicus' book for his own extensive library. Dee lived at Mortlake, then on the outskirts of London near Richmond Park, where he received frequent visitors. He made his library available to other gentlemen to use and his residence provided a meeting place for intellectual discussions. Thus, the idea of Copernicus' solar system would have been fairly well broadcast amongst the academics living around London at that time. What many of Dee's colleagues didn't know was that he was also a secret agent for Elizabeth I, keeping a look-out for threats to England from the continent. Dee's code number was ōō7, the number symbolising his hand-shaded eyes as he looked around for trouble. The fictional 007, James Bond, came 400 years later.

If the Earth was not the centre of the Universe it called into question Aristotle's view of gravity and his idea about the properties of falling bodies. In 1586 the Belgian-Dutch mathematician Simon Stevin introduced the simple, but clever, idea of slowing down the rate of fall of a ball under gravity by using an inclined plane. In this way, only a component of gravity acted on the ball and the ball rolled down more slowly. Galileo Galilei, the young Professor of Mathematics at Pisa University in Italy, made use of the inclined plane technique. Using a water clock, Galileo was able to measure the distance a ball rolled down the plane in equal units of time. He discovered that consecutive distances got longer, showing that the ball was increasing its speed. The ball was accelerating. Galileo tried the experiment with balls of different mass. It made no difference; they all accelerated at the same rate. The acceleration due to gravity was the same, whatever the mass size.

What was true for the inclined plane must also be true for free fall. Earlier, Simon Stevin had dropped bodies of different masses from a tower and noted that they fell together. Legend has it that in 1590 Galileo repeated this experiment by dropping a cannon ball and a musket ball from the Leaning Tower of Pisa. A large crowd of on-lookers confirmed that they hit the ground together. So Aristotle was wrong! Those Italian academics responsible for teaching physics had great difficulty in coming to terms with the fact that Aristotle was wrong. It also called into question Aristotle's view that the Earth was the centre of the Universe, as espoused by the Church and as taught at the universities. Instead, it gave support to the Copernican hypothesis that the Sun was the centre of the Universe. The idea of a solar system, with planets revolving about the Sun at the centre, began to gain ground. Some academics sought support from the Church authorities to refute the claim that Aristotle was wrong. The ensuing uproar resulted in Galileo being chastised by the Church at home but his predicament and his scientific observations being widely publicised abroad. For other scientists, particularly in northern Europe, an important principle had been established: no theory should ever be accepted until it had been tested by experiment. The Royal Society, founded in London in 1660, took as its motto, *Nullius in verba*, meaning 'Don't take anyone's word for it; do the experiment'.

While attending mass in Pisa cathedral, in 1593, Galileo had been fascinated by a swinging lamp suspended from the high ceiling by a long rod. What he had noticed was that for small oscillations the amplitude of the lamp swing (the to and fro of the lamp) made no difference to the time taken to complete a swing (the period). This led Galileo to suggest that a pendulum started swinging could be used to measure regular intervals of time because, although the amplitude of oscillations naturally reduced, the period stayed the same.

In 1597 Johannes Kepler, a German mathematician, wrote to Galileo advocating the Copernican system of a Sun-centred Universe with circular planetary orbits, rather than an Earth-centred Universe where the planets moved in circles within circles. Galileo replied that he was already persuaded to that view. However, it wasn't until 1609, when Galileo used a telescope to see the moons orbiting the planet Jupiter, that he finally had some evidence to support his view since, in a like fashion, the planets surely orbited around the Sun.

In the meantime, working from astronomical data collected by the Danish astronomer Tycho Brahe, Kepler tried to fit the planetary orbits into Sun-centred circular orbits, but failed. Eventually, he deduced three empirical laws for the motion of the planets from Brahe's data. The first two laws were published in 1609. The first law stated that the planetary orbits were ellipses, not circles, with the Sun as a focus. The second law stated that the radius of each planet joining it to the Sun swept out equal area triangles in equal times. What led Kepler to make this observation? Perhaps it was because originally he expected the orbits to be circular and knew of Archimedes' method (Fig. 2.1) to calculate the enclosed area by summing a large number of small equal area triangles. The third law was given in 1619 and concerned the orbital periods (time for one complete circuit) of the planets. Kepler went on to speculate that the Earth possessed a force which kept the Moon trapped in its orbit and that, by extension, the Sun must possess a force, too, which kept the planets in their orbits. Kepler had a copy of William Gilbert's book on magnetism (Chapter 6) and knew that the Earth possessed a magnetic field. He wondered whether the force holding the Moon to the Earth and the planets to the Sun might be magnetic. He even went so far as to consider that the force might diminish in strength with the inverse square of distance, analogous to the intensity of a point light source. But he rejected this idea in favour of a force confined to the plane of the ecliptic containing all the planets.

In 1632 the Church in Rome relented and granted Galileo permission to publish a book which discussed the Aristotelian and Copernican systems, entitled *Dialogue Concerning the Two Chief World Systems*. Much to the Church's dismay, the book popularised interest in the Copernican system of planetary motion throughout Europe, rather like Stephen Hawking's book, *A Brief History of Time*, which has popularised interest in gravitational research throughout the world today. However, Galileo still thought in terms of circular planetary orbits and didn't believe in Kepler's elliptical orbits.

In 1639 Jeremiah Horrocks, a young graduate of Cambridge University, showed that Kepler's laws applied to the Moon orbiting the Earth. From astronomical observations, that he made at Much Hoole in Lancashire (near to British Aerospace's military aircraft site at Warton), he showed that the Moon moved in a slightly squashed circular, or elliptical, orbit around the Earth. In squashing a circle the centre point splits into two points, each called a focus. For the Moon's orbit the Earth was one focus. The broad width of an ellipse is called its major axis. Horrocks showed that the major axis of the Moon's elliptical orbit rotated, so that its perihelion (point of nearest approach to the Earth) precessed, meaning that it circled around the Earth.

In 1656 Giovanni Alfonso Borelli became Professor of Mathematics at Pisa University. He was well versed in Galileo's work on gravity and Galileo's discovery of the moons of Jupiter. In 1666 Borelli extended Kepler's idea of a force emanting from a body attracting a satellite moving in a circular orbit by balancing it with the satellite's centrifugal force. His paper, *Theoricae mediceorum planetarum,* was an attempt to model the motion of the moons of Jupiter. But he didn't extend the idea to elliptical orbits. He had almost explained the reasoning behind the laws of planetary motion, but not quite. Borelli is better known for his groundbreaking studies of biomechanics, showing that animal movements are linked with machines, via muscle contractions and levers.

Christiaan Huygens was a famous Dutch mathematical physicist who was a contemporary of Newton and, like many philosophers at that time, was interested in optics, Galileo's experiments with gravity and with time. In 1673 Huygens published his book, *Horologium Oscillatorium*, which provided the theory for Galileo's simple pendulum and described models for other more

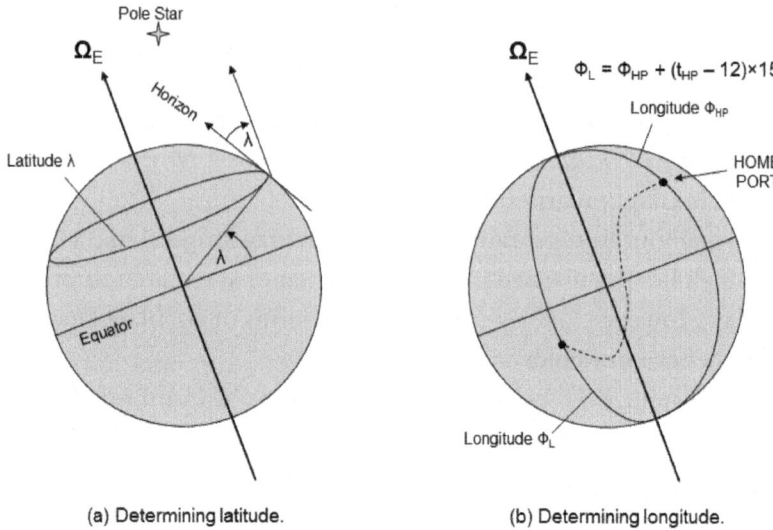

Fig. 2.2 Measuring latitude and longitude.

complex oscillating systems. Huygens was particularly active in the development of pendulum clocks.

Mechanical clocks had first appeared in Europe during the 13th century, their development largely being championed by the Church, as it enabled better organisation of each day's religious activities. The central mechanism of the first clocks was merely a horizontal spindle attached to a rope. The rope was wound around the spindle and a heavy weight was attached at the free end. As the weight fell in the Earth's gravitational field the rope unwound, thereby turning the spindle, with the rate of rotation being controlled by gear wheels. Thus, the first crude clocks were naturally placed at the top of towers. Initially, such clocks only had an hour hand, attached to the spindle end, marking the passing of time by the distance the hand moved around a circular dial. Based on Galileo's earlier proposal to use a pendulum to regulate time intervals, Huygens developed the first weight-driven clock with a pendulum in 1656. After 1680 pendulum clocks generally featured both hour and minute hands.

One particular application for time keeping was in maritime navigation, especially from the 16th century onwards as the European countries sent out ships across the vast oceans on their voyages of discovery. Navigation on the high sea is difficult as there are no geographical features to help you locate your position. Knowing where you are on the Earth's surface boils down to knowing your latitude and longitude. Latitude is your distance north, or south, of the Equator. In the northern hemisphere at any time on a clear night the latitude can be determined exactly by measuring the elevation of the Pole Star above the horizon (Fig. 2.2(a)). Although not so straightforward, the latitude can be determined during the day by measuring the Sun's elevation above the horizon at noon. By and large, maritime measurement of latitude was not deemed to be too difficult.

Determining the change in longitude, due to moving eastwards, or westwards, from a starting position, depends on being able to measure the passage of time. Based on the notion that there are 360 degrees in a circle (introduced by the Ancient Babylonians), the Earth's angular velocity Ω_E, or rate of rotation, is

$$\Omega_E = \frac{360°}{24\,\text{hr}} = 15°\,\text{per hour} \tag{2.1}$$

This means that every 15° change in longitude eastwards, or westwards, of a starting position results in a change in time of one hour ahead, or behind, that of the starting position. Suppose when a ship left its home port, with known longitude Φ_{HP}, the ship's clock was set with the local time t_{HP}. Then at noon (determined by the Sun at its zenith, or maximum height above the horizon), during a long sea voyage, noting the time difference in hours with that at the home port t_{HP} and multiplying the difference by 15 gave the ship's longitude Φ_L (Fig. 2.2(b)).

Improved accuracy in determining longitude at sea (for trade and military purposes as well as for avoiding dangerous areas at sea) meant improving the accuracy of the ship's clock. In 1714 the British Government felt that this was so important that they announced a prize of £20,000 to be paid to anyone who could make an instrument capable of determining longitude at sea to within half a degree. John Harrison, an English clock maker from Yorkshire, took up the challenge and over a period of thirty years submitted four special clocks for testing. They were called chronometers and were almost unaffected by violent movement and great changes in temperature and provided extremely accurate time measurement. The last chronometer, made in 1760, clearly satisfied all the

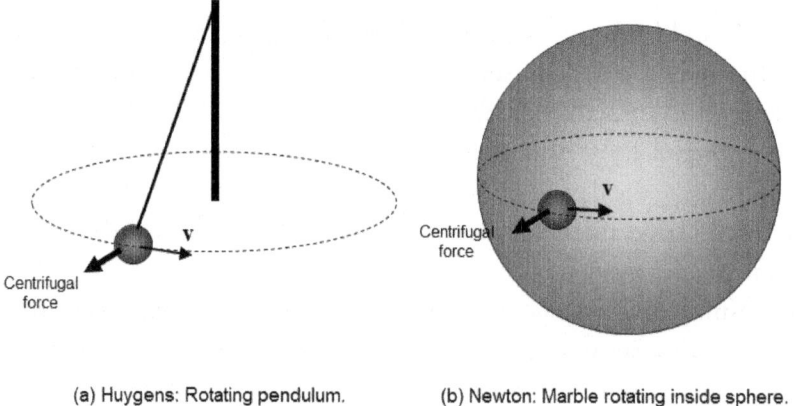

(a) Huygens: Rotating pendulum.   (b) Newton: Marble rotating inside sphere.

Fig. 2.3 Circular motion and centrifugal force.

Government's requirements and, after much wrangling and the intervention of King George III, Harrison was eventually awarded the prize money in 1773.

In the appendix of his book on the theory of oscillating systems and clocks, published in 1673, Huygens gave details of the centrifugal force which arises when a pendulum swings in a circular motion (Fig. 2.3(a)). As Borelli had suggested earlier, Huygens also concluded that centrifugal force must have some connection with gravitational force. At about the same time Isaac Newton, in England, was also considering circular motion, in the form of a marble running round the inside of a hollow sphere (Fig. 2.3(b)). Newton, who was aware of Borelli's work, was also interested in centrifugal force and its invisible counter-force, linked with centripetal acceleration, which seemed very much like a force of gravity.

Newton pondered on Borelli's idea and Galileo's observations about bodies in gravitational free fall accelerating towards the Earth. According to Newton, an apple falling to the Earth gave him the clue which led to his breakthrough in understanding. Under gravitation, the Moon was falling towards the Earth, too! Likewise, the Sun must have a gravitational field and all the planets were in free fall, accelerating towards it. From Archimedes, Newton knew that he could replace a mass by its centre of gravity, so gravitational force must act between the centres of gravity of masses. It was only the fact that the planets were in orbit and that their centrifugal forces exactly balanced the Sun's gravitational force of attraction that stopped them falling into the Sun. Robert Hooke, a philosopher and the Royal Society's leading experimenter, was of the same opinion and said so in several letters to Newton, further suggesting that the Sun's gravitational attraction was inversely dependent on the square of a planet's distance from the Sun. Synchronicity is a well-known phenomenon in scientific progress. This is when two or more scientists, or research teams, working separately on the same problem, make a breakthrough in understanding at almost exactly the same time. However, Hooke lacked Newton's mathematical skills and was not able express his ideas in mathematical terms.

The modern method of explaining planetary motion, called central orbit theory, is based on an algebraic approach, but Newton solved the problem geometrically. As an alternative, suppose we visualise a planet orbiting around an ellipse with speed, v. For a small time t a segment of the curve is of length vt so that the triangular area swept out by the planet's radius r is ½rvt. From Kepler's second law this means that although r and v change with time t their product rv is constant. As the time interval shrinks we get a smooth curve. If the planet's mass is m then its momentum is defined as mv. Now the momentum of a mass about a point distance r away is its angular momentum, defined as mrv (actually the velocity v is the component perpendicular to the radius). So, the angular momentum of each planet is seen to be constant in its elliptical orbit around the Sun. Therefore, the rate of change of angular momentum is zero, which means that there is no transverse force perpendicular to the radius pushing each planet along. In other words, the force experienced by a planet is only in the Sun's radial direction. Circular motion is a special case, being where the radius and the orbital velocity are both constant.

If you sit in a nearly frictionless swivel chair, holding a 2kg bag of sugar, and get someone to rotate you and then leave you alone, you will have nearly constant angular momentum. As you spin you can feel a force on the bag of sugar pulling it radially away from you. This is called centrifugal force. If you think of the bag of sugar as a planet and yourself as the Sun, then the force provided by your arms holding onto the bag of sugar is akin to the mysterious force of gravity holding on to the planet. When you hold the bag of sugar way out in front of you, you slow down. When you clutch the bag of sugar tightly to your chest you speed up. The planets do the same thing. As they get further from the Sun they slow down; as they get nearer they speed up.

All that remained for Newton to do was to determine the radial force between the Sun and a planet. Again, he used a geometrical method. Instead, suppose we consider a mass M as a gravitational source with a radial gravitational field, or region of influence around it, just like a light source. Suppose the mass is concentrated at a point in space and that we draw a sphere of radius r around it (Fig. 2.4). The sphere has a surface area of $4\pi r^2$. The gravitational influence of the mass M at this distance r from it has to be shared over the whole sphere. So at one particular point on the sphere the gravitational influence of M is proportional to $M/4\pi r^2$. We say that the gravitational field intensity g at a distance r from a point mass M is given by

$$g = \frac{1}{\gamma}\left(\frac{M}{4\pi r^2}\right) \qquad (2.2)$$

As the gravitational field of a mass permeates the space around it, its influence is moderated by the factor γ, which we may call the gravitational permeability of space.

If an electric bell is placed in an airtight glass container and switched on, we can hear it ringing and see the clanger vibrating. The sound of the bell

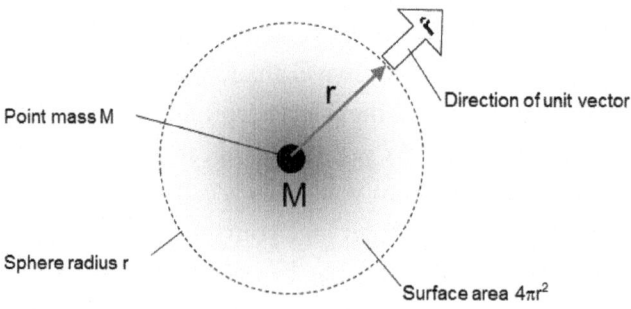

Fig. 2.4 Sphere of influence.

permeates through the medium, air in this case. If some of the air is removed from the container, then the sound that we hear is reduced. The sound permeability has been reduced. If all the air is removed, we can't hear the bell, even though we can see the clanger moving. The permeability of sound in the container is zero.

In the SI (International Standard) units of the kilogramme (kg), the metre (m) and the second (s), the gravitational permeability of the vacuum is large, being

$$\gamma = 1193000000 = 1.193 \times 10^9 \text{ kg m}^{-3} \text{ s}^2 \qquad (2.3)$$

So, gravitational fields are sustained in space and cross the Universe. Originally it was assumed that, because the gravitational permeability of space is non-zero, there must be some medium in space which allows gravitational influence to be broadcast. This medium was called the ether. Whether the ether exists, or whether space is empty, remains a modern-day controversy.

Newton chose to combine the constant γ with the term 4π into a factor G, where G = $1/(4\pi\gamma)$, so that his law of the gravitational influence of a mass M is

$$g = G\left(\frac{M}{r^2}\right) \qquad (2.4)$$

G is called the Universal Gravitational Constant, as Newton assumed that it applied across the whole Universe. Scientists usually refer to it as Big G.

In SI units Big G is actually very small.

$$G = \frac{1}{4\pi\gamma} = 0.00000000006672 = 6.672 \times 10^{-11} \text{ kg}^{-1} \text{ m}^3 \text{ s}^{-2} \qquad (2.5)$$

A quantity which just has a magnitude is called a scalar. A quantity which also has a direction associated with it is called a vector. To distinguish between the two in the following pages, bold print is used for vectors. We can also think of scalar and vector fields, where a field encompasses a region of space. In the former, every point of the field has a magnitude associated with it. In the latter, every point of the field has a magnitude and a direction associated with it.

Consider the following two examples. Suppose a body's position x changes with time t, then the rate of change of position is dx/dt = v is called the speed of the body. Speed is a magnitude measured, say, in miles per hour (mph) or metres per second (m/s). But if the speed is also in a certain direction, then **v** is a vector called the velocity. Suppose an angle θ rotates with time t. The magnitude of the rate of rotation dθ/dt = Ω may be expressed in revolutions per minute (rpm), or radians per second (rads/s). But the rotation takes place about a certain axis, too, so **Ω** is a vector called the angular velocity.

The magnitude of the influence of gravity is usually referred to as the gravitational intensity g. But gravity also pulls in a certain direction, so we talk about the gravitational field of strength **g** at a point in space.

A unit vector has a magnitude of one unit and points in a particular direction. It is a vector signpost. In the case of the point source of gravitation M, we choose ř as a unit vector pointing in a radial direction outwards from M.

Because a mass gravitationally attracts other masses towards it, we have to introduce a negative sign to get the vector direction of **g** right.

$$\mathbf{g} = -\frac{1}{\gamma}\left(\frac{M}{4\pi r^2}\right)\hat{\mathbf{r}} \quad m/s^2 \tag{2.6}$$

The gravitational influence (g) depends on the square of the distance (r^2) from the centre of the source of gravity M. But it's an inverse relationship, because g decreases as r^2 increases and vice versa. For this reason, Newton's law of gravitational attraction is called an inverse square law.

As an approximation, we can treat the Earth as a sphere of radius R_E = 6370 km, with a mass M_E = 59.65 × 10^{23} kg. For these values we find that at the Earth's surface

$$\mathbf{g} = -\frac{1}{\gamma}\left(\frac{M_E}{4\pi R_E^2}\right)\hat{\mathbf{r}} = -9.81 \, (m/s^2)\,\hat{\mathbf{r}} \tag{2.7}$$

Because a mass m has the potential to fall in Earth's gravity field and do some work it has gravitational potential energy E. A mass placed on a concentric

GRAVITY AND TIME

sphere surrounding the Earth's spherical surface has a fixed energy level. Such a surface on which the potential energy is fixed is called a potential surface. We can introduce any number of these surfaces (Fig. 2.5), distinguishing each of them by a different scalar value ϕ, called the gravitational potential. Note that the gravitational field **g** cuts any potential surface at right angles. On any potential surface the gravitational potential energy E of a mass m is given by

$$E = m\phi + \text{constant} \tag{2.8}$$

In SI units force is measured in Newtons, giving a free mass of 1kg an acceleration of 1m per second per second. The force of gravitational attraction experienced by mass m is given by

$$\text{Force F} = (\text{mass}).(\text{strength of gravitational field}) \tag{2.9}$$

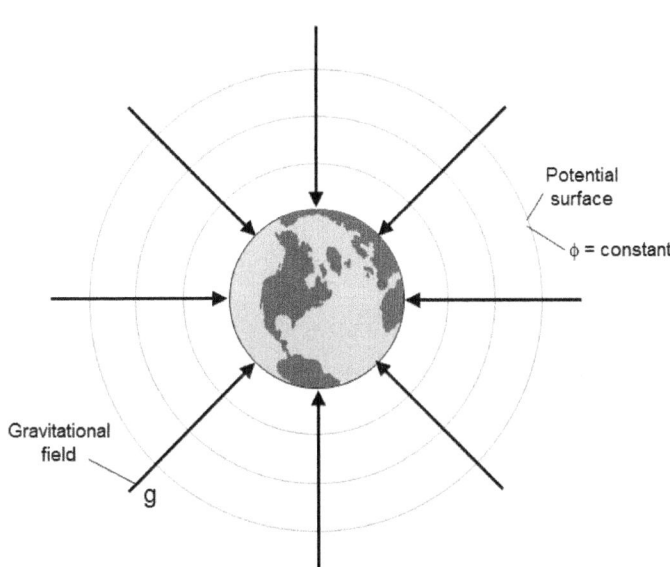

Fig. 2.5 Earth's gravitational field and potential surfaces.

So, from equation 2.6

$$\mathbf{F} = m\mathbf{g} = m\left[\frac{-1}{\gamma}\left(\frac{M}{4\pi r^2}\right)\hat{\mathbf{r}}\right] = \frac{-1}{\gamma}\left(\frac{mM}{4\pi r^2}\right)\hat{\mathbf{r}} \quad \text{Newtons} \quad (2.10)$$

However, force can also be defined as being equal to the negative gradient of potential energy. Introducing the gradient operator ∇ this can be written symbolically as

$$\mathbf{F} = -\nabla E = -\nabla(m\phi) = -m\nabla\phi. \quad (2.11)$$

From equations 2.10 and 2.11 we see that

$$\mathbf{g} = -\nabla\phi. \quad (2.12)$$

In words, this says that the gravitational field is equal to the negative gradient of gravitational potential. The symbol ∇ is the mathematician's shorthand for obtaining the gradient, or slope, in any direction.

To satisfy equation 2.12, the gravitational potential of a mass M is

$$\phi = -\frac{M}{4\pi\gamma r} \quad (2.13)$$

By convention it is accepted that the constant in equation 2.8 is zero, so that the gravitational potential energy is always negative, or zero at infinity. The advantage of using scalar potentials is that for any number of discrete masses, the individual potentials can just be added to get the total potential.

On Earth, the weight of a mass m held in check is the force that gravity exerts on it.

$$\text{Weight} = m.\mathbf{g} \quad (2.14)$$

If free, the mass m will accelerate towards the Earth's centre, until it is stopped by the Earth's surface. Just before hitting the surface it will have developed an acceleration of about 9.8 m/s².

Newton showed, theoretically, that if the Sun attracted a planet with a gravitational force that obeyed the inverse square law, then the orbiting planet would follow an elliptical path with constant angular momentum. However, Newton's theory only applies to two bodies at a time, say the Sun and one planet, or a planet and one moon. In trying to model the motion of all the planets at once, using Newton's inverse square law, the interactions between the gravitational fields of the planets become exceedingly complex. The resulting model is non-linear and the predicted planetary orbits become chaotic. In 1886 King Oscar II of Sweden and Norway offered a prize for the first scientist who could analytically solve the model for the motion of the whole solar system. This is known as the n-body problem. However, in 1906 the French mathematician Jules Henri Poincaré proved theoretically that for three bodies the Newtonian model must give chaotic orbits, never mind any more bodies. In Poincaré's view, Newton's two-body gravitational model was a special case. And yet, the solar system seems to be stable and regular. Is there something missing from Newton's gravitational theory? Or could a slight perturbation in one of the planetary orbits cause the solar system to go unstable and break up? At present we are stuck with Newton's linearised two-body model. But what a model it proved to be! With it, Newton was able to explain the reasoning behind Kepler's empirical laws. However, to demonstrate conclusively that his theory fitted with natural observations, he needed some real astronomical data.

Newton already had access to Jeremiah Horrocks' data of the Moon's motion but he needed more comprehensive data. This he finally obtained from John Flamsteed, the first Astronomer Royal, which showed that the Moon moved in a slightly elliptical orbit around the Earth with a period of 27.3 days. In fact, the Moon's orbit was almost circular with a radius of about sixty times the Earth's radius. Assuming that the Moon obeyed an inverse square law of gravitational attraction by the Earth which was exactly countered by centrifugal force, Newton deduced from the data that the gravitational intensity at the Earth's surface would be about 9.8 m/s². Since the result agreed pretty well with the measured acceleration due to gravity, Newton assumed that his assumption of an inverse square law was justified.

What was true for the Earth and the Moon applied to other pairs of masses in the Universe. Newton's theory of gravity was a fundamental theory from which other observations could be explained. Not just the orbits of the planets, but the tides, too.

In 1687 Newton, with support from his friend Edmund Halley, published his book, *Philosophiae naturalis principia mathematica*, better known as the *Principia*, which included a description of his theory of gravity. It also contained details of Newton's three laws of motion, partly based on Galileo's earlier observations.

The first law stated that a body moving with uniform speed in a straight line would continue to do so, unless acted upon by a force. This is sometimes called the law of inertia. Inertia is the force that we experience when we try to start a body moving, or try to stop a moving body. Inertia arises during periods of acceleration or deceleration.

The second law defines force. In mathematical form this can be written as

$$\mathbf{F} = m.\mathbf{a} \qquad (2.15)$$

Where **F** stands for force, m stands for mass and **a** stands for acceleration, or the rate of change of velocity. The law probably stemmed from the observation that weight can be separated into mass times the acceleration due to gravity. However, in the case of weight (2.14) the stationary mass m is called a gravitational mass, while in Newton's second law (2.15) the accelerating mass m is called an inertial mass. Scientists have long pondered on the observation that gravitational mass is equivalent to inertial mass. It's why, at any point in space, we can't tell the difference between gravity and acceleration.

The third law was that action and reaction are equal and opposite. That is, to every force there is an equal and opposite force.

Newtonian mechanics has evolved from Newton's laws. When combined with the inverse square law for gravity, the motion of the planets in the solar system can be predicted with remarkable accuracy. We can use Newtonian mechanics to predict the motion of bodies in flight, such as cannon balls, aircraft, satellites or even the fall of an apple. But although the secret of gravity's universal influence had been discovered by Newton, gravity still remained an immutable force to him and his contemporaries and to us, too, more than 300 years later. We aren't able to manipulate gravitation, yet, in the way that we can manipulate electromagnetism.

3
ANALOGUES, POLES AND DIPOLES

It was soon realised that other phenomena with sources capable of exerting influence at a distance all gave rise to similar force-field patterns, as well as obeying the inverse square law of force. Repeated patterns in nature are called analogues.

In 1785 Charles Coulomb, a French scientist, demonstrated that the force between two stationary electric charges obeyed an inverse square law. But there was a difference when compared with the gravitational case. There is only one known source of gravity, namely mass. Masses attract each other. There are two sources of electricity, namely positive electric charge and negative electric charge. Two like electric charges repel each other, while two unlike charges attract each other. However, the force of repulsion, or attraction, between stationary charges still obeys an inverse square law of distance.

Following the pattern with gravity (eqn 2.10) we can write

$$\mathbf{F} = \frac{1}{\varepsilon}\left(\frac{Q_1 Q_2}{4\pi r^2}\right)\hat{\mathbf{r}} \qquad (3.1)$$

Q_1 and Q_2 are the strengths of two electric charges measured in SI units of Coulombs. The parameter ε defines the electric permeability, or permittivity, of the medium through which the electric influence spreads. For free space ε is:

$$\varepsilon = 8.854 \times 10^{-12} \text{ F/m}. \qquad (3.2)$$

The farad (F) is the SI unit of capacitance. So, by comparison with gravitational permeability γ, the electric permittivity ε appears small, but this depends on the units chosen.

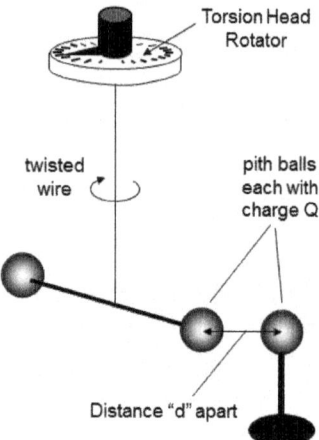

Fig. 3.1 The torsion balance.

To demonstrate that the inverse square law applied to electric charges, Coulomb used a torsion balance (Fig. 3.1). Torsion is associated with rotation, or twist. The torsion balance is a detector of torsion, or twist. It is merely a balanced lightweight horizontal bar which is suspended from its centre by a vertical wire. The top of the wire is attached to a device which allows the wire to be turned. The important characteristic of the balance is the resistance to twisting of the suspension wire, defined as the torsional stiffness.

Coulomb attached a small gilded pith ball to each end of the torsion balance bar and charged them with static electricity. Then he introduced an uncharged gilded pith ball near to one end of the bar. He rotated the wire until one of the charged pith balls just touched the uncharged one. The charge was then shared between both pith balls. Since like charges repel, the pith balls were forced apart and the wire twisted. Coulomb measured the distance between the two repelled pith balls and determined the angle through which the wire was twisted. With this information, and knowing the torsional stiffness of the wire, he was able to show that the force between the two electrically charged pith balls obeyed an inverse square law.

Following the analogue with gravitation (eqn 2.6), we can write the electric field $\mathbf{E}_1$ surrounding a point charge Q_1 as

$$\mathbf{E}_1 = \frac{1}{\varepsilon}\left(\frac{Q_1}{4\pi r^2}\right)\hat{\mathbf{r}} \quad \text{Volts/metre} \quad (3.3)$$

The force experienced by a charge Q_2 in the electric field $\mathbf{E}_1$ is

$$\mathbf{F} = Q_2.\mathbf{E}_1 \quad \text{Newtons} \quad (3.4)$$

In Britain, at about the same time that Coulomb was carrying out his electrical experiment, the Reverend John Michell, a retired Cambridge professor of geology, was also experimenting with a torsion balance with identical masses at each end of the bar with the intention of demonstrating the inverse square law for gravity in a laboratory on Earth. Newton had assumed that, given the smallness of Big G, the inverse square law for masses could only be demonstrated using astronomical bodies, but he was wrong. Unfortunately, Michell died in 1793 before completing his experiment. The apparatus was eventually passed to Henry Cavendish, an extremely rich reclusive gentleman scientist, who carried out an improved version of Michell's proposed experiment in 1798. He was the first scientist to measure Big G using small masses.

Incidentally, the Reverend John Michell was the first person to predict the existence of dark stars. In a paper read to the Royal Society in 1783, the Reverend Michell said that massive stars might exist which generate enough gravity to prevent corpuscles of light (photons) escaping from them so that they would be invisible. Completely independently, the French mathematician Pierre Laplace also predicted the existence of invisible stars in 1796. We now call dark, or invisible, stars black holes, a term introduced by Professor John Wheeler in 1960.

Sources of the magnetic field, called monopoles, only occur in pairs of north-seeking and south-seeking poles, called dipoles. The magnet is a natural magnetic dipole. When a magnet is placed in an external magnetic field $\mathbf{H}$ it will tend to line up with the field. This is the basis of the magnetic compass, where the magnetised needle lines up with the Earth's magnetic field.

It is found that like poles of different magnets repel, while unlike poles attract. For a very long magnet, where the two opposite poles are widely spaced apart, each pole approximates to that of a single pole, or monopole. Using two such long magnets, experiment shows that the force arising between two poles of different magnets seems to obey the inverse square law of distance.

For theoretical purposes it is assumed that magnetic monopoles do obey the inverse square law. Thus, the force law between monopoles m_1 and m_2 a distance r apart is

$$\mathbf{F} = \frac{1}{\mu}\left(\frac{m_1 m_2}{4\pi r^2}\right)\hat{\mathbf{r}} \qquad (3.5)$$

μ is the magnetic permeability of the medium. For the vacuum of space

$$\mu = 1.257 \times 10^{-6} \text{ H/m}, \qquad (3.6)$$

so it is fairly small. The henry (H) is the SI unit of inductance.

In 1738 the Dutch-Swiss mathematician Daniel Bernoulli published his book *Hydrodynamics*. This was the first major work on the motion of fluids, which includes liquids and gases. For simplicity, an ideal fluid is introduced as a fluid in which the density ρ remains constant and where there is no viscosity, or stickiness. Based on the analogue with the point mass m as the source of the gravitational field **g**, the point source q is introduced as the point from which an ideal fluid flows outwards in 3-D to create a velocity field **v** (Fig. 4.1), emitting q kg of fluid per unit time. Its counterpart is the point sink which sucks fluid in. Due to the analogue approach, we expect pairs of sources, q_1 and q_2, to obey an inverse square law of force. As is the case with pairs of masses, the interaction is one of attraction. This is due to a peculiarity of nature, known as the Bernoulli effect (Chapter 5), where, as a fluid speeds up, so the pressure drops and vice versa. The inverse square law of force for sources in an ideal fluid is given by

$$\mathbf{F} = -\frac{1}{\rho}\left(\frac{q_1 q_2}{4\pi r^2}\right)\hat{\mathbf{r}} \qquad (3.7)$$

The density ρ of the fluid is the permeability of the medium to fluid velocity v and q_1 and q_2 are the source strengths. Similarly, pairs of sinks attract.

The idea of continental drift was hypothesised by the German scientist Alfred Wegener in 1911. His idea that a supercontinent, called Pangea, had split up and drifted apart across the surface of the Earth to form the continents of today was treated as nonsense by many scientists of the time. What possible mechanism was there for such movement? As an answer to this question, in 1930 the German scientist Ott Christoph Hilgenberg postulated that the Earth was expanding, albeit at an extremely slow rate, so that over millions of years the surface of the Earth ballooned and the supercontinent naturally broke apart. This was not seen as such an outrageous idea at the time because, only the year before, Edwin Hubble had just discovered that the Universe itself was expanding. To explain his idea, Hilgenberg made use of the ether, an exceedingly fine fluid filling the entire Universe. He proposed that ordinary matter was made of ether sinks and that, over eons of time, as matter swallowed more ether so it became more massive. Since sinks attract sinks the inverse square law for matter applied, while the gravitational constant was dependent on the ether density and rate of ether absorption.

It wasn't until the 1970s that Wegener's hypothesis of continental drift was finally accepted, while Hilgenberg's postulate of an expanding Earth as the cause of the drift was rejected in favour of plate tectonics, an idea backed by experimental evidence.

We have now established some patterns and can see that nature repeats itself, although under different guises. So, when examining a new phenomenon with similar features to an understood phenomenon it is natural to enquire whether the same theoretical model can be used again, merely re-designating the various terms. Moreover, one is led to question whether phenomena sharing similar patterns do so because they are further linked in some way. But the judge of whether a particular theory is applicable depends on experimental data.

As we saw in equation 2.12, under static, or steady, conditions, the gravitational intensity **g** is equal to the negative gradient of gravitational potential ϕ. We have $\mathbf{g} = -\nabla \phi$. Following the same pattern we can introduce an electric potential ϕ, so that for the electric field **E**, under static conditions,

$$\mathbf{E} = -\nabla \phi. \tag{3.8}$$

We can do the same thing for the magnetic field **H** and for the fluid velocity field **v**, each having its own particular potential.

Heat sources also obey the inverse square law, where the permeability term of the medium is the thermal conductivity, κ. In the case of the thermal conduction field τ we can write

$$\tau = -\nabla\phi. \tag{3.9}$$

The potential ϕ now has its own name and is called the temperature. So, if we replace ϕ with T we have $\tau = -\nabla T$. From the analogue with gravity (eqn 2.13), the temperature T at a distance r from a heat source Q in a medium will be

$$T = \frac{Q}{4\pi\kappa r} \tag{3.10}$$

Note that the temperature T is positive. We can apply the same approach to potential energy as we did for gravity. The potential energy E of a heat source q in a thermal field τ created by the heat source of strength Q is given by

$$E = qT \tag{3.11}$$

Equation 3.11 is the thermal analogue of the potential energy of a mass in a gravitational field, given in equation 2.8. In equation 2.11 we defined force as being the negative gradient of potential energy. So, for the thermal case we can write

$$\mathbf{F} = -\nabla E = -\nabla(qT) = -q\nabla T = q\tau \tag{3.12}$$

This is the thermal analogue of $\mathbf{F} = m\mathbf{g}$. It is also a form of the second law of thermodynamics. It shows that if the gradient ∇T in a thermal system is zero (i.e. the system is in temperature equilibrium), then the force **F** is zero and, consequently, no work can be done.

In the mid-19th century scientists learnt how to make an artificial magnet, using a coil of wire, or solenoid, fed with an electric current. This was the electromagnet. This form of magnet was a magnetic dipole with virtual, or imaginary, poles.

We can make an electret, or electric dipole, too. Electrets were discovered in 1925 and involve placing certain molten ceramics and plastics in a very strong electric field of about 1Megavolt/m and allowing them to cool. Once solidified, the electret has unlike electric charges at opposite ends and it will align with an electric **E**-field in the same way that a magnetic dipole aligns with a magnetic **H**-field.

Using a water table, called the Hele-Shaw apparatus, we can demonstrate some basic fluid flows in two dimensions (2-D). A horizontal metal plate has a glass plate fixed just above it, allowing water to flow in the sandwich. Water introduced at one edge and removed at the opposite edge gives rise to a uniform stream which can be observed through the glass from above. Holes in the metal back plate allow more water to be injected into the layer, or sucked out of it, to form a 2-D source, or a 2-D sink. If a source and a sink are located in close proximity, then a dipole (sometimes called a doublet) forms. If the source is on the upstream side of the stream, a separate oval of fluid is created, separating the fluid dipole flow from the free stream (Fig. 3.2). In three dimensions (3-D) this would form a bubble.

If we could create a 3-D fluid dipole in water, in which the dipole was free to move, then it would propel itself. The sink is repelled by the source, which

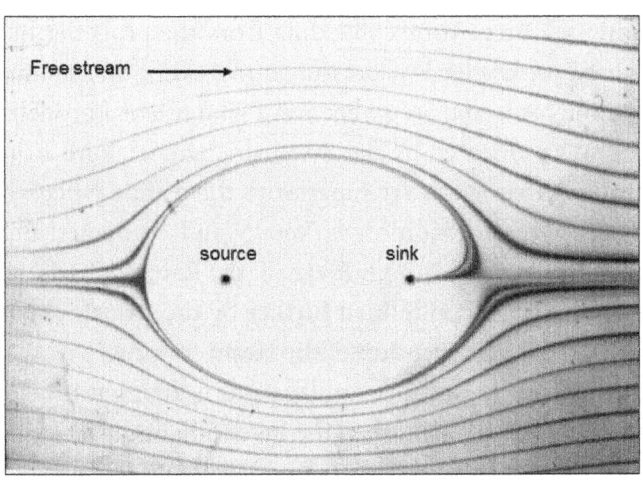

Fig. 3.2 Streamlines for a source and a sink in a uniform flow.

Courtesy of Dr. Derek Brighton, DRA Marine Division, ARE Haslar, Gosport, Hants.

Fig. 3.3 The ship's screw, or propeller.

moves away but attracts the source after it so the two unlike poles move off together with constant velocity.

The forerunner of the ship's screw, or propeller, was the Archimedean screw, which was invented about 3000 years ago. The propeller creates a dipole in the water (Fig. 3.3). Water is drawn in by the screw, forming a sink, and then ejected, forming a source. Vortex shedding from the propeller blades is made visible by cavitations. Under low pressure a myriad of tiny bubbles of boiling steam form on the blade and are swept away. The use of a propeller on steam-powered ships was pioneered by the Swedish engineer John Ericsson during the 1830s, showing that it was far superior to the paddle wheel. Prior to that, Ericsson built the steam locomotive *Novelty* and competed unsuccessfully against George Stephenson, with his *Rocket*, at the Rainhill Trials near Liverpool in 1829. Ship propulsion was refined further by the British engineer Charles Parsons, in 1884, with his invention of the steam turbine.

Aircraft propulsion followed a similar pattern, first with the development of the air screw, or propeller, and then the invention of the jet engine, based on the gas turbine. At the front of the jet engine (Fig. 3.4) is the intake, which forms a sink, while at the rear is the exhaust, which forms a source.

The flight of the toy balloon is a good example of the propulsive property of

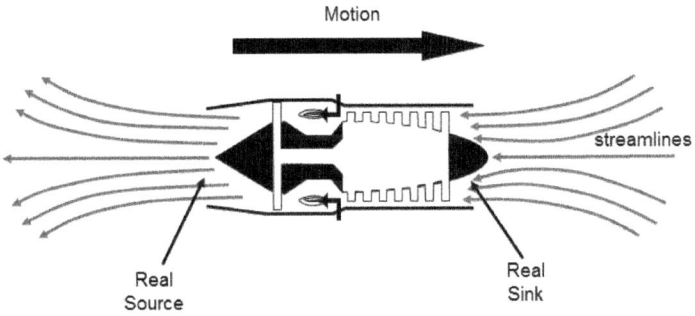

Fig. 3.4 The jet engine as a flying dipole.

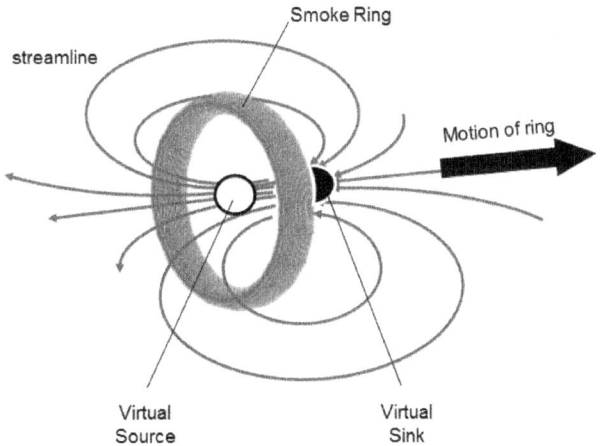

Fig. 3.5 Vortex ring.

the fluid dipole, which emulates the jet engine and the rocket motor. For the best result a nozzle, or rigid tube, is fitted into the neck of the balloon to keep the neck open. The pressurised air from the balloon flowing into the nozzle entry forms the sink and the air flowing out from the nozzle exit forms the source.

The vortex ring is a natural jet-propelled phenomenon (Fig. 3.5). In air, a smoke ring sucks air in at the front and blows it out the back as it jets along. The vortex ring forms a fluid dipole with a virtual source and a virtual sink.

From a different viewpoint, Figure 3.3 shows a vortex solenoid, akin to the electric solenoid which creates a magnetic dipole. However, in the case of magnetism, where unlike poles attract, the magnetic dipole does not form a

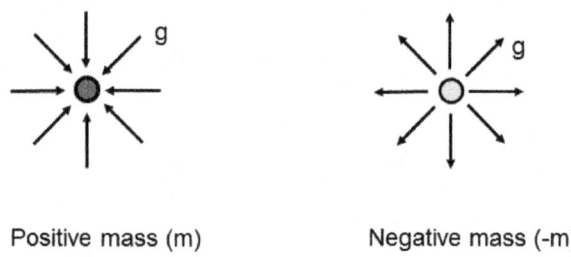

Fig. 3.6 The gravitational fields of positive and negative masses.

propulsive device. The same goes for the electric dipole. These are examples where the analogue with fluids does not read across.

The patterns for gravity, electricity, magnetism, fluid flow and heat flow are very similar, but not exactly the same. Electricity, magnetism and fluids have positive and negative sources. For heat flow positive (hot) and negative (cold) sources may exist in a medium, but are relative phenomena, being dependent on the background temperature of the medium. For gravity, we only know of positive sources, or ordinary mass.

In 1957 Professor Sir Herman Bondi, of King's College London, pointed out that there is nothing in physics that says negative mass can't exist. Since the gravitational field pattern for an ordinary point mass is the analogue of a sink, we assume that the gravitational field pattern for a negative point mass is the analogue of a source (Fig. 3.6). We might think that if negative mass does exist then if equal amounts of positive mass and negative mass come into contact they will merely annihilate each other, with no violent burst of energy to mark their disappearances. However, without experimental evidence, we can't be certain that this is so. Indeed, in a contrary view, John Mike, an ex-particle physicist from MIT, has suggested that negative mass might be present on Earth in extremely small quantities such that we are completely unaware of its presence. Perhaps it is the existence of negative mass deep within the heart of the atom that is the underlying cause of radioactivity.

It is assumed that gravity is the force that holds the stars together in galaxies. However, examining the rotation of certain spiral galaxies shows that the stars in such galaxies do not possess enough gravity to hold themselves together and that stars in the outer reaches of such galaxies move too fast. It is postulated that the source of the missing gravity comes from an invisible halo of matter surrounding these galaxies. Since the matter in the halo is not visible it is known as dark matter, although that is a misnomer; transparent matter would be more correct. Amazingly, scientists now think that ordinary visible matter only accounts for about 5% of the matter in the Universe while 20% is invisible dark matter, leaving a staggering 75% of unknown matter being associated with something called dark energy. Where does negative matter fit into this picture? And what happens to light entering a region of space containing negative mass?

To get some idea of how non-touching positive and negative masses interact we must be guided by theory, while noting that experiment is the final arbiter. From equations 2.6 and 2.10 we have the inverse square law for gravity.

Fixed mass	Acceleration due to gravity	Free mass	Force of gravity	Acceleration of free mass	Newton's second law	Reference & movement of M_2 relative to M_1
M_1	$\mathbf{g}_1 = \left[\dfrac{-M_1}{4\pi\gamma r^2}\right]\hat{\mathbf{r}}$	M_2	$\mathbf{F} = M_2 \mathbf{g}_1 \hat{\mathbf{r}}$	$a_2 \hat{\mathbf{r}}$	$\mathbf{F} = M_2 a_2 \hat{\mathbf{r}}$	
+	− Towards M_1	+	− Attraction	− Towards M_1	− Attraction	3.14(a) Towards
+	− Towards M_1	−	+ Repulsion	− Towards M_1	+ Repulsion	3.14(b) Towards
−	+ Away from M_1	+	+ Repulsion	+ Away from M_1	+ Repulsion	3.14(c) Away from
−	+ Away from M_1	−	− Attraction	+ Away from M_1	− Attraction	3.14(d) Away from

Table 3.14

$$\mathbf{F} = M_2 \left[\frac{-M_1}{4\pi\gamma r^2} \right] \hat{\mathbf{r}} \qquad (3.13)$$

The fixed mass M_1 creates the gravitational field in which the inertial mass M_2 is free to move. The predicted interactions between the positive and negative masses, a distance r apart, are given above. Some of the results seem counter-intuitive.

As we might expect, like masses attract, while unlike masses repel. But the overall effects are more confusing. Free positive masses and free negative masses both accelerate towards a fixed positive mass, but accelerate away from a fixed negative mass.

From 3.14(b) we see that negative mass alone cannot be used for anti-gravitational purposes as it would accelerate towards the Earth. In fact, equal amounts of positive and negative matter, with equal but opposite densities, would weigh the same on the surface of the Earth. Nor can a centrifuge be used to separate them.

From 3.14(c) we can see that if we plunge our hand into the gravitational field of a fixed negative mass our hand would be forced away, just like the force experienced when trying to push two magnetic like-poles together. However, until it has been established what happens when positive and negative mass come into contact this simple experiment must be regarded as hazardous.

From 3.14(d) we see that negative masses do not naturally congregate together to form lumps, unlike ordinary matter. Instead they would form clouds trapping, or swallowing up, any ordinary matter which strayed into them.

In 1990 Dr Robert Forward, at one time a senior scientist at the Hughes Aircraft Research Centre in Malibu, California, suggested that a gravitational dipole might form a propulsion device. Suppose we could, somehow (perhaps by some form of electromagnetic force), gang together a positive mass (M) and a negative mass (-M), a small distance d apart to make a gravitational dipole (Fig. 3.7).

If we use equation 2.11 to examine the dipole it correctly gives the force directions experienced by the positive and negative masses (3.14(b) and 3.14(c)), but it doesn't give the whole story and is misleading. A closer examination shows that the positive mass experiences a force of repulsion which causes it to accelerate away from a negative mass (3.14(c)) with acceleration GM/d^2. But, the negative mass is attracted by the positive mass (3.14(b)) and moves towards it with an acceleration GM/d^2. So, the two unlike masses act in unison and accelerate in the

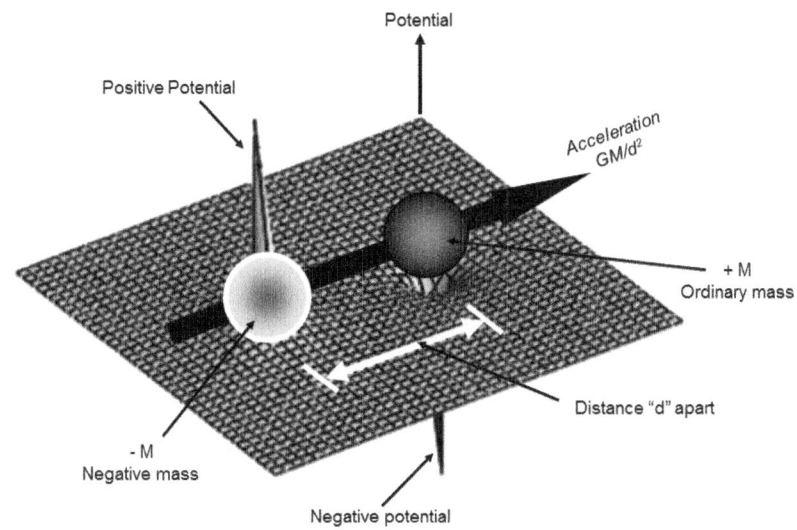

Fig. 3.7 The gravitational dipole.

same direction. The acceleration vector **a** points in the direction from the negative to the positive mass. The magnitude of the acceleration **a** is given by

$$a = \frac{GM}{d^2}. \qquad (3.15)$$

Thus, a gravitational dipole would work as a propulsive device. It is the analogue of the fluid dipole.

Earlier, in 1964, Banesh Hoffmann, a collaborator of Einstein's while at the Advanced Studies Institute at Princeton University, had invoked the idea of the gravitational dipole in his attempt to explain the mechanism underlying the observed stellar bipolar jets of matter generated by quasars, or radio stars.

The combined energy of the gravitational dipole is always zero, so that the law of conservation of energy applies. Similarly, the combined linear momentum of the gravitational dipole is always zero, so that the law of conservation of momentum also applies. Since there is no change of momentum, from Newton's second law of motion, we see that the accelerating dipole is not subject to inertial force (Fig. 8.3(e)).

If we can't have negative mass, then we must think of a way of artificially generating its properties. Then we can build a gravitational engine, being the analogue of the jet engine. The medium for the jet engine is air; perhaps the medium for the gravitational engine is the ether!

4

DIVERGING AND CONVERGING FLOWS

During the 18th and 19th centuries the phenomena of flowing fluids, electricity and magnetism were much explored but, curiously, not much happened with gravity. The first to develop in any depth was the study of fluid motion, or fluid dynamics. So let us start by having a closer look at the basics of this subject, which includes the flow patterns of treacle, water and air.

The first thing that scientists do, when confronted with a new phenomenon to model mathematically, is to make some very simple assumptions about the phenomenon. So, to model the flow of a fluid they began by assuming that they had an ideal fluid, one that had constant density and was, therefore, incompressible (couldn't be squeezed) and that possessed no fluid friction (zero viscosity). In doing this, it meant that they couldn't model some well-known effects of fluids, such as the generation of shock waves (which are a compressibility effect) and the generation of vortices (which depend on viscosity). But a new mathematical model is just like a baby. It can't do much to begin with but, as it develops, it is able to do more and more.

Leonhard Euler was a prolific 18th century Swiss mathematician who spent most of his professional life either at the Academy in St Petersburg, in Russia, or at the Academy in Berlin, in Prussian Germany. One of Euler's interests was in the gravitational interactions between the Sun, the Moon and the Earth and their effect on the Earth's tides. This led to his general interest in the flow of fluids. In 1755, using Newton's laws of motion and the simple assumptions for an ideal fluid, Euler derived the basic equation for fluid motion.

Just like a point mass is a source of the gravitational field **g** and a point charge is a source of the electrical field **E**, so in fluids we have a point source which creates a fluid dynamic field, where at any point in the field the fluid flow has a velocity **v**. It's easy to demonstrate a fluid source in two dimensions (2-D). A downward jet of water, on hitting the floor, forms a source centre from which the fluid spreads out radially in 2-D. By introducing ink, or milk, at a

DIVERGING AND CONVERGING FLOWS

point into the flow a radial streamline of the 2-D velocity field is made visible. A 3-D source, which is easier to imagine than it is to demonstrate, emits fluid from a centre point which spreads out radially, in all directions. That is, at any point in the fluid flow, the velocity vectors all point outwards in a radial direction.

The ideal 3-D fluid source obeys an inverse square law of influence. At any point a radial distance r from a point source, of strength q, the fluid velocity v is given by

$$v = \frac{q}{4\pi\rho r^2} \tag{4.1}$$

In one second q kg of fluid emitted from the source centre flows across the spherical surface with radius r. The radial streamlines from the source centre diverge outwards.

The 3-D sink, of strength -q, also obeys the inverse square law, but it sucks the fluid in. The streamlines bunch together as they converge towards the hole forming the sink centre.

In a fluid containing sources, or sinks, we can detect their presence by revealing, say with ink, or milk, regions of divergent, or convergent, streamlines.

The observation that fluids appear to be 'continuous' plays an important part in developing the mathematical model to describe their flowing properties. We have assumed that the fluid density ρ is constant, so that the fluid can't be squeezed, or compressed. For this simple case, if we concentrate on a fixed

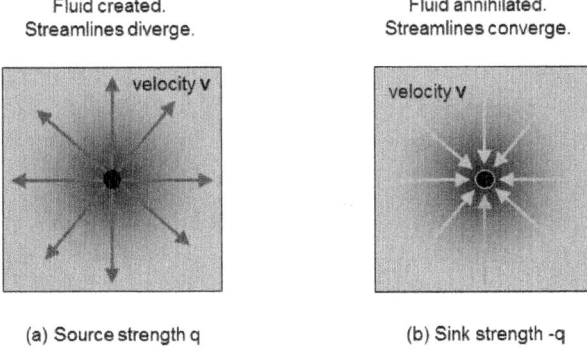

Fig. 4.1 *Source and sink.*

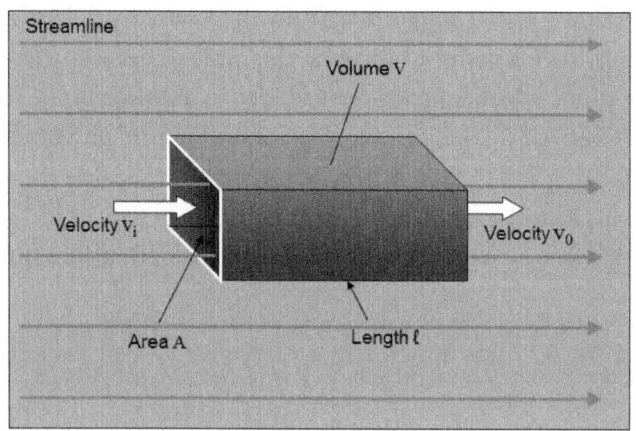

Fig. 4.2 Velocity flux.

volume in space and let the fluid flow through it, then what goes into the volume over a certain time must equal what comes out in the same time. This is what is meant by continuity.

But, suppose the volume that we are looking at contains a source, or a sink, creating more fluid, or annihilating some of it. Then what goes in might not equal what comes out.

If we want to monitor what is flowing into, or out of, a volume V we use a quantity called the flux.

Let us consider the very simple idea of placing an empty matchbox of length ℓ into a fluid flow, such that it doesn't disturb the fluid. If A is the surface area of the open ends of the box through which the fluid flows, then the box volume $V = A\ell$. Suppose that v_i is the input velocity of the fluid, which crosses the entry area A perpendicularly as it flows into the box. Then, in mathematical terms, the entry velocity flux value is just $v_i A$. Similarly, $v_o A$ is the exit velocity flux of the flow out of the box.

Since $V = A\ell$, we can rearrange the velocity flux difference as

$$(v_o - v_i)A = \left[\frac{(v_o - v_i)}{\ell}\right] V \qquad (4.2)$$

Now $(v_o - v_i)/\ell$ is the change in velocity v over distance ℓ and in mathematical shorthand this is written as ($\nabla \bullet \mathbf{v}$). It is called the divergence of the velocity and is a scalar value. So we can write the velocity flux difference as

$$(v_o - v_i)A = (\nabla \bullet \mathbf{v})V. \qquad (4.3)$$

This equation is a simplified form of what, nowadays, is called Gauss's divergence theorem, named after the famous German mathematician Carl Friedrich Gauss, whose life spanned the late 18th and early 19th centuries. This theorem will help us detect whether there are any sources, or sinks, within the volume V causing the streamlines to diverge (spread out), or converge (bunch together). Although we have derived the result for flow through a rectangular box, in fact it applies to any shaped volume in 3-D which doesn't disturb the flow.

The divergence theorem has an interesting history in that it was first derived by George Green, a Nottinghamshire miller who lived at the turn of the 18th into the 19th century. Just prior to Green's birth in 1793, Charles Coulomb, a French experimenter, had demonstrated that static electric charges obeyed an inverse square law of influence – the same law of influence for masses in gravitation. There was a great deal of public interest in the physics of electrostatics at that time, rather like there is in gravitation today.

Although Green only had a basic education, he developed a keen interest in mathematics, perhaps stimulated by the need to be good at arithmetic for keeping accounts for the mill. Working in the mill, he would have been quite used to the idea of flow. He could probably visualise the number of bags of corn pouring into the volume of the hopper, containing the grinding wheels, and how many bags of flour were produced as a result.

In his leisure time Green joined a subscription library and began reading about the latest developments in electrostatics being studied in France. Sitting in his study at the top of his windmill, Green began developing a mathematical model for the flow of electrical influence (the electric field) $\mathbf{E}$, based on the inverse square law of electric charge. Perhaps he viewed the flow of electrical influence as being analogous to the flow of corn down a chute.

Green's final result, linking the flux of electric field $\mathbf{E}$ across an area A with the divergence that occurred over a length surrounded by a volume V, was none other than

$$(E_0 - E_i)A = (\nabla \bullet \mathbf{E})V \qquad (4.4)$$

It's the same equation that we obtained for ideal fluid flow, only **E** has replaced **v**. It is the divergence theorem later made famous by Gauss.

When Green thought about trying to get his work published in a learned journal, he felt that it might be rejected, since he was not an academic. So, in 1828, to avoid any embarrassment, he decided to have his work printed privately in Nottingham.

Green's exceptional mathematical ability eventually came to the notice of Cambridge University who took him in as a mature student. At the age of forty he gained his degree and was subsequently elected to the fellowship of his college. Nearly two decades later, after Green's death, another Cambridge graduate, the Irish physicist William Thomson (better known as Lord Kelvin), discovered Green's privately printed book and was amazed at its contents. It didn't only contain the divergence theorem but other mathematical ideas, too. Today, Green is better known through the use of 'Green's functions', a technique for solving differential equations. These find use in subjects as far afield as biology and quantum mechanics.

We can now derive the equation of continuity for the flow of a fluid. Suppose we have a 3-D point source of strength q, which we surround by an imaginary sphere, or potential surface, of radius r. The surface area of the sphere is $A = 4\pi r^2$. The outward velocity flux crossing the area A is $v_0 A$, where the velocity $v_0 = q/(4\pi\rho r^2)$, due to the inverse square law for a fluid source (eqn 4.1).

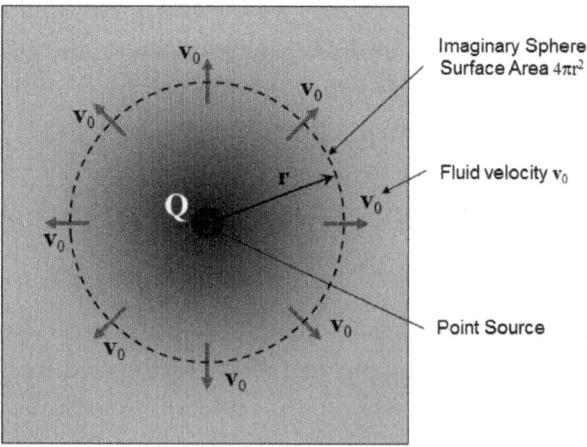

Fig. 4.3 Point source surrounded by an imaginary sphere.

Substitution shows that

$$v_o A = \left[\frac{q}{4\pi\rho r^2}\right]\left[4\pi r^2\right] = \frac{q}{\rho} \qquad (4.5)$$

Since $v_i A = 0$, on substituting for $v_o A$ in the divergence theorem (eqn 4.3) we get

$$(\nabla \bullet \mathbf{v})V = \frac{q}{\rho} \qquad (4.6)$$

This can be rewritten as

$$\nabla \bullet \mathbf{v} = \frac{q}{\rho V} \qquad (4.7)$$

Suppose we introduce σ as the source density inside the volume V, so that $\sigma = q/V$. Then, finally, we have the equation of continuity for fluids:

$$\nabla \bullet \mathbf{v} = \frac{\sigma}{\rho} \qquad (4.8)$$

Here ρ is the density of the fluid medium, or the permeability of the medium to velocity **v**. If there are no sources or sinks inside the volume V, so that $\sigma = 0$, then the ideal flow is continuous and $\nabla \bullet \mathbf{v} = 0$. No sources or sinks inside the volume V means that the streamlines don't diverge or converge, but remain evenly spread.

Nature's patterns repeat themselves. So, with hardly any effort, we can write down the analogue forms of equation 4.8 for a number of other phenomena, using the constants that appear in their own inverse square laws.

For electrostatics $\nabla \bullet \mathbf{E} = \dfrac{\rho}{\varepsilon} \qquad (4.9)$

The parameter ρ is the charge density and ε is the permittivity of the medium to the electric field **E**.

For magnetism $\nabla \cdot \mathbf{H} = \dfrac{(\rho' - \rho')}{\mu} = 0$ (4.10)

Theoretically, the density of the north poles is +ρ' and the density of the south poles is -ρ', but they always appear together. The term μ is the permeability of the medium to the magnetic field **H**.

For gravitation $\nabla \cdot \mathbf{g} = -\dfrac{\rho}{\gamma}$ (4.11)

The parameter ρ is the density of the sources of mass. γ is the permeability of the medium to the gravitational field **g**. The negative sign occurs because the inverse square law is one of attraction between like masses.

Courtesy of Nottingham City Council

Fig. 4.4 George Green's windmill.

Incidentally it was George Green who introduced the idea of potentials that we explored in Chapters 2 and 3. Surprisingly, Green is often omitted from books describing the history of mathematics. His windmill at Sneinton (now absorbed by the City of Nottingham) is still there and has been transformed into a mathematics centre. It reminds us that occasionally gifted amateurs play an important part in developing a subject.

Another point is that although experiment usually leads theory, help with the insight provided by analogues is often valuable. Nevertheless, it was by trial and error that engineers first learnt how to build windmills, as no proper theory of aerodynamics (a branch of fluid dynamics) existed. We now know that as the wind blows, each sail, or blade, of the windmill develops a vortex around it, which is responsible for the turning effect. George Green, sitting in his study at the top of his windmill watching the sails rotate past the window was unaware of this. As we will discover later, the vortex has an analogue in the phenomena of electricity and magnetism, too. Unknown to many people, but as we will show, the vortex also has an analogue in gravitation.

5
THE VORTEX PHENOMENON

We are all familiar with the vortex, that swirling core of water, often with a hollow tubular centre, that we see when we empty the bath. But, when the plug is first pulled the initial flow of water draining away down the plug hole follows a radial sink pattern. However, the proximity of the plug hole to the bath wall means that viscous effects interfere with the flow causing it to rotate and so a vortex appears, too. Whenever a real fluid rotates, then a vortex appears. When air blows past a sharp edge, like the corner of a building, then vortices or dust eddies often appear, made visible by the swirling dust and leaves that they pick up.

Like the genie escaping from a bottle, the tornado is a huge version of the small street-corner dust eddy. It is caused by rising hot air which spirals as it goes up. A narrow tubular funnel extends from the clouds down to the ground. It sweeps across the land at speeds of about 60 km/hr (40 mph), sucking up everything in its path. Thankfully, most of us are not plagued by these yearly, deadly and extremely destructive, colossal whirlwinds, which occur mostly in the mid-western states of the USA.

Over the sea the tornado sucks up water and is transformed into a waterspout. Sometimes only a small column of water is sucked up, like a stalagmite. It may look like the head and neck of a sea serpent emerging from the deep. Every year, hundreds of waterspouts form in the sea near to the US coast at the Florida Keys. There has been some speculation that they are responsible for the mysterious losses of ships and aircraft in the Bermuda Triangle.

On a much vaster scale, the Atlantic hurricane, the Pacific typhoon and the Indian cyclone are formed from sink and vortex combinations and can be many hundreds of miles across.

Around the Earth's equatorial belt the heat from the Sun warms the oceans, and the moist air above the sea surface, being less dense, starts to rise and form thick clouds. A central region develops where the air up-current is the most pronounced and the surrounding air is sucked radially in, across the sea surface, towards it. This

THE VORTEX PHENOMENON

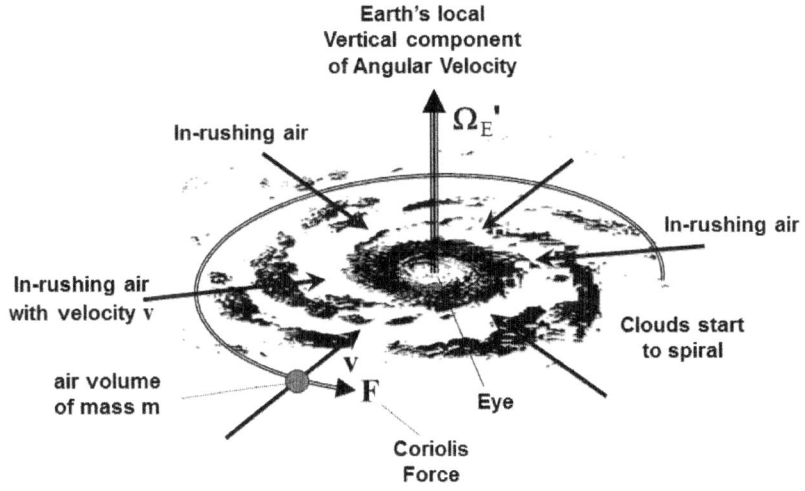

Fig. 5.1 Hurricane in the northern hemisphere.

is the eye of the developing storm. The drop in atmospheric pressure is most noticeable at the eye and on an isobar chart the region of depression is labelled as a *LOW*. Without the Earth's rotation, the clouds streaming towards the centre of the depression would mark out a 2-D sink flow on the face of the Earth as seen from space. Adding the effect of the Earth's rotation, or angular velocity Ω_E, the in-rushing air rotates around the eye forming a combined sink and vortex flow which moves slowly along. Satellite photos (Fig. 5.1) show that the air rotates in an anticlockwise direction in the northern hemisphere and in a clockwise direction in the southern hemisphere. These gigantic circulating movements of air, made visible by clouds, coupled with invisible gigantic circulating currents of water in the oceans, are largely responsible for the weather patterns on Earth.

The force on the moving air due to the Earth's rotation is called the Coriolis force. Gaspard de Coriolis was a French professor of mathematics who studied rotating flows during the early 19th century.

In mathematical symbols the Coriolis force **F** is written as

$$\mathbf{F} = m(\mathbf{v} \times 2\Omega_E'). \tag{5.1}$$

Here, m is a typical mass of moist air (Fig. 5.1) moving with velocity **v**, and $\Omega_E' = \Omega_E \cdot \sin\lambda$ is the vertical component of Earth's rotation, at latitude λ, at the centre of the low pressure region.

Giant vortices, or hurricanes, form around really strong depressions. We know that hurricanes appear in the atmospheres of other planets, too. The giant red spot on Jupiter is a huge hurricane.

Swirling galaxies show that there are also gigantic vortices in space. Looking out in any direction into space there are countless spiral galaxies. One of the nearest to the Earth is the Whirlpool Galaxy, M51. Just as the spiral vortex in air breaks up into small turbulent spots, so too does the rotating spiral galaxy. Nature repeats itself; it's just a matter of scale. In the case of galaxy formation the gas is hydrogen and in the rotational process the turbulent spots condense to become stars. Occasionally, one of these stars explodes, forming a supernova, creating new elements in the process and showering the whole of the Universe with dust. Professor Fred Hoyle of Cambridge University was the first to point out that we are all made of stardust, but it was Carl Sagan's *Cosmos* TV series that publicised this fact. Our bodies (and those of all creatures) are composed of elements which have come from dying stars, from all corners of the Universe. From the utter chaos of a supernova explosion has come intelligent life-forms. We are star children.

In Homer's *Odyssey* it is recorded that Odysseus and his crew were caught up in a large whirlpool, or vortex, called Charybdis which formed in the Strait of Messina, between Italy and Sicily. The crew was able to look down the hollow core of the vortex and see rocks on the seabed below being ground together. It was a truly terrifying experience from which they were lucky to escape. In the world of the Ancient Greeks, being sucked into Charybdis was the ultimate disaster, like today's fear of the crew of a spaceship being sucked into a black hole.

We even have our own tidal whirlpool in the UK. It is called Corryvreckan and it forms in the channel between the islands of Scarba and Jura, in the Inner Hebrides. Ships are warned to keep well clear of it. Much more dangerous is the Maelstrom, or great whirlpool, off the coast of Norway between the islands of Moskenës and Mosken, in the Lofoten group.

If we are unfortunate enough to get caught up in a large whirlpool, then we will be swept round and round on a streamline, circling the vortex core, with tangential velocity **v**. Imagine that you could grip the axis of the vortex core with your right hand and twist it so that your curled fingers pointed in the direction that you are being swept round in (Fig. 5.2). Your thumb would then point in the direction of the whirlpool's vorticity vector ζ.

Using symbols we can express the curling effect in mathematical shorthand as

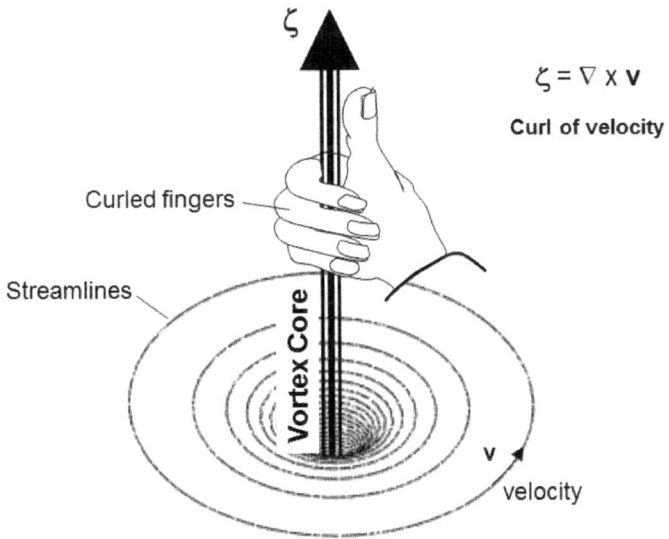

Fig. 5.2 The vortex.

$$\zeta = \nabla \times \mathbf{v}, \tag{5.2}$$

In words this says that the vorticity ζ equals the curl of the velocity $\mathbf{v}$. It relates the speed of the rotating, or curling, flow with the strength of the vortex.

There is a duality between a vortex core and streamlines:

(i) A straight vortex core has circular streamlines around it.
(ii) The core of a vortex ring (e.g. a smoke ring) has a straight streamline through it.

If you drag a spoon sideways through the surface of your cup of tea or coffee, you will notice that a small vortex is shed from it. But according to the laws of vorticity expressed by the German scientist, Hermann von Helmholtz, in the mid-19th century, "vorticity cannot be created or destroyed". So, what about the vortex in the cup? The answer is that an equal and opposite vortex has been created around the spoon. Adding the vorticity of both vortices together we see that the total vorticity created in the cup is zero. Similarly, when an aeroplane takes off from an airfield a vortex is developed around its wings and another vortex, called the starting vortex, is left behind at the airfield. The unseen starting vortices left behind by jumbo jets take some time to disperse and can be very dangerous to small aircraft caught up in them.

In the 18th century Daniel Bernoulli, a Dutch-Swiss scientist who taught Leonhard Euler, noticed an effect in fluid flow that is now named after him. Where a fluid speeds up there is a corresponding drop in pressure. If streamlines are made visible with dye, then in a region where the flow speeds up the streamlines are seen to bunch together. The Bernoulli effect is used in many devices and also explains some weird effects. What causes the raspberry noise from an evacuating balloon, or 'whoopee cushion'? As the high speed air passes through the narrow rubber neck, the drop in pressure causes the neck to suck together, shutting off the air flow. The pressure then rises forcing the neck to open again, the cycle repeating over and over again until all the air has gone. The constant opening and shutting of the rubber neck causes the raspberry sound.

Consider a uniform stream of air flowing past a vortex of rotating air (Fig. 5.3). Where both air flows are in the same direction, then the mixed air flow will speed up, otherwise it will slow down. Where the air speeds up there is a drop in pressure, while where it slows down there is an increase in pressure. If the vortex is formed around a solid body, say a spinning ball, or a rotating cylinder, then the body will be sucked and pushed at right angles to the flow of air. A note written by Isaac Newton in 1672 recorded that he observed such an effect on a spinning tennis ball. However, it was Professor Heinrich Magnus, a Prussian German scientist at the University of Berlin, who first studied the phenomenon in detail. In 1852 he named the force which caused a spinning body moving through the air to veer off-course as the Magnus effect.

When an aircraft moves forwards a vortex is formed around the wing and the air streams past it. The result is that the wing is sucked and pushed upwards.

The sycamore seed is a passive propeller which rotates as the air flows past it, in the same way that the sails turn on a windmill. The active propeller is a driven rotating wing which gets sucked forwards due to the vortices which develop around its blades (Fig. 3.3). The propeller is an ideal device to provide thrust for an aircraft, but what is needed is a means of turning the propeller fast enough to create enough thrust and forward motion through the air for the main wing to develop a strong enough vortex to generate lift. In Britain during the late 1840s, John Stringfellow and William Henson used a lightweight steam

THE VORTEX PHENOMENON

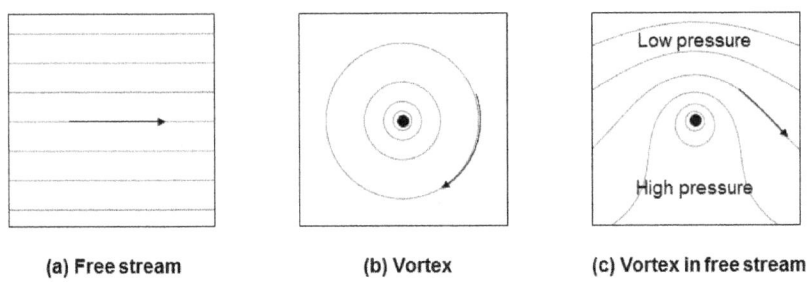

(a) Free stream   (b) Vortex   (c) Vortex in free stream

Fig. 5.3 Vortex in a uniform flow.

engine to drive two pusher propellers on a small unmanned aircraft which they called their *Aerial Steamer*. Powered flight was achieved, providing a glimpse of the future, but the thrust was extremely limited and there was no chance that their aircraft could carry a man. Further development had to wait until 1885, when the German engineers Gottlieb Daimler and Wilhelm Maybach developed the internal combustion engine. After that, the 'Age of Steam' went into gradual decline.

In the USA the brothers Orville and Wilbur Wright built a lightweight petrol engine and fitted it to their box-kite-cum-aircraft, *Flyer 1*. In December 1903 they made the world's first powered manned flight. The world's press didn't believe it, though, and thought that photographs showing one of the Wright brothers airborne were part of a huge hoax. It was only after a public flying demonstration by the Wright brothers in 1908 that everyone, including the press, realised that the breakthrough in manned flight had been made.

Vortex creation and control plays a major role in the mechanics of flight. As we will see when we investigate the idea of gravitational propulsion, a vortex-type phenomenon plays a major role too.

6
ELECTRICITY AND MAGNETISM

The phenomenon of magnetism was known to the Ancient Chinese as far back as 2500 BC and they were the first to use the naturally occurring black ore of iron to make magnetic compasses. Much later, around 600 BC, the Ancient Greeks also discovered deposits of the same magnetic mineral near the Ancient Greek city of Magnesia, in Asia Minor, giving it the name magnetite, and so began the study of magnetism in Europe. No doubt the Ancient Greeks were fascinated by the strange attractive and repulsive powers of magnets, just as we are today. In their jewellery the Ancient Greeks used amber, a yellowish fossilised resin of coniferous trees. They noticed that when amber was rubbed it would sometimes attract lightweight objects and repel others, a property similar to that of magnets. How many of us, as school children, have run a plastic comb through our hair and then used it to pick up small pieces of paper, through electrostatic attraction?

By the 13th century the use of the magnetic compass for direction-finding had spread throughout Asia, Arabia and Europe. The 13th century French scholar and soldier, Peter Peregrinus, published his book, *Epistola de Magnete*, which was the first serious book about magnetism. In it he described how unlike magnetic poles (a north-seeking pole and a south-seeking pole) attracted each other, while two like poles repelled each other.

In the 15th century Christopher Columbus used a direction-finding magnet on his voyage of discovery, leaving Europe to cross the Atlantic Ocean in the search for a new route to India. "Was the venture a success, since the original goal was not achieved?" is a much asked question. Like scientific research, it is the setting out on a journey of discovery that is the important thing. Who knows what will be found along the way and where the journey will lead to?

The first in-depth scientific study of both magnetism and electrostatics, collecting together all the previous knowledge on these subjects, was done by William Gilbert, a Cambridge University graduate and Court Physician to Queen Elizabeth I of England. His book, *De Magnete*, was published in 1600 and became a best seller in scientific circles all over Europe. It was Gilbert who correctly explained why the compass needle pointed north-south, because the Earth was a giant magnet, with a south-seeking pole at the North Pole and a north-seeking pole at the South Pole. Gilbert also noted that the Earth's magnetic axis roughly aligned with the Earth's rotational axis, suggesting that rotation and magnetism were linked together. The word electricity was introduced into the English language by Gilbert. During his study of electrostatics he found that, as well as amber, other substances could be electrified by friction, such as rubbing an ebony rod with fur, or a glass rod with silk. But, apparently, metal could not be electrified.

Otto von Guericke, a German scientist, lawyer and mayor of the city of Magdeburg, was also interested in Gilbert's observation about frictionally produced electricity. In 1663 he built a continuous frictional-rubbing machine, based on a rotating sphere of sulphur. Placing one's hand on the rotating surface could produce small sparks, seen in a darkened room. Henry Oldenburg, a German diplomat and philosopher living in London, became the first secretary of the Royal Society, newly founded in 1660. He maintained an extensive correspondence with scientists throughout Europe and, consequently, von Guericke's scientific investigations were closely monitored. Oldenburg was also the founding editor of the *Philosophical Transactions of the Royal Society*, which was the world's first scientific journal.

Francis Hauksbee, an instrument maker employed by the Royal Society and a former student of Robert Boyle's, was one of the earliest investigators of electrical phenomena in England. He replaced von Guericke's rotating sphere of sulphur with a rotating sphere of glass and showed that by rubbing it with a silk cloth an electrical charge was built up. He then went a step further and, in 1702, using an air pump to evacuate his glass sphere, he showed that the remaining low pressure air developed a ghostly glow when it was excited by the frictionally produced electrical charge. Electrical experiments such as these fascinated the Fellows of the Royal Society and did much to stimulate further research in the subject.

During the late 1720s and early 1730s, Stephen Gray, a Fellow of the Royal Society, carried out experiments which showed that electrical effects could be transmitted along some substances, but not others. For example, he showed that when an ivory ball was suspended from a glass rod it became electrified when the glass rod was electrified if the suspension was a brass wire, but not if it was a silk thread. The electrical effect seemed to flow along the wire connection, just like a fluid.

Gray carried out further tests and showed that the electrical effect, generated by a continuously rotating glass cylinder and rubbing silk cloth (Fig. 6.1), could be conducted along quite long distances, of several hundred feet, using wet packthread or string. Later, such materials became known as conductors, while the non-conducting materials were called insulators. Wire and wet string were found to be good conductors, while glass and silk were insulators. Some of the insulators could be electrically charged by friction.

Gray's experiments were repeated almost immediately by Charles du Fay, a French scientist. Du Fay was actually the Chief Gardener to the French King Louis XV and, like other gentlemen at that time, he carried out electrical experiments for his own amusement. It was he who discovered that metal substances could be charged, providing that they were insulated and held in an amber, or glass, handle.

Du Fay found that charged substances of the same type, such as glass rods (charged by rubbing with silk), would repel each other, but that they would attract some other charged substances, such as an ebony rod (charged by rubbing with fur). He concluded that there were two types of electrical fluid which he called vitreous and resinous. Vitreous electrical fluid in one substance

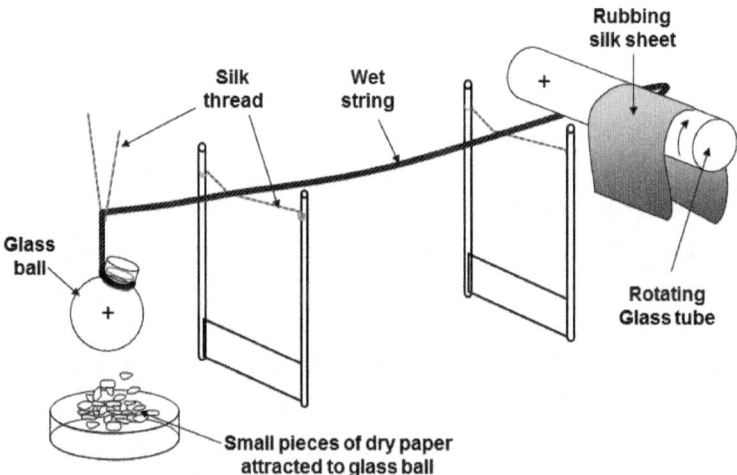

Fig. 6.1 Gray's wet string conductor.

would repel another also containing vitreous fluid, but would attract it if it contained resinous fluid. We would say that like charges repel, whereas unlike charges attract. So, electric charges are analogous to magnetic poles. The repulsive property of like electric charges could be simply demonstrated with the gold-leaf electroscope, invented by Abraham Bennet, a Derbyshire clergyman and pioneer of electrical science, in 1789.

Benjamin Franklin was a British subject born in Boston, Massachusetts, in 1706, in pre-revolutionary America. His parents and older siblings had emigrated from Banbury (the town made famous by the poem about a fine lady upon a white horse), in Oxfordshire, to the New England state in 1685.

Franklin began his career as a natural philosopher, or scientist, in 1748. Most famous philosophers begin in their early twenties, so Franklin, who was forty-two years old at the time, was rather a late starter. However, he had already made his fortune in printing and his interest in science, particularly electricity, really began as a fascinating pastime. Franklin is particularly famous for his dangerous experiment, carried out in 1752, of flying a kite into a thundercloud. Electric charge in the cloud was conducted down the wet string to an iron key and then transferred to a Leyden jar. The Leyden jar was the world's first capacitor used for storing electrical charge. It was independently invented by Ewald von Kleist, the Dean of Kanim Cathedral in German Pomerania, in 1745, and by Pieter van Musschenbroek, the Professor of Physics at Leyden University in the Netherlands, in 1746. What is amazing is the speed at which scientific information travelled around Europe and North America during the 18th century and that Franklin had a Leyden jar. Franklin showed experimentally that the electric charge that he had collected from the clouds was the same as that produced by frictional methods. The kite experiment quickly led to the use of metal lightning rods on tall buildings to provide a preferred conducting path for atmospheric electric charge to flow to the ground, thereby reducing the possibility of lightning strikes causing structural damage. The Royal Navy also used lightning conductors on the masts of their sailing ships, particularly vulnerable to lightning strikes at sea, to prevent lightning reaching the magazine and blowing their ships apart. The same principle is used to protect modern aircraft from lightning strikes.

Contrary to Du Fay's view, Franklin was convinced that there was only one type of electrical fluid and that its transfer left substances either positively (vitreous), or negatively (resinous), charged. He guessed that the current of

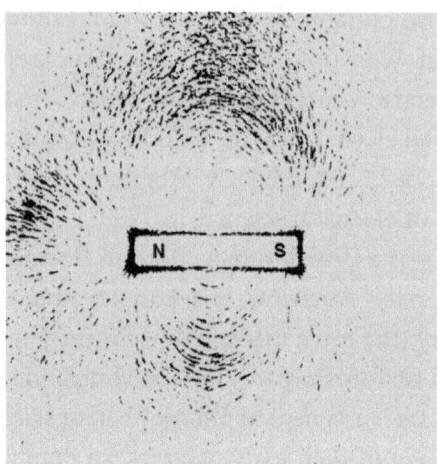

Fig. 6.2 Magnetic field around bar magnet.

electrical fluid was the flow of positive electrical charges which flowed from positive to negative (remember that unlike charges attract). Although his idea was right, he guessed wrong. About 150 years later, the electrical fluid was traced to the flow of electrons. But the electron has a negative charge and flows from negative to positive. So the electron current flows in the opposite direction to Franklin's current, or what we now call the conventional electric current.

With hindsight, we know that by rubbing glass with silk, some of the electrons from the surface atoms of the glass are removed, leaving it positively charged. On rubbing ebony, or amber, with fur some of the electrons from the fur are transferred to ebony, or amber, leaving it negatively charged.

During the mid-1740s the study of magnetism was continued by Gowin Knight, a British scientist, who used iron filings to map out the magnetic fields around magnets (Fig. 6.2).

The famous Lunar Society of Birmingham was formed by Matthew Boulton, who was a far-sighted manufacturer and pioneer of new technologies, being particularly associated with the development of steam power in collaboration with James Watt. The Lunar Society brought together a number of leading scientists and engineers and, perhaps, did more to stimulate the Industrial Revolution than did the more prestigious Royal Society. During an early meeting of the society, in 1765, Joseph Priestley (who discovered oxygen and initiated the fizzy drinks industry) met Benjamin Franklin, on one of the latter's frequent stays in England. Franklin, with his deep interest in electricity, encouraged Priestley to write a history of the subject, which resulted in Priestley's *History of Electricity*, being published in 1767. Based on an analogue with gravitation, Priestley predicted that the force between static electric charges would obey an inverse square law, a view endorsed by a certain Reverend John Michell. But it was the French experimental scientist, Charles Coulomb, who was the first to publically demonstrate (Fig. 3.1) the truth of this prediction in 1785. Strangely, as it may now seem to us, Coulomb was of the opinion that although the phenomena of electricity, magnetism and gravity were analogous, they were totally separate phenomena which could not interact. Although, today, we are not aware of any interaction between gravity and electromagnetism, most scientists are convinced that a link exists and will eventually be discovered.

Until nearly the end of the 18th century, only frictional electricity had been generated, giving rise to short discharge currents. From the time of Gray, it was known that electrical discharges to a body caused muscles to contract. In 1791 Luigi Galvani, an Italian professor of anatomy, was investigating whether the muscles of a frog's legs were a source of 'animal electricity'. He prepared the frog's legs for dissection and hung them from a copper hook over an iron table top. Galvani probably used a frictional electrical apparatus to make the frog's legs twitch. Galvani also noticed that the frog's legs twitched of their own accord whenever the frog's feet touched the iron table top. Galvani carried out further experiments and observed that if the exposed nerve at the top of the frog's legs was connected to a metal plate and the feet to another, dissimilar, metal plate, then the legs kicked convulsively. Galvani thought that he had proved that muscles were a source of 'animal electricity'.

Alessandro Volta, an Italian physicist, investigated Galvani's findings and showed that the frog's legs were only acting as conductors of electricity. The essential fact was that dissimilar metals separated by a membrane of some form created a current of electricity when connected. In 1799 Volta described his 'voltaic pile', a pile consisting of alternate layers of zinc and copper discs interspersed with paste board wetted with brine to make a good contact. Each sandwich of zinc and copper discs formed a cell, so the 'pile' was made of a column of cells. On touching the top and bottom discs of the pile he received an electrical discharge, or shock. Volta's pile was the first means of generating a continuous electric current, although what it was a current of no one knew.

Volta went to Paris in 1801 to demonstrate his voltaic pile. The French Emperor Napoléon Bonaparte was very interested in science, being a mathematical member of the Institute of France, the highest academic body in the land. Napoléon was intrigued by electricity and ordered that an enormous voltaic pile should be built in Paris. The disadvantage of a tall pile was that the weight of it squeezed out all the brine and the pile ceased to function. Instead, Volta laid out 600 cells horizontally and connected them in series. This was the first battery. Developments led on to the accumulator and, later, the car battery. The dry cells that we use in our Walkmans and mobile phones are based on Robert Bunsen's development of Volta's original idea. (Yes, it's the same Bunsen, a German scientist, of Bunsen burner fame.) And to think that the idea had come from studying twitching frog's legs!

Once the secret of how to make a voltaic pile had been revealed it was fairly simple for other scientists to copy the idea. In Britain, Humphry Davy, newly appointed as the Professor of Chemistry at the Royal Institution in London, was very quick in following up all of Volta's discoveries. Not to be outdone by the massive voltaic battery in Paris, Davy arranged for a battery to be assembled in the cellar of the Royal Institution with twice the number of cells.

In 1784 Joseph Priestley and Henry Cavendish, working together, had discovered that a mixture of oxygen and hydrogen, when ignited by a spark, produced water. While experimenting with an early voltaic pile, in 1801, Sir Anthony Carlisle accidently discovered the reverse effect: that water could be decomposed into hydrogen and oxygen by passing an electric current through it. Working with William Nicholson, the process of electrolysis was rapidly developed. Later, Sir Humphry Davy used the technique to decompose certain chemical solutions to isolate the metallic elements sodium and potassium.

In 1812 Michael Faraday, previously a London bookbinder's apprentice, was taken on to work in the laboratory at the Royal Institution. Soon, by keenness and ability, he had moved on to become Sir Humphry Davy's personal laboratory assistant. Incredibly, during the final stages of the Napoleonic Wars, Faraday accompanied Davy on a scientific journey through Europe, with a safe passage being granted by Napoléon.

Earlier, during the 18th century, Immanuel Kant, the famous German philosopher of Königsberg in East Prussia, had declared that there was a unity between the attractive and repulsive forces of nature and that it should be possible to convert one force into another. Hans Christian Oersted, a Danish scientist, was much influenced by Kant (his PhD thesis was on Kantian philosophy) and in 1810 he wrote, "It is my firm conviction that a fundamental unity permeates all Nature." In 1812 he went further, predicting that electricity and magnetism were connected. However, it wasn't until 1820 that he finally discovered the link that he'd been searching for. Even then, as he admitted later, it had been a chance observation made during a laboratory demonstration. What he noticed was that when a current produced by a twenty-cell voltaic battery flowed along a wire it caused a nearby magnetic compass needle to move. Carrying out further experiments he discovered that there was a circular magnetic field around the wire.

Oersted's major discovery came as a surprise to many scientists, especially those influenced by Charles Coulomb's view that electricity and magnetism were separate phenomena which couldn't interact. Within a few months the German scientist Johann Schweigger, professor of mathematics at the University of Halle, had shown that by forming a current carrying wire coil Oersted's magnetic effect could be magnified, or 'multiplied' within the coil. This led on to the development of the galvanometer, a device used to detect electric currents. One of the first scientists in France to read about Oersted's discovery was Dominique François Arago. Based on Schweigger's discovery, Arago showed experimentally that an iron bar placed along the axis inside a current carrying wire coil became magnetised. Arago went to Paris in the autumn of 1820 and reported to the French Academy on Oersted's work and his own observation that iron could be magnetised by placing it in a live coil. Later, in 1823, William Sturgeon, an English scientist and friend of Michael Faraday, developed Arago's observation to make the first useable electromagnet.

Oersted's amazing breakthrough set off a flurry of activity amongst scientists in Europe, especially in France where scientists were no longer blinkered by Coulomb's view. As well as repeating Oersted's experiment for themselves, several French physicists, led by André Ampère, developed a mathematical model of the magnetic effect associated with an electric current. An electric current i is defined as the rate of flow of electric charges Q past a fixed point.

$$i = \frac{dQ}{dt} \qquad (6.1)$$

If we consider a length ℓ of a thick wire of cross-sectional area A (Fig. 6.3), containing a total charge $Q = \rho V$, where ρ is the charge density and $V = A\ell$ is the volume, then

$$i = \frac{\rho A \ell}{t} = \rho A v \qquad (6.2)$$

where v is the average, or drift, velocity of the individual charges along the wire. Typically, drift velocities are a few metres per hour.

For a long straight wire, carrying a current i, the strength of the circular magnetic field H_o outside the wire is

$$H_o = \frac{i}{2\pi R} \qquad (6.3)$$

The variable R is the radial distance from the centre of the wire. The formula is known as Biot-Savart's law, named after the two French scientists most associated with its derivation.

The magnetic field exists inside the wire, too. Theory shows that the strength H_i of the internal circular magnetic field is given by

$$H_i = \frac{iR}{2\pi a^2} \qquad (6.4)$$

The cross-sectional area of the wire $A = \pi a^2$. As we will see later, the gravitational analogues of both of the Biot-Savart formulae are of interest.

In Germany, Oersted's discovery prompted the scientist Georg Ohm to begin his investigation of the electrical properties of different types of wire. He searched for an experimental relationship between the current i, which he measured through Oersted's magnetic effect, and the resistance R (not to be confused with radial distance), dependent on the length ℓ and inversely dependent on the thickness, or cross-sectional area A, of the wire under test, while keeping the voltage V constant.

The voltage V is the difference in electric potential $\Delta\phi$ between two points

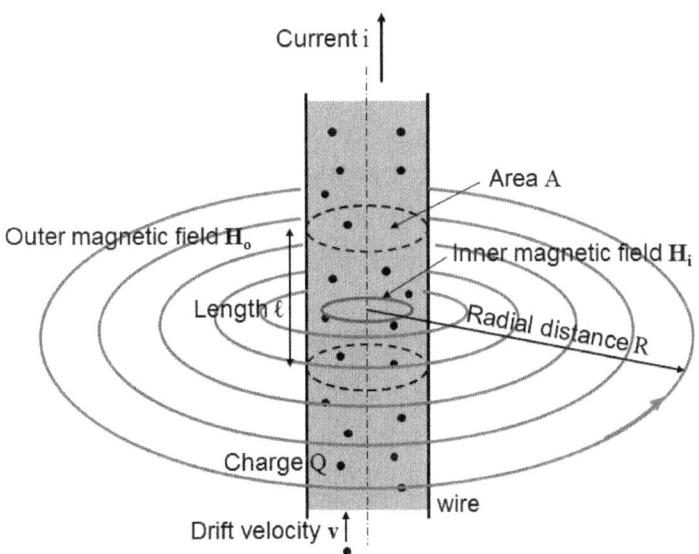

Fig. 6.3 Biot-Savart's law for a linear current.

along the wire. Over a length ℓ of the wire the potential gradient $\nabla\phi = V/\ell$ is due to an electric field $\mathbf{E} = -\nabla\phi$ (eqn 3.8) which forces free charges Q to flow along the wire (eqn 3.4). However, Ohm did not have this insight, which only came later. Instead, he based his idea on an analogue of the flow of heat, or caloric fluid, along a body, described in Jean Baptiste Fourier's famous French textbook, *The Analytical Theory of Heat*, published in 1822. Ohm assumed that the electrical conduction, or current i, was analogous to the thermal conduction, or flow of heat. Thus, the electric potential gradient $\nabla\phi$ was the analogue of the temperature gradient ∇T and the constant of electrical conductivity σ was the analogue of the constant of thermal conductivity κ. Thus, Ohm sought to determine the electrical conductivity σ of different types of metals in his experiments, defining the resistance R for particular wires as

$$R = \frac{\ell}{A\sigma} \tag{6.5}$$

Based on his experiments, Ohm published his law in 1827:

$$V = iR \tag{6.6}$$

Although this empirically derived law is now famous, at first it was ignored. It wasn't until 1841, when scientists at the Royal Society realised the importance of Ohm's law and made Ohm a member of the Society, that other scientists began to take notice.

We can rewrite Ohm's law in a less familiar vector form. The magnitude of the voltage V is Eℓ. This, together with the current i (eqn 6.2) and the resistance R (eqn 6.5) can be substituted into equation 6.6 to give

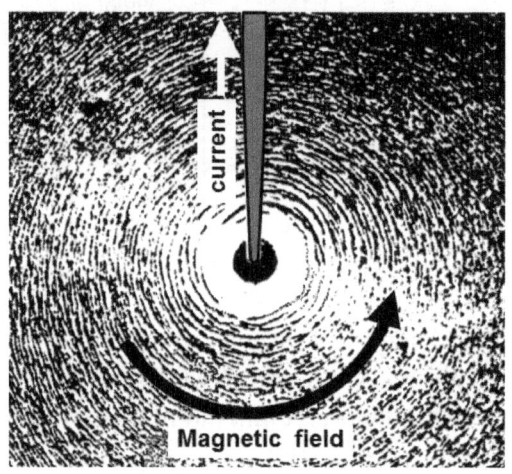

Fig. 6.4 Magnetic field around a linear current.

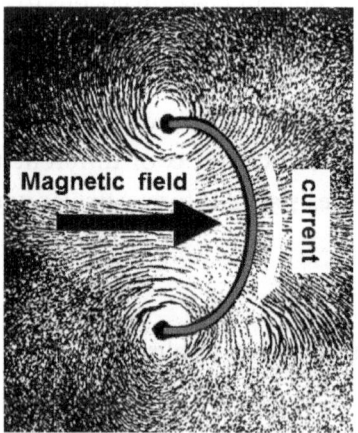

Fig. 6.5 Magnetic field through ring current.

$$\rho \mathbf{v} = \sigma \mathbf{E} \tag{6.7}$$

Like his colleagues in the Royal Institution, Faraday was fascinated by Oersted's discovery. Using Gowin Knight's technique, Faraday used iron filings to make the magnetic field visible around wires carrying a current. He noticed that there was a duality in the patterns.

(i) The current produced by an electric E-field along a straight conductor caused a circular magnetic H-field around it (Fig. 6.4).
(ii) The current produced by an electric E-field around a conducting ring caused a straight magnetic H-field along its centre line (Fig. 6.5).

We can see that the pattern is the same as that for the straight vortex core and the circulating streamlines and the vortex ring and the streamlines through it. The vortex core is analogous to a current carrying wire and the streamlines are analogous to the magnetic field. But this wasn't realised until later.

The link between the circular streamlines and the vorticity (eqn 5.2) can be written as

$$\nabla \times \mathbf{v} = \zeta, \tag{6.8}$$

The similarity in patterns suggests that the link between the circular, or curling, magnetic field lines and the electric field can be written as

$$\nabla \times \mathbf{H} = \sigma \mathbf{E}. \tag{6.9}$$

Using the vector form of Ohm's law (eqn 6.6), this can be written as

$$\nabla \times \mathbf{H} = \rho \mathbf{v}. \tag{6.10}$$

The curl of **H** describes Oersted's discovery exactly. For a steady current, where electric charge of density ρ flows with constant velocity **v** along a conductor, a circular magnetic field **H** is formed which curls around the path taken by the moving charge.

In 1821 Faraday made a novel discovery which, eventually, led to the invention

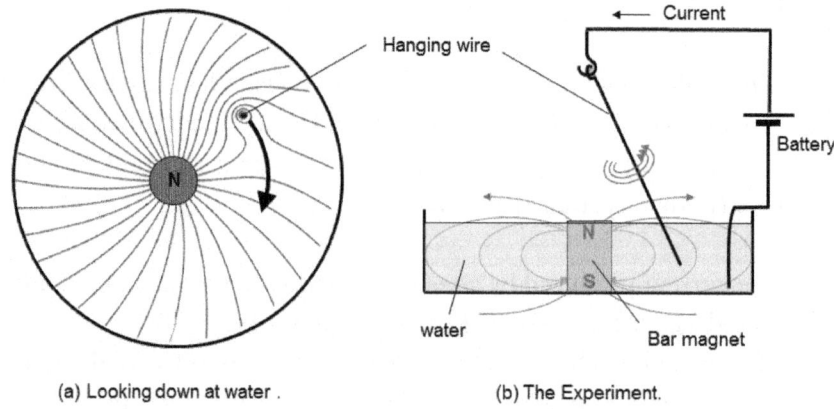

(a) Looking down at water.   (b) The Experiment.

Fig. 6.6 Faraday's simple motor.

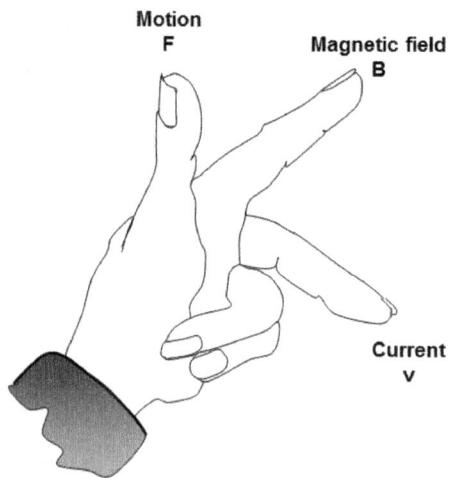

Fig. 6.7 Fleming's left hand rule.

of the electric motor. Through experiment (Fig. 6.6), he found that a dangling live wire would circle around the pole of a magnet. This is a fascinating experiment to repeat. A short bar magnet is stuck with its north-south poles upright in the centre of a Petri dish that is filled with salty water to make it conducting. A length of copper wire is freely suspended from a hook above the face of the magnet so that its other end just enters the water to the side of the magnet. Another piece of copper wire is bent over the side of the dish and into the water. When the two ends of the copper wires out of the water are connected to a battery, then the length of wire dangling down begins to rotate around the magnetic pole.

Faraday visualised the effect in terms of interacting magnetic fields. Near the top magnetic pole the magnetic field in a horizontal plane is roughly radial, while the magnetic field around the live wire is circular. The interaction of the two magnetic fields results in the wire moving perpendicularly (Fig. 6.6) to the radial field of the top magnetic pole. From our analogue with the vortex, we see that the movement of the wire is analogous to the movement of the vortex in the Magnus effect in fluids (Fig. 5.3).

Much later, Professor Sir John Ambrose Fleming of Imperial College, who invented the diode valve, introduced 'Fleming's left hand rule' to predict the direction of the force moving the live wire (Fig. 6.7). With the thumb and first and second fingers all at right angles we learn, "thu**M**b is **M**otion, **F**irst finger is **F**ield and se**C**ond finger is **C**urrent".

If we write the result in mathematical symbols we have

$$\mathbf{F} = Q(\mathbf{v} \times \mathbf{B}). \tag{6.11}$$

F is the force which gives rise to the motion of the wire, Q is the positive charge flowing along the wire, **v** is the velocity of the current of charge and **B** is the approximately radial magnetic field induced by the pole of the magnet. The pattern formed by the interacting magnetic fields is analogous to the pattern for the Magnus force in fluids.

Adding this force to the electric force, the total force experienced by a charge in electric and magnetic fields is called the Lorentz force **F** and is given by

$$\mathbf{F} = Q\mathbf{E} + Q(\mathbf{v} \times \mathbf{B}) \tag{6.12}$$

Sir Humphry Davy died in 1828 and, eventually, in 1833 Michael Faraday was chosen to fill the vacant post of Professor of Chemistry. Among a whole range of electrical studies, Faraday also continued the investigation of electrochemistry, so brilliantly initiated by Davy. In 1834 Faraday published his laws of electrolysis.

But there was a much bigger discovery waiting to be made. If there is a unity in nature, as all the leading scientists of Faraday's time felt that there should be, it seemed reasonable to suppose that if an electric current could cause a magnetic field then, somehow, it ought to be possible to reverse the effect and convert magnetism into electricity. But no one knew how to do it.

7
ELECTROMAGNETISM

Michael Faraday was mesmerised by the idea of turning magnetism into electricity. In 1825 he began a series of experiments aimed at finding an answer. He began by looking for a current induced in a wire by a neighbouring steady current. But he found nothing and had to put the work aside.

In 1831 Faraday decided to have another look. What prompted him to continue with his experiments may have been work with Charles Wheatstone (Wheatstone was the first Professor of Physics at King's College London) on the acoustical vibration patterns of Chladni plates covered in light sand. Faraday noticed that one plate set vibrating would induce a nearby plate to vibrate, too. Faraday looked on the magnetic field around a live wire (one with a current) as an electrical vibration effect so he may have reasoned that, if the analogue held true, the electrical vibration ought to induce an effect in a nearby dead wire (one not connected to a battery, so without a current).

He made a wooden frame and coiled a live wire on it, thus creating an electromagnet. On top of this he coiled a dead wire connected in series, in a closed circuit, with a galvanometer. As before, he began by experimenting with a steady current, but there was no effect. However, on switching the current in the live circuit on, or off, he happened to notice that the galvanometer needle in the dead circuit momentarily flicked from its central position, indicating a transient current. By chance, he had stumbled on the breakthrough that many other contemporary scientists were searching for. As the magnetic field grew, or collapsed, inside the live coil it induced a current in the otherwise dead coil connected to the galvanometer. He had made the discovery! How excited he must have been!

Faraday knew, from Arago's discovery, that he could make the magnetic field much stronger by placing a length of soft iron inside the coil. He made a soft iron ring and coiled the live wire around one side and coiled the wire connected to the galvanometer around the other. Whenever he switched the current on, or off, the changing magnetic field through the iron ring caused a transient current to flow in the galvanometer circuit. Once steady current was achieved the galvanometer needle settled back to zero, at the centre of the scale. Next he made a paper tube and coiled both the live wire and the wire to the galvanometer around it. He then placed a length of soft iron inside the tube. The effect was still there; changing the current in the live circuit produced a changing current in the otherwise dead circuit.

Finally, he took out the soft iron core and disconnected the live wire. Then he plunged a strong magnet into the paper tube, into the centre of the wire coils. Sure enough, as the magnetic field changed, the galvanometer needle momentarily flicked into motion. He pulled the magnet out and the galvanometer needle momentarily moved across the scale in the other direction.

Changing the magnetic field though the coil created a voltage in the wire of the coil, which induced a current to flow around the coil, creating a changing magnetic field exactly in opposition to the moving magnetic field (Fig. 7.1). In 1834 the Russian scientist Heinrich Lenz, professor of physics at the University of St Petersburg, explained that the meaning of Faraday's discovery was all to do with satisfying Newton's third law of action and reaction. If the north pole of a magnet approaches the coil, the induced current creates a north pole in the end of the coil facing the magnet, in order to oppose it, since like poles repel. When it's

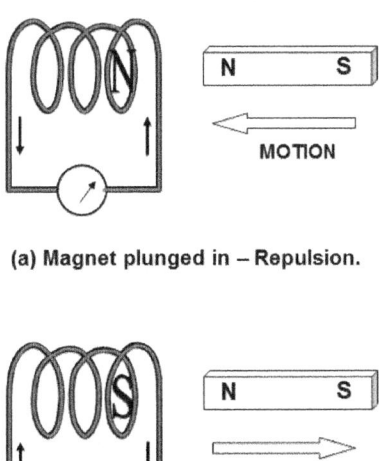

Fig. 7.1 *Relative motion between coil and magnet.*

the north pole end of a magnet being withdrawn from the coil, then the induced current creates a south pole, which attracts the magnet and opposes its motion.

So Oersted was right. There is a unity between electricity and magnetism. A steady current creates a magnetic field, but it needed a magnetic field changing with time to create an electric current.

Faraday now turned his attention to a variation of Arago's disc experiment. In 1825 Arago had discovered that when a magnetic compass was centrally placed on a glass plate held just above a rotating copper disc then the compass needle was deflected. When the copper disc had a high angular velocity, the compass needle rotated continuously in the same direction as the disc. The phenomenon puzzled many famous scientists, including Faraday.

Faraday arranged for the outer periphery of a revolving copper disc to pass between the poles of a very strong magnet (Fig. 7.2). Conducting strips in contact with the perimeter of the disc and the axle of the disc were connected to either side of a galvanometer, thus forming a closed circuit. As the disc rotated the galvanometer showed a constant deflection, indicating the generation of a constant current. Rotary mechanical motion had been converted in the flow of electricity. This was the world's first dynamo and it is now on display in the Science Museum in London.

With hindsight, we know that the movement of the magnetic needle seen in Arago's experiment was due to magnetic vortices, or eddy currents, created in the rotating copper disc. The magnetic fields they generated opposed the motion

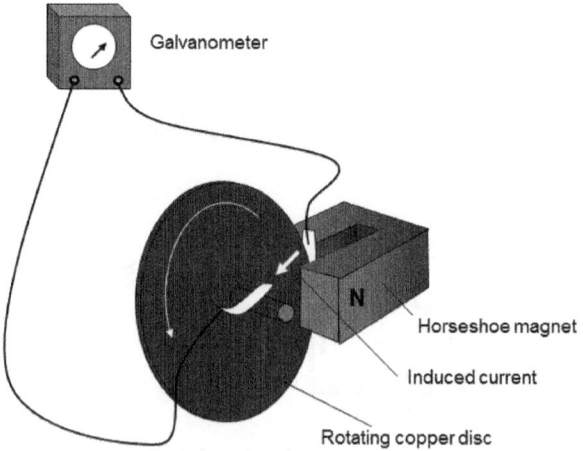

Fig. 7.2 Rotating copper disc between poles of magnet.

of the disc relative to the magnet. Faraday had tapped into these eddy currents.

In the USA, Joseph Henry, a professor of mathematics, was also conducting electromagnetic experiments. He had already greatly improved the design of the electromagnet, simply by using insulated wire in the coils. He had also developed the electromagnetic motor. While Faraday was carrying out speculative experiments on electromagnetic induction, so too was Henry. They both seemed to have discovered the secret at the same time – another example of synchronicity.

Very quickly, following the revelation of Faraday's and Henry's breakthrough, several practical electric generators were built. The first was made by the French scientist, Hippolyte Pixii, in 1832. It was capable of generating alternating current (ac) and direct current (dc), the latter requiring commutators, an idea devised by William Sturgeon. The combination of the steam engine, ideally suited to provide the power needed for rotary motion, and the electrical generator heralded the birth of the electrical power generation industry. This was the start of the Electromagnetic Revolution that has changed the world that we live in!

In 1855 the young Scottish mathematician, James Clerk Maxwell, had just graduated from Cambridge University. He was fascinated by Faraday's discoveries in electromagnetism. At that time there was still no complete mathematical model describing Faraday's results. Maxwell was very taken with the 'iron filings' magnetic field patterns made by Faraday and was aware of the analogy between vortex cores and electric field lines and between streamlines and the magnetic field lines. But the analogy is not exact, as there are no 'vortons' moving along a vortex core analogous to electrons moving along a wire.

For steady conditions the two analogies were

$$\nabla \times \mathbf{v} = \zeta, \qquad (7.1)$$

$$\nabla \times \mathbf{H} = \rho \mathbf{v} = \sigma \mathbf{E}. \qquad (7.2)$$

The electric vortex is shown in Figure 7.3(a), where a steady magnetic field $\mathbf{H}$ curls around a steady electric field $\mathbf{E}$ in a conductor. This was Oersted's discovery. Maxwell considered the duality between electric $\mathbf{E}$-fields and magnetic $\mathbf{H}$-fields. Perhaps there was a vortex-type analogue for $\mathbf{H}$, too. Was it possible that an $\mathbf{E}$-field curled around a steady $\mathbf{H}$-field, giving the following equation?

$$\nabla \times \mathbf{E} = \text{constant}.\mathbf{H} \tag{7.3}$$

But does this equation model nature? Faraday's experiments had shown that if the magnetic **H**-field remained steady then there was no effect. So the equation couldn't be true.

Maxwell thought about Faraday's experiment to induce a current in a conductor by locally changing the magnetic field. In mathematical terms, while keeping the position fixed, the magnetic field **H** changing with time t can be written as $\partial \mathbf{H}/\partial t$. Mathematicians use the curly ∂, rather than the straight d, to indicate that the function being considered is dependent on more than one variable. In the present case, the magnetic field strength **H** depends on the position (x,y,z) in space and time t. Now the induced electric field **E**, responsible for the induced current in the coil, curls around the opposite direction of the changing magnetic field (Fig. 7.3(b)). In mathematical form this is expressed as

$$\nabla \times \mathbf{E} = -\mu \frac{\partial \mathbf{H}}{\partial t} \tag{7.4}$$

So, there was a magnetic vortex! The constant μ is the magnetic permeability. The minus sign occurs because the induced current creates a magnetic field to oppose the motion of the magnet, as explained by Lenz.

Maxwell then realised why there was no steady term for **H**. Magnetic monopoles always appeared in pairs, one positive, the other negative, their sum being zero. There couldn't be a $\rho'\mathbf{v}$-type term. But, duality in pattern suggested that there might be an effect as follows:

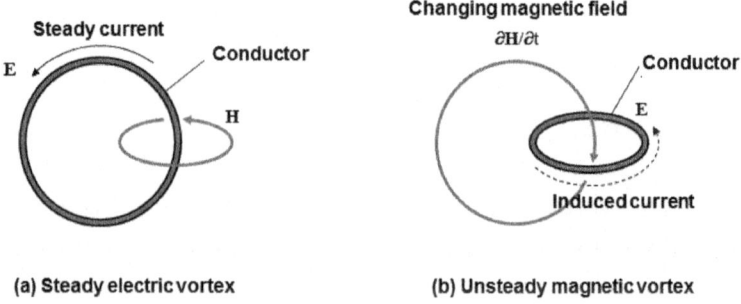

Fig. 7.3 *Electric and magnetic curls.*

$$\nabla \times \mathbf{H} = \rho \mathbf{v} + \varepsilon \frac{\partial \mathbf{E}}{\partial t} \qquad (7.5)$$

Indeed there was. Since the ρ**v** term represented the flow of charges, or the conduction current, then the ε ∂**E**/∂t term must represent some other form of current that was time dependent.

Today, we are familiar with the idea that atoms consist of a nucleus, containing neutrons and positively charged protons, around which orbit negatively charged electrons. Of course, this idea is just based on an analogue with the solar system, but it has proved to be a useful model.

Maxwell noted that some materials, called dielectrics, when placed in an electric field became slightly stretched, storing electric energy like a spring stores elastic energy. These materials are used in capacitors which store electric charge. In terms of our simple model of the atom based on a mini-solar system, we can see that within each atom the positively and negatively charged constituents are pulled in opposite directions by the external electric **E**-field and are slightly displaced from their natural position. Under this condition dielectrics form chains of atomic electric dipoles and the material is said to be polarised. During the stretching process, the slight movement of the atomic charges means that tiny currents occur. Maxwell called these tiny, short-lived currents displacement currents and identified them with the stretching caused by the changing electric field ∂**E**/∂t.

Maxwell assumed that the ether filling space was also a dielectric material which could be stressed, or polarised, by electric and magnetic fields.

Equations 7.4 and 7.5, when combined with equations 4.9 and 4.10, form Maxwell's famous set of equations for electromagnetic phenomena. Maxwell's theory was published in his *Electromagnetic Treatise* in 1864, using Cartesian components (the x, y and z spatial components of a vector) and quaternions, a rather obscure mathematical form particularly useful for dealing with rotations, as well as his curl notation. Oliver Heaviside, a self-taught English mathematical genius, converted Maxwell's theory into the simpler vector form, retaining the curl term. And it's this form of Maxwell's equations that engineers and scientists the world over have used ever since.

Coupling the curl equation 7.4 for **H** with the curl equation 7.5 for **E**, Maxwell derived an equation which described the propagation of electromagnetic waves (waves that had not been seen) travelling through the ether with speed c, given by

$$c = \frac{1}{\sqrt{\varepsilon\mu}} = 3 \times 10^8 \text{ m/s} \qquad (7.6)$$

Since c was the same as the known speed of light, Maxwell concluded that light was an electromagnetic wave and that there might be other forms of electromagnetic waves, too. Light is an extremely narrow band of waves in the electromagnetic spectrum, but the eyes of many animals are sensitive to the fluctuations in this band allowing us 'to see'. Perhaps there is a gravitational analogue of light in the gravitational spectrum to which some animals are sensitive. Are there any indications?

Probably the first scientist to detect an electromagnetic wave was Professor David E. Hughes, an Anglo-American scientist. Like a number of leading scientists, Hughes was a gifted musician, being a good pianist. Hughes was also a successful businessman, being the inventor and manufacturer of early electric telegraph printing machines. The teleprinter and the modern computer keyboard can be traced back to his inventions, which appeared before the invention of the typewriter.

In 1878 Hughes, living at the time in London, invented the carbon granule microphone. While experimenting with the device, he heard 'clicks' on it which he traced back to a loose connection on an induction coil used to provide a high voltage current. Hughes completely disconnected his microphone, but still heard clicks on it when he caused a spark to appear at the loose connection of the coil. Knowing of Maxwell's prediction of electromagnetic waves, Hughes thought that he might possibly have detected them with his carbon granule microphone. In 1880 he demonstrated his findings at a meeting of the Royal Society. However, several distinguished scientists in the audience, including the President, Sir George Stokes, did not agree with Hughes' view. This discouraged Hughes who didn't pursue his research any further. With hindsight, we can be fairly sure that Hughes had been right, but unfortunately for him he was not sufficiently confident enough to be able to grasp the offering made by serendipity.

In 1884, with the search for Maxwell's predicted waves still ongoing, the English theoretical physicist Professor John Henry Poynting derived a formula for the power P, or rate of flow of electromagnetic energy dE/dt, crossing a surface area S (Fig. 7.4) perpendicular to an advancing electromagnetic wave.

$$P = \frac{dE}{dt} = (\mathbf{E} \times \mathbf{H}) \cdot \mathbf{S} \quad \text{watts} \qquad (7.7)$$

ELECTROMAGNETISM

The story of how, in 1888, the German physicist Professor Heinrich Hertz detected Maxwell's electromagnetic waves is one of the high points of 19th-century experimental science. Let us start by considering waves that we can see in the form of ripples on the surface of water. If we poke a stick into the water and draw it up then, due to surface tension, some of the water will be drawn up, too. If we rapidly oscillate the stick vertically, transverse water waves will radiate out in a circular pattern from the stick. By transverse we mean that the waves, moving up and down, are perpendicular to the outward direction in which they travel. Likewise, if we rapidly oscillate a current in a wire then, above a certain frequency, transverse electromagnetic waves will radiate out from the wire. In this case, the electric E-wave and the magnetic H-wave are both transverse to the direction of travel and are perpendicular to each other (Fig. 7.4).

To create a high frequency oscillating current i, Hertz used a giant capacitor, with two large flat metal plates, A and B, arranged vertically with one above the other (Fig. 7.5). The two plates were oppositely charged using an induction coil and the vertical connecting wire separated by a spark gap C. As the plates were charged, at a certain voltage a spark appeared across the gap and the current surged from one plate to the other, reversing the plate voltage, only to surge back in the opposite direction. The current oscillations continued until the

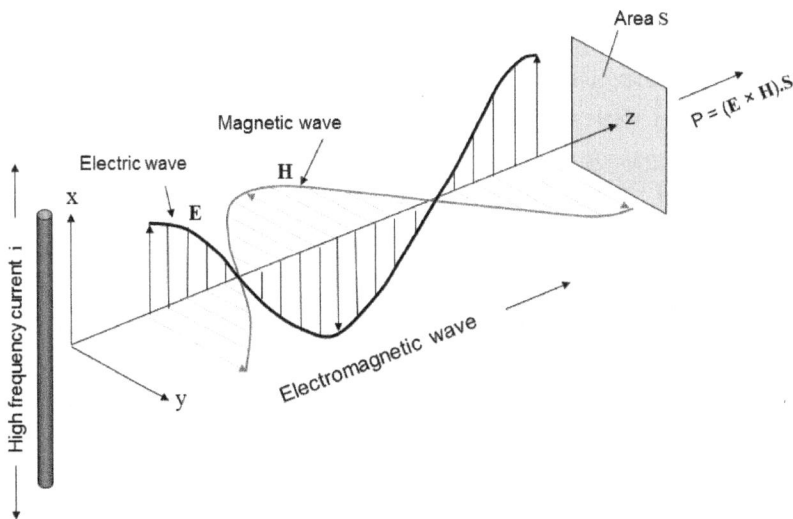

Fig. 7.4 The electromagnetic wave.

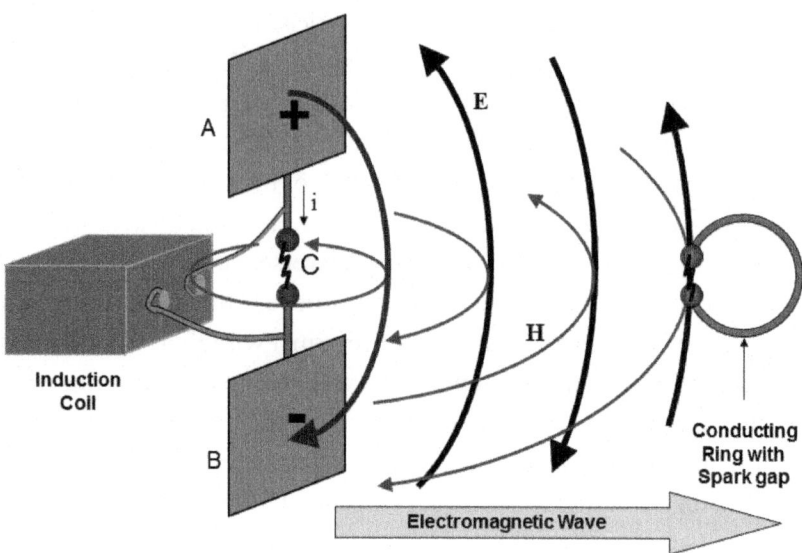

Fig. 7.5 Hertz's wireless waves.

voltage dropped to a level below which the spark could not form across the gap. During the high frequency current oscillations in the connecting wire electromagnetic waves were transmitted.

Hertz reflected the waves back from a flat metal surface, so that the outward waves interfered with the reflected waves to form standing waves, so-called because they are frozen in position (Chapter 10). Finally, to detect the presence of the waves, Hertz used a conducting ring with a spark gap, rather like an open copper bracelet with knobs on the ends. In his darkened laboratory, Hertz moved the ring along the path of the waves searching for a position where a spark appeared. At the anti-nodes of the standing waves, where the waves reinforced each other, a current was induced in the ring. By using a ring circumference with the right dimensions a resonance effect was realised, akin to tuning-in, thereby maximising the induced oscillating ring current and resulting in a tiny spark appearing across the gap. Finding adjacent anti-nodes, Hertz was able to determine the wavelength of the electromagnetic waves, which were microwaves, later associated with radar technology.

Today, technologies based on the use of electromagnetic waves across a wide spectrum of frequencies have created huge new markets and global business empires. This pattern is likely to repeat itself when the breakthrough in gravitational physics is made.

8

GRAVITY AND THE MISSING CURLS

Michael Faraday was convinced that gravity was related to electricity and magnetism and even, possibly, to heat. In his laboratory diary, Faraday's entry for 19 March 1849 (paragraph 10018) reads, "Gravity. Surely this force must be capable of an experimental relation to electricity, magnetism and the other forces, so as to bind it up with them in reciprocal action and equivalent effect. Consider for a moment how to set about touching this matter by facts and trial". Over a period of ten years, between 1849 and 1859, he carried out a series of experiments searching for a connection between these force fields. However, Faraday's research into unifying gravity with electricity, magnetism and heat was not successful, unlike the dazzling successes he achieved in other areas of experimental science.

Although most scientists continue to believe that it should be possible to convert one force field into another, the riddle of unification has proved to be such a difficult subject to unravel that very few scientists have continued the attempt to find an answer. However, rather than tackling the mystery of force-field unification head-on, some scientists have investigated the force of gravity alone, to see whether it is open to being manipulated in a way similar to that for electromagnetism. But they, too, have hit a brick wall and gravity remains as inscrutable as ever.

Nevertheless, as a first step, it would seem sensible to concentrate on the gravity problem first. One factor to bear in mind is that in previous experimental studies detector sensitivity might not have been sufficient to detect effects that were looked for. Or, maybe, a new detection technique is needed. In planning a programme of research we must have a simple model, or models, in mind to investigate. A model with a theoretical basis is even better. This is where analogues are particularly useful. In an experimental programme to test the validity of a proposed model it is most unlikely that it will survive in its original form, but the history of speculative scientific investigation shows that serendipity often lends a hand, with an unpredicted experimental result throwing new light on a phenomenon which then leads on to a breakthrough in understanding.

If we place a force field source or sink at a point in the ether, or the vacuum of space, the local conditions are disturbed from the uniform condition of emptiness. The presence of the source or sink effectively stresses the surrounding region and the field contains potential energy to do work. The effect of a source is rather like driving a round peg into a block of rubber, where work is done to displace and locally compress the rubber. If the peg is removed the stored energy of the compression is released and work is done to return the rubber to its original condition. The first person to think about turning Maxwell's equations for electromagnetism into an analogous set of equations for extending Newton's theory of gravity was James Clerk Maxwell himself. Maxwell assumed that electric, magnetic and gravitational force fields all stored potential energy in the ether.

Energy density is the energy contained in a volume of space divided by the volume. In the case of static electrical **E**-fields created by distributions of positive and negative charges the energy density is positive.

$$\text{Electrical energy density} = \frac{\varepsilon}{2} E^2 \text{ Joules/m}^3 \qquad (8.1)$$

In the case of static and steady magnetic **H**-fields the energy density is also positive.

$$\text{Magnetic energy density} = \frac{\mu}{2} H^2 \text{ Joules/m}^3 \qquad (8.2)$$

In the case of periodic electromagnetic waves, anticipated by Maxwell and discovered by Hertz, the moving energy density is also positive, being the sum of the averaged energy densities over a wavelength for each time-cycling **E**- and **H**-field.

However, for static gravitational **g**-fields created by distributions of positive masses and negative masses (if such phenomena exist) the energy density is negative. This is because like masses attract (Table 3.14).

$$\text{Gravitational energy density} = -\frac{\gamma}{2} g^2 \text{ Joules/m}^3 \qquad (8.3)$$

Although Maxwell did not concern himself with the possible existence of negative mass, he was confronted with the notion that the vast ethereal region of empty space between the stars must contain a huge store of negative energy. This perplexed Maxwell, who felt that energy needed to be a positive quantity. To keep the overall gravitational energy content of space positive, or zero, Maxwell wondered whether the ether was endowed (somewhere, somehow) with an enormous amount of hidden positive energy. But this thought was too much for Maxwell to contemplate, so he didn't pursue the idea of an electromagnetic analogy for gravitation very far. Anyway, he had other important matters to deal with, namely making the Cavendish Laboratory at Cambridge University a world-class centre of electromagnetic research, which he succeeded in doing before dying at the early age of 48, in 1879.

In 1905 Einstein published a short paper with the title 'Does the inertia of a body depend upon its energy content?' Beginning with the energy in a beam of light, Einstein derived an equation linking energy with mass (eqn 8.8), which he then extended to all forms of energy.

Figure 8.1 shows the main points of a rough method used to obtain the mass-energy relationship.

Let us begin with the observation that light is energy in motion. From experiment, we know that if a beam of light illuminates a surface area A the surface experiences a slight force F, referred to as radiation pressure $p = F/A$.

From Newton's second law (eqn 2.15) we know that a force F is associated with a mass m, so we conclude that the incident beam of light giving rise to

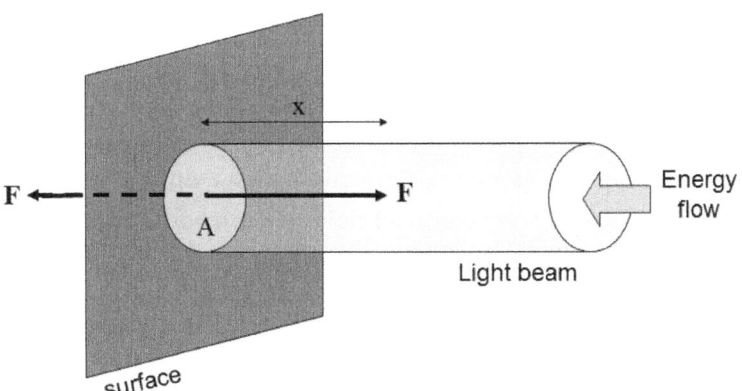

Fig. 8.1 Inertia of a light beam and radiation pressure.

radiation pressure possesses effective mass. (Note: There are other ways of creating force so we can't be certain that real mass is involved.) By continuity, we then conclude that the whole beam of moving energy (now thought of in terms of photons) contains effective mass.

The velocity of the beam of light moving in the x-direction, say, is

$$c = \text{speed of light} = \frac{dx}{dt} \quad (8.4)$$

But, Einstein asserted that the speed of light is constant, so in applying Newton's second law, to determine the force F exerted by the incident light beam, it must be the effective mass m that changes with time t, not the velocity c.

$$F = \text{rate of change of momentum with time} = \frac{d(mc)}{dt} = \frac{dm}{dt} c \quad (8.5)$$

In words this says that the force experienced by the illuminated surface is equal to the rate at which it receives momentum mc from the beam of light.

From Newton's third law the force experienced by the illuminated surface is equal and opposite to the force experienced by the incident beam. Now a gradient in energy E gives rise to a force (eqn 2.11). So, using equation 8.5, over the short length dx of the incident beam, we can write

$$\nabla E = \frac{dE}{dx} = F = \frac{dm}{dt} c \quad (8.6)$$

On rearranging equation 8.6 and substituting from equation 8.4 we get

$$dE = dm \frac{dx}{dt} c = dm c^2 \quad (8.7)$$

Now, providing the elemental effective masses dm (think in terms of photons) do not possess a gravitational field, we can add a collection of such masses together to get Einstein's famous equation

$$E = mc^2 \quad (8.8)$$

This result applies to the whole beam. Well away from the surface the effective mass of the beam remains unchanged. Note that if the elemental effective masses do have a gravitational field, then we cannot just sum the energies of a collection of them to get the total energy, as negative energy (eqn 8.3) is also involved.

Einstein later argued that since solid bodies could lose energy by radiation (including radioactivity) then inert solid bodies could be treated as lumps of frozen energy. So, without any proof, Einstein declared that the mass of a solid body determined its energy content, via equation 8.8. Experiments have confirmed the inverse of Einstein's assumption, namely that increasing the energy of a body results in it increasing its mass (eqn 9.6). Thus, it seems, energy E and mass m are interchangeable. It is claimed that the mass-energy relationship explains the source of the Sun's radiant energy.

Finally, Einstein claimed that all forms of energy, including the energy contained in static electric, magnetic and gravitational fields all obeyed equation 8.8.

We must be careful how we interpret Einstein's mass-energy equation. It implies that a region containing a distribution of positive or negative energy density may be thought of as containing a distribution of positive or negative effective mass. But whether an effective mass creates its own gravitational field is unproven. In Einstein's theory of gravity it is assumed that "gravity gravitates". By this is meant that since a static gravity field surrounding a mass contains negative energy then, by equation 8.8, the surrounding region contains effective negative mass which creates its own gravitational field, thereby altering the fundamental gravity field around the mass. Allowing for this makes Einstein's gravitational theory non-linear and rather complicated.

The model of gravity explored in this book is a linear theory. Following from equation 8.8, an energy field of any form is still assumed to give rise to an equivalent, or effective, mass distribution, but since it is not a distribution of real mass it does not create its own gravitational field. So, in the linear theory of gravity, "gravity does not gravitate". However, effective mass does respond to the influence of gravity. For example, the moving energy in an electromagnetic field has an effective mass which will be influenced by gravity.

The English mathematician, Oliver Heaviside, a nephew of Charles Wheatstone (he who developed the Wheatstone bridge, in 1843, to measure electrical resistances very accurately), was less inhibited by the idea of negative gravitational energy. In a short comment in Appendix B in Volume 1 of his book

Electromagnetic Theory, published in 1893, Heaviside noted, "Now what is there analogous to magnetic force in the gravitational case? And if it has an analogue, what is there to correspond with electric current? At first glance it might seem that the whole of the magnetic side of electromagnetism was absent in the gravitational analogy. But this is not true".

Based on our analogue approach we have treated the steady current of electrical charge along a wire, with velocity **v** and density ρ, as though it was a vortex core, creating a magnetic force field **H** around the wire like streamlines.

In mathematical terms we wrote

$$\nabla \times \mathbf{H} = \rho \mathbf{v} \tag{8.9}$$

Extending the vortex analogue into the realm of gravitation suggests that for a current of mass flowing along a pipe we can write

$$\nabla \times \mathbf{h} = -\rho \mathbf{v} \tag{8.10}$$

In equation 8.10 the parameter ρ is now the mass density, **v** is the mass velocity and **h** is a force field around the piped flow of mass (Fig. 8.2). The force field **h** is the absent part of the electromagnetic analogue of gravity that Heaviside referred to. The negative (-) sign occurs because gravitation is an attractive force between like masses (eqns 3.14(a) and 3.14(d)). Since **h** is based on an analogue with the magnetic field, it is called the gravitomagnetic field (note the use of Arial font for the gravitomagnetic field). This curl term is

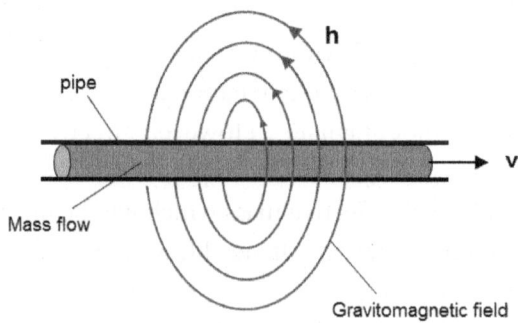

Fig. 8.2 Gravitomagnetic field around a mass current.

missing in Newtonian gravitational theory. But is **h** a real force field or is it merely a theoretical construction?

We are not aware of any natural sources, or monopoles, of gravitomagnetism **h**. But, assuming the **h**-field is real, the analogue with magnetism suggests that only gravitomagnetic dipoles can exist, although we are not aware of any such objects at the macroscopic scale (Chapter 16). For theoretical purposes, however, we will assume that gravitomagnetic monopoles satisfy the inverse square law. Following the pattern with the other inverse square laws, the field strength of a gravitomagnetic monopole p (the analogue of the theoretical magnetic monopole) is

$$\mathbf{h} = -\frac{1}{\eta}\frac{p}{4\pi r^2}\hat{r} \text{ kg/ms} \qquad (8.11)$$

The factor η is called the gravitomagnetic permeability or, as suggested by Professor Eric Laithwaite of Imperial College London, the inertial permeability. The SI units of η are m/kg. The negative (-) sign is introduced because we suspect that gravitomagnetic monopoles will follow the pattern of like poles attracting each other and unlike poles repelling each other.

But what is this field **h** that we have introduced? At this stage, all we can do is guess, again. From equation 8.10, using the dimensional terms of mass M, length L and time T, the dimensions of **h** are $ML^{-1}T^{-1}$. But these are the dimensions of angular momentum density. This suggests that the **h**-field has something to do with rotation, or torsion. Let us assume that at any point in space the field intensity **h** represents the angular momentum there of a moving mass occupying a volume V. As an example, consider a distributed mass m moving with speed v, then at some point of space a distance r away the magnitude of the gravitomagnetic field is given by

$$\mathbf{h} = \frac{mrv}{V} \qquad (8.12)$$

Like gravity, the energy density of a static gravitomagnetic field **h** is assumed to be

$$\text{Gravitomagnetic energy density} = -\frac{\eta}{2}h^2 \text{ Joules/m}^3 \qquad (8.13)$$

Field theory allows for action at a distance. Gravitational effects extend across space through a **g**-field. So, we assume that angular momentum-type effects can be transmitted across space in the form of an **h**-field. To detect the presence of a magnetic field **H** which permeates through space it must induce an effect in a sensing device, say a magnetic compass. The strength of the induced magnetic field is **B** = μ **H**, where μ is the magnetic permeability. The analogue with magnetism suggests that an induced gravitomagnetic field can be written as **b** = η **h**, where η is the gravitomagnetic permeability. The induced gravitomagnetic field **b** has dimensions of T^{-1}, the same as those of angular velocity, or of spin, with units of radians/second.

Newton's model of gravity assumes that disturbances in the gravitational field are broadcast instantaneously to every part of the Universe, implying an infinite speed. This doesn't accord with our natural experience, which shows that the effect of a disturbance takes time to travel outward from the source of the disturbance. In the electromagnetic case, disturbances in the electric field are transmitted as electromagnetic waves which travel at the speed of light c (eqn 7.6), where $c = 1/\sqrt{\epsilon\mu}$. By analogy with electromagnetism, our simple extended model of gravitation suggests that disturbances in the gravitational field are transmitted as gravitoelectric-gravitomagnetic (shortened to gravitational) waves which also travel at the speed of light, where

$$c = \frac{1}{\sqrt{\gamma\eta}} \quad (8.14)$$

This is an extension of the Newtonian model.

But what is the value of the gravitomagnetic permeability η in free space? We know (from eqn 2.3) that $\gamma = 1.193 \times 10^9$ kg m^{-3} s^2 and that $c = 3.0 \times 10^8$ m/s in space, which means that

$$\eta = \frac{1}{\gamma c^2} = 0.74 \times 10^{-27} \text{ m/kg} \quad (8.15)$$

To determine the induced effect that an **h**-field has on a body, across space we must multiply the **h**-field by 0.74×10^{-27} m/kg. The resulting induced gravitomagnetic field **b** = η**h** is miniscule, which explains why we don't see any strange effects on bodies caused by ordinary-sized bodies moving slowly nearby. It's just possible that planetary-sized moving bodies might have an effect.

However, we should note that although the magnetic permeability μ of the vacuum of space is small, within some bodies the magnetic permeability can be quite large, perhaps 1000 times greater than the free space value. Maybe something similar is true for the gravitomagnetic permeability η, and its magnitude inside some bodies may be much greater than the free space value.

Based on the analogue with electromagnetism (eqn 7.4 and 7.5) we can expect that for unsteady conditions the following two equations with curls arise in our modified Newtonian gravitational theory:

$$\nabla \times \mathbf{h} = -\rho \mathbf{v} - \gamma \frac{\partial \mathbf{g}}{\partial t} \qquad (8.16)$$

$$\nabla \times \mathbf{g} = = \eta \frac{\partial \mathbf{h}}{\partial t} \qquad (8.17)$$

In the first equation (eqn 8.16) if gravity **g** is constant, so that $\partial \mathbf{g}/\partial t = 0$, then we have equation 8.10 which models the gravitational analogue of Oersted's observation that a circular magnetic field surrounds a steady electric current. On the other hand, $\partial \mathbf{g}/\partial t$ is equivalent to jerk, the rate of change of acceleration, so even if there is no continuous mass current a jerked mass should induce a curled gravitomagnetic field **h**. Due to the magnitude of γ (eqn 2.3), the induced gravitomagnetic field may be large, but only short-lived.

The second equation (eqn 8.17) is the gravitational analogue of Faraday's law of induction. It tells us that if we knew how to change the **h**-field then we could create a curled gravitational field **g**. In other words, this equation is the key to the technology of creating gravitational fields! We can also write the equation as

$$\nabla \times \mathbf{g} = \frac{\partial \mathbf{b}}{\partial t} \quad \text{since } \mathbf{b} = \eta \mathbf{h}. \qquad (8.18)$$

From equation 8.17 and Table 3.14 the following observations are made for the inertial forces involving accelerating point masses. In the case of the dipole, the forces experienced by the positive and negative masses (tending towards coincidence in position) are in opposite directions, so that the overall inertial force experienced by the dipole is zero.

An accelerating mass m experiences an inertial force $F = mg_i$. (8.19a)
A mass m falling in a gravity field **g** experiences no inertia. (8.19b)
An accelerating mass -m experiences an inertial force $F = (-m)(-g_i)$. (8.19c)
A mass -m falling in a gravity field **g** experiences no inertia. (8.19d)
An accelerating gravitational dipole experiences no inertia. (8.19e)

For a mass m in a gravitational field we write the force equation as $\mathbf{F} = m\mathbf{g}$. But, if a body accelerates we can also write $\mathbf{F} = m\mathbf{a}$, where **a** is the acceleration. At any moment at any point in space, we can't say whether our feeling of weight is due to our being in a gravity field or to our being accelerated. In fact, acceleration can be used locally to transform away a gravity field. So gravity and acceleration are equivalent. But, they are not the same phenomenon!

In the same way that magnetism **H** is the dual of electricity **E**, gravitomagnetism **h** is the dual of gravitation **g**. Therefore, since **g** has an equivalence property, we can expect that **h** has one, too. But what is it? Dimensional analysis suggested that **h** might be equivalent to angular momentum density (eqn 8.12). This hints at a link with Planck's constant h in quantum mechanics, which may be a clue to a theory of quantum gravity. Dimensional analysis also showed that the induced gravitomagnetic field **b** had dimensions of T^{-1}, which is the same as that for angular velocity Ω. With hindsight, we can guess that induced gravitomagnetism is equivalent (the sign ≡ signifies equivalence) to angular velocity, given by

$$\mathbf{b} \equiv -2\Omega. \qquad (8.20)$$

Consider a torsion balance with equal masses at each end of the rigid lightweight cross-bar (Fig. 8.4), immersed in a vertical gravitomagnetic field **h**. If the gravitomagnetic field changes with time, $\partial \mathbf{h}/\partial t$, a gravitational field **g** is induced (eqn 8.17), which curls around the vertical and forces the two masses to rotate causing the suspension wire to twist, generating a reaction torque in the wire.

Now let's consider the same situation in terms of our equivalence with the gravitomagnetic field. Suppose we suspend the torsion balance from a frame mounted on a turntable (Fig. 8.5).

Under steady conditions, while the angular velocity Ω remains constant, there is no twist in the suspension wire (ignoring any wind resistance). The masses at each end of the rigid lightweight bar both rotate with angular velocity

Ω and have tangential velocity **v**. In mathematical terms the equation describing the motion is

$$\nabla \times \mathbf{v} = 2\Omega \qquad (8.21)$$

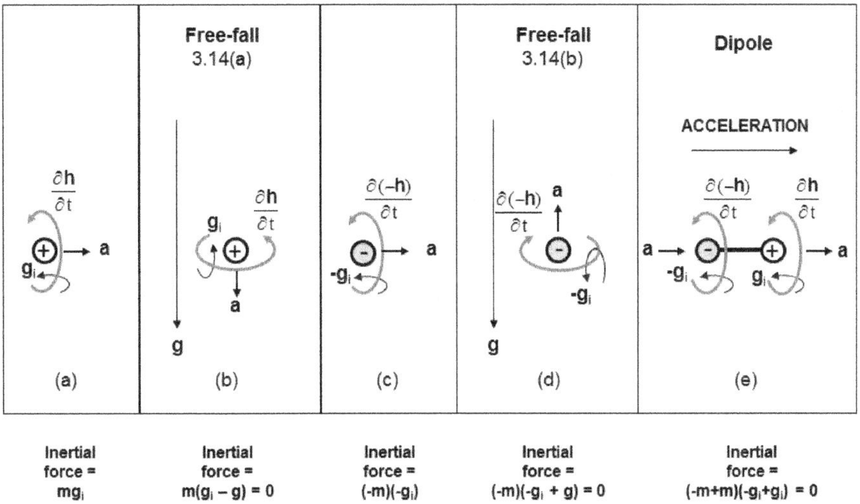

Fig. 8.3 *The force of inertia for point masses.*

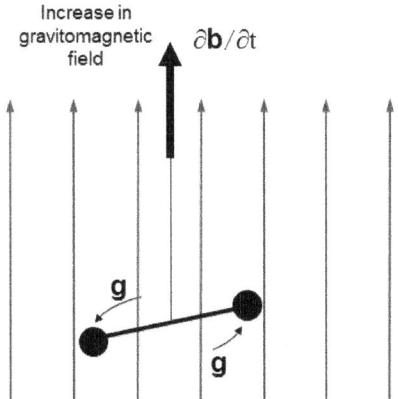

Fig. 8.4 *Torsion balance in a changing gravitomagnetic field.*

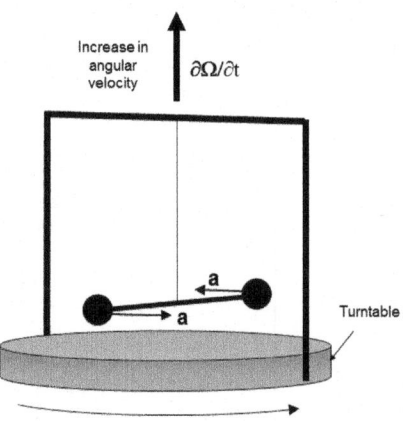

Fig. 8.5 Torsion balance mounted on a turntable.

Suppose that the angular velocity Ω of the turntable, is made to increase with time, as $\partial\Omega/\partial t$. The resulting twist of the suspension wire is communicated to the two masses causing them to increase their tangential velocity **v** with acceleration **a** = $\partial\mathbf{v}/\partial t$ (Fig. 8.5).

Thus equation 8.21 becomes

$$\nabla \times \mathbf{a} = \frac{\partial(2\Omega)}{\partial t} \tag{8.22}$$

If we replace -**a** with **g** and -2Ω with **b** we get the equation for the equivalent motion

$$\nabla \times \mathbf{g} = \frac{\partial \mathbf{b}}{\partial t} \tag{8.23}$$

This is the inertial analogue of Faraday's law of induction (eqn 8.18), which gives support to our guess (eqn 8.20) that angular velocity -2Ω is equivalent to induced gravitomagnetism **b**.

There is a difference between the two cases. In the gravitomagnetic case the masses, under the influence of a force field, accelerate, causing the suspension wire to twist. In the equivalent angular velocity case, the suspension wire twists, causing the masses to accelerate.

We must remember that -2Ω and **b** are only equivalent; they are not the same phenomenon. There is no influence field for -2Ω, the same as there is no influence field for acceleration **a**. We can't project acceleration or angular velocity across space. The equivalent effects can only be incurred by direct contact with a mass.

Let's explore the equivalent form of this gravitational analogue a bit further. A free electric charge moving with velocity **v** in a magnetic field **B** experiences a continuous sideways force $\mathbf{F} = Q(\mathbf{v} \times \mathbf{B})$, causing the charge to circle around a **B** field line (eqn 6.11). This is the magnetic part of the Lorentz force.

The gravitational analogue of the Lorentz force suggests that a free mass m, moving with velocity **v** across a gravitomagnetic field, should experience a force **F** given by

$$\mathbf{F} = m(\mathbf{v} \times \mathbf{b}). \qquad (8.24)$$

In which case, a free-moving mass should circle around a **b** field line. Well, we know that this is true in the equivalent case! This is just the Coriolis force $\mathbf{F} = m(\mathbf{v} \times 2\Omega)$ (eqn 5.1) that causes hurricanes to form (Fig. 5.1).

During the 1974 Royal Institution Christmas Lecture, Professor Laithwaite pondered on the possibility that a precessing flywheel might lose weight, arousing much controversy at the time. In the Professor's demonstration, shown in Figure

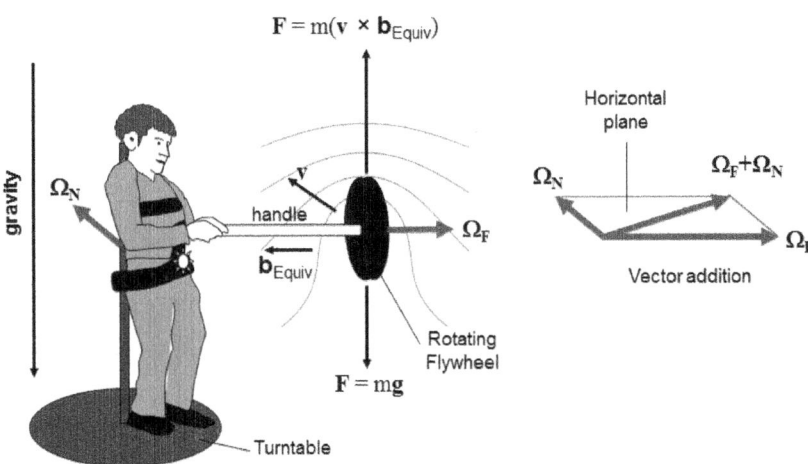

Fig. 8.6 *Professor Laithwaite's demonstration at the Royal Institution.*

8.6, a boy (actually Professor Laithwaite's son) was stood on a turntable, strapped to a vertical support and given a flywheel attached to a long handle. When the flywheel was not spinning, it was far too heavy for the boy to hold it out horizontally. Yet, when the flywheel was spinning and precessing horizontally the boy had no difficulty in holding it out in front of him. Had the flywheel lost weight? The Professor's demonstration was a large-scale version of the spinning toy gyroscope, with its axle horizontal and one end located on the top of a miniature Eiffel Tower, which rotates round in a horizontal circle, apparently defying gravity. The key principle to explaining the phenomenon is that the rate of change of angular momentum of the flywheel is equal to the moment of the applied forces acting on the flywheel. The applied force is the weight of the flywheel, while the moment results from the point of application of the force on the flywheel being transferred a distance along the axle (called the moment arm) to the top of the tower. The reaction, or rate of change of angular momentum, always takes place at right angles to the plane containing the moment arm and the force. If left undisturbed this is naturally horizontal. If the overall force is increased by pushing the gyro axle in the horizontal direction of motion, the summation of forces moves out of the horizontal plane with the curious result that the gyro rises.

In more simple terms, since angular velocity Ω has both magnitude (rpm) and direction (axis of spin) it is a vector. Suppose the flywheel's angular velocity is Ω_F and that we start with the axle horizontal. In the natural case, due to gravity, the flywheel attempts to fall. In doing so it pivots about its axle end and thereby develops an instantaneous angular velocity, say Ω_N, with its axis in the horizontal plane and at right angles to Ω_F. From vector addition (Fig. 8.6) of Ω_F and Ω_N, since both angular velocities lie in the horizontal plane the resultant angular velocity $\Omega_F + \Omega_N$ will also lie in the horizontal plane, and the flywheel's axle (or handle, in Laithwaite's demonstration) will instantly rotate around horizontally to its new position. Thus, the flywheel never actually gets the chance to fall, always moving on. But has the spinning flywheel lost weight in the process?

The basic weight of the flywheel is m**g**. Controversy arises if an attempt is made to use the gravitational analogue of the Lorentz force to argue that it modifies the flywheel's weight. From equation 8.24, suppose we assume that the velocity **v** is the horizontal precessional speed of the flywheel and that the equivalent induced gravitomagnetic field of the flywheel is $\mathbf{b}_{Equiv} = -2\Omega_F$. We might then attribute the flywheel's loss of weight to its moving through its own gravitomagnetic field. Compare this with the analogue of the Magnus effect (Fig. 5.3(c)). So, was the Professor right all those years ago? Gyro experts are adamant that he was not. The velocity **v** must be measured relative to the Ω_F axis, rather than relative to a stationary set of coordinates, so that **v** is zero and there is no weight change.

Although we can't generate a detectable gravitomagnetic, or torsion, field across space, from equivalence we do know what sort of effects to expect.

In 1916, only a year after Einstein had published his theory of general relativity, he derived a solution which showed that gravitational waves could exist in space-time. At first, some of the scientists following Einstein's theory were sceptical and suggested that the solution arose due to the co-ordinate system and was merely a mathematical artifice rather than a model of a real phenomenon. Indeed, in 1936 even Einstein expressed his own doubts about the existence of such waves. Since then there have been numerous disagreements between the experts about how the theory should be interpreted, although they now generally agree that gravitational waves should exist. Because of the complexity of Einstein's theory it seems a reasonable idea to investigate the linearised version of gravitational theory, based on the electromagnetic analogue, to try to gain some insight into the possible phenomenon of gravitational waves in three-dimensional space (not four-dimensional space-time!).

Following the pattern set by electromagnetism, if the curled equations for gravitation (eqn 8.16) and gravitomagnetism (eqn 8.17) are coupled together they lead to transverse wave equations for **g** and **h**. With this model, a periodic gravitational wave is a combination of **g** and **h** waves in planes at right angles to one another. Figure 8.7 shows a vertically polarised linear gravitational wave, with a **g**-wave in the vertical xz-plane, with amplitude g_0, and an **h**-wave in the horizontal yz-plane, with amplitude h_0. The gravitational wave travels in the z-direction and, at first glance (based on eqns 8.3 and 8.13), it looks as though the energy transported by the wave is negative but this is not so. According to equation 8.20, we can treat the induced gravitomagnetic component as being equivalent to a negative angular velocity, say -2Ω. In the gravitational wave case, $\Omega \equiv \omega$ is the angular frequency of the wave. Then the time-averaged energy density of the gravitational wave is $+h_0\omega$. It may seem rather strange but when a gravity field is disturbed the redistribution of the negative energy is achieved with the movement of positive energy.

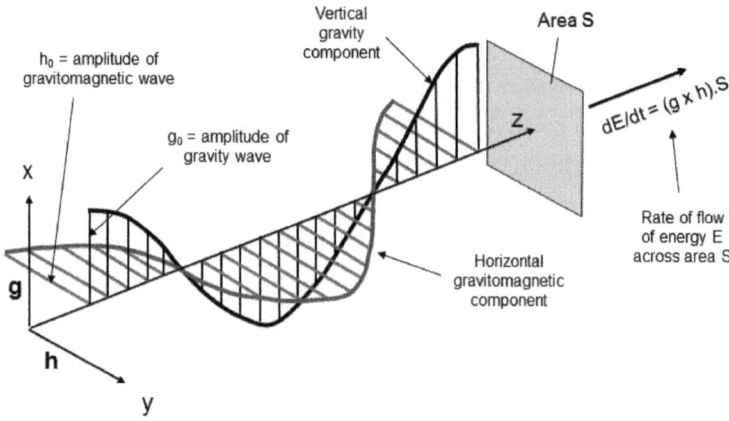

Fig. 8.7 Vertically polarised linear gravitational wave.

To clarify the above, consider the following. At any point in the gravity field surrounding a mass the gravitational potential is negative and inversely related to the distance from the mass (eqn 2.13). The gravitational energy density in any region surrounding the mass is negative, becoming more negative in the region closer to the mass. Now suppose that we have two masses. As they move closer together the gravitational energy density of the region between them becomes more negative. To achieve this, positive energy must be displaced away from the inner region, making the gravitational energy density of the outer region less negative. The process of displacing the positive gravitational energy may be thought of in terms of a wave. It's reminiscent of hole conduction in semiconductor theory.

Suppose a test mass with finite dimensions is placed in the path of a gravitational wave. As the wave passes by, the mass will be subjected to two orthogonal force components (the gravitational Lorentz force). There will be an oscillating gravitational force, given by equation 2.9, in the vertical direction. This can't be detected since the effect would cause any mass, including the test mass, the sensing apparatus, any observers, the laboratory etc. to oscillate together in free fall, with no relative motion allowing detection. But, due to the combined action of the gravity and gravitomagnetic fields, there will also be an osillating force in the direction of wave motion, given by equation 8.24, causing the mass to be squeezed and stretched by an incredibly tiny amount. These oscillatory forces may cause the test mass to resonate, offering one means of detecting the presence of a passing gravitational wave. Another possibility, untried as far as I know, is to try to detect the rotation of space caused by the h-field component of the gravitational wave. The plane of polarisation of a laser beam parallel to the h-field of a gravitational wave may oscillate, thereby revealing the presence of such a wave via Faraday rotation (Fig. 20.5). But, again, any effect is expected to be exceedingly tiny.

In developing the linearised model of a gravitational wave, we have based it on the analogue with an electromagnetic wave and dipolar radiation. It is the disturbance caused by the relative oscillations of neighbouring positive and negative masses which gives rise to the radiation. So, the existence of dipolar gravitational radiation depends upon the existence of negative mass.

Many scientists don't accept the existence of negative mass, so dismiss the idea of dipolar radiation and the associated form of linear gravitational waves. When positive masses oscillate relative to one another then quadrupolar gravitational radiation is predicted to occur.

Later, in Chapter 12, we will look at the idea of replacing waves with particles. Electromagnetic waves may be replaced with positive energy particles called photons. Photons are said to have energy $\hbar\omega$ (eqn 19.1) and spin 1. That is, as the photon moves forward a wavelength it spins once around its axis, given by the direction of propagation. Gravitational waves, assuming they exist, may be replaced with positive energy particles called gravitons. Gravitons also have energy $\hbar\omega$ and travel at the speed of light in the direction of the wave. With the linearised 3-D model developed here, based on dipolar produced gravitational waves, the graviton has spin 1. However, in the academically accepted view, where gravitational waves are only generated by quadrupolar radiation, then such waves may be replaced with positive energy gravitons, with spin 2. The difference between the linearised version and Einstein's model would show up as a difference in the way that a mass would be disturbed by a passing gravitational wave.

Naturally occurring quadrupolar gravitational waves are predicted to emanate continuously from a pair of closely orbiting stars, or binary stars, with a frequency range 10 – 300 Hz. Although gravitational waves have still not been detected, side effects appear to have confirmed their existence. In 1974 radio telescope measurements of a particular binary star system showed an increase in the orbital speed of one of the stars around its unseen neutron star companion, indicating a decaying orbit which implied that energy was being radiated away from the binary pair in the form of gravitational waves. In my view, the orbital decay implies that the star was moving through a resisting medium, causing it to fall in a spiral towards its companion star. But the star was not in free fall, so, as a consequence, gravitational energy was being radiated away (eqn 20.15). The US physicists Russell Hulse and Joseph Taylor received the Nobel Prize in Physics in 1993 for their observation.

In the case of cataclysmic changes in astronomical masses, such as exploding stars, or supernovae, the gravitational waves are expected to be more like blast waves, having a massive leading edge, containing gravitational waves of many frequencies predicted to be in the range 1 – 10 kHz, which tail away to nothing. Since the 1960s there has been an increasing effort to try to detect gravitational waves from space. Professor Joseph Weber, of the University of Maryland in the USA, was the first scientist to try to detect gravitational waves using a bar antenna, rather like trying to detect the electromagnetic waves generated by lightning using a copper rod as an aerial.

Weber's gravitational receiver antenna was a cylindrical bar of aluminium, 2 m in length and 0.5 m in diameter. Bonded around the circumference of the bar, near to its centre, was a band of piezo-electric transducers which could detect any circumferential strain. The bar was suspended by wires inside a vacuum chamber to isolate it from seismic and sound vibrations and to keep the bar at a uniform temperature (Fig. 8.8). Based on the radial resonance of the bar, Weber hoped to detect gravitational radiation having a frequency around

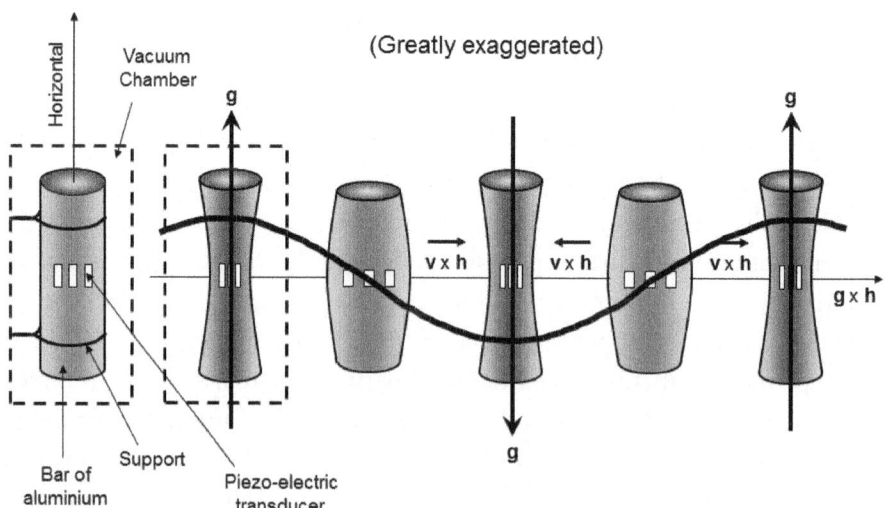

Fig. 8.8 Gravitational wave passing through Weber's bar.

1 kHz. The world's first gravitational wave antenna was, in fact, operated by Bob Forward, Professor Weber's postgraduate researcher from Hughes Aircraft Company. This apparatus is now in the Smithsonian.

Weber had two bar antennae for comparison purposes: one at Maryland University and the other at the Argonne National Accelerator Laboratory, near Chicago, nearly 1000 km away. Filtering out known spurious effects, the two bars initially seemed to confirm that gravitational waves were arriving from space. But experiments must be repeatable by other researchers. However, attempts by other groups to repeat Weber's experiment all failed. It has since been concluded that bar antennae are not sensitive enough to detect gravitational waves.

To improve sensitivity, Dr Forward and Professor Weber considered using interferometry (Fig. 10.4) since changes in mass deformation could then be measured in terms of a fraction of the wavelength of light. That is, inter-atomic changes in distance of order 10^{-18}m can be detected. This idea led to the development of the laser interferometer gravity wave detector. The apparatus consisted of a pair of suspended test masses fitted with mirrors (Fig. 8.9). A laser beam was split and caused to bounce back and forth between the mirrors with the resulting interference pattern being detected by a photodiode. Any subsequent change in the interference pattern signalled a difference in the

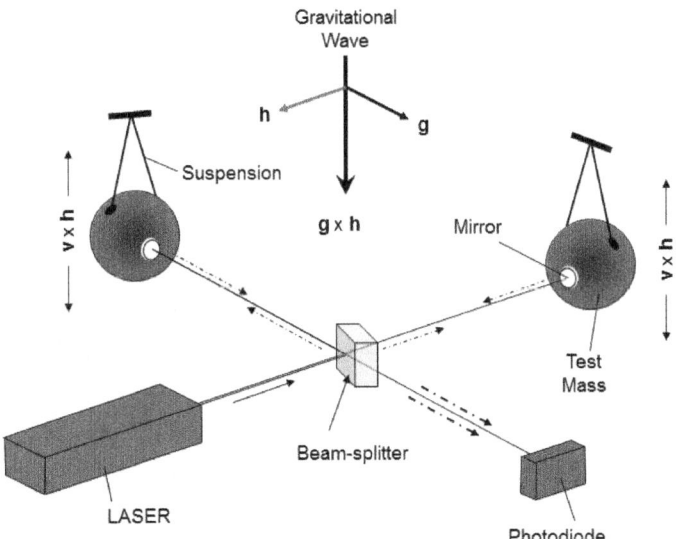

Fig. 8.9 Laser gravitational wave interferometer.

distance between the test masses, possibly resulting from the distortion caused by the passage of a gravitational wave through the apparatus. Dr Forward built the first such device, in 1973, at the Hughes Aircraft Company site at Malibu in California. To detect gravitational waves from a supernova explosion, with wavelengths of several kilometres, the distance between the mirrors has to be commensurate. Since gravitational waves can be elliptically polarised, just like electromagnetic waves, right-angled arms are used (Fig. 8.9). Consequently, the pairs of masses are spaced several kilometres apart with the laser beams being housed within evacuated pipes. This greatly improved sensitivity has meant that gravitational bar antennae experiments have, mostly, been superseded by laser interferometer gravitational wave experiments.

In January 1988 I visited Professor Jim Hough and Dr Norna Robertson, in the Department of Physics and Astronomy at Glasgow University, to see their prototype gravity wave interferometer. I was interested to learn that the optical quality mirrors used in the prototype interferometer were made by British Aerospace. Due to the high costs involved, British and German scientists have now collaborated and jointly run the GEO 600 gravity wave interferometer at a site near Hannover. Over the last decade, about half a dozen large gravitational wave interferometers have become operational around the world. Of these, the two laser interferometer gravitational wave observatories (LIGO) in the USA, based 3000 km apart, are the biggest. But despite the very large amount of dollars, euros, yens and pounds that have been spent on expensive apparatus, no gravitational waves have yet been detected in Earth-based experiments. If such waves really do exist, the sensitivity of the detector systems will have to be increased still further.

My original interest in gravitational wave detection was stimulated by the possibility that a gravitational radar, or gradar, might be developed to detect objects from their mass signatures, thereby defeating our radar and infra-red stealth measures. Existing gradars are passive devices limited to sensing large objects over a range of just a few tens of metres, so are not a threat. The academic opinion in the early 1990s was that an active gradar system, able to transmit and receive gravitational waves, was a very long way off. However, in April 2002 Professor Raymond Chiao, of the University of California, revealed that he'd been experimenting with the idea of a gravity radio, able to transmit and receive gravitational waves, although his work had not been successful. A breakthrough in this area of research could happen at any time.

9

FRAMES OF REFERENCE

With a 3-D frame of reference any point in space can be uniquely defined. Such a framework has a common origin O with three axes at right angles, say Ox, Oy and Oz.

Following Albert Einstein's paper on special relativity, in 1905, scientists realised the importance of the frame of reference from which an observation is made and whether it is an inertial, or a non-inertial, frame. An inertial frame is one in which the origin is either stationary or moves with constant speed in a fixed direction, without the axes rotating. A non-inertial frame is one where the framework introduces accelerations. This includes linearly accelerating and rotating frameworks.

In 1875, long before the paper on special relativity was published by Einstein, the US physicist Henry Rowland showed that an electrically charged disc produced a magnetic field when it was set rotating. In other words the moving charges produced a magnetic field. Since the magnetic field surrounding his rotating charged disc was similar to the magnetic field surrounding a live coil of wire, Rowland concluded that the current in the coil must consist of moving charged particles. We now know these charged particles are electrons, but the existence of the electron wasn't discovered until 1897, by the English physicist Professor Joseph J. Thomson, then Head of the Cavendish Laboratory.

Suppose we have a long straight length of conducting wire. If we pass a current through the wire we know that we will get a curled **H** field around the wire which can be detected as an induced **B** field by a stationary observer. Actually, the magnetic field is created by the flow of electrons moving, in the opposite direction to the conventional current, with an average velocity **v**, of a few metres per hour, called the drift velocity. If an observer moves in the same direction and with the same speed v as the electron current then, as observed in their moving frame of reference, there is no current of electrons, so the magnetic field should vanish. It is as though the current has been switched off.

To detect a magnetic field there has to be a relative movement between the electric charge and the observer with the detector.

We expect something similar to occur for mass. Let us observe a point mass under the following situations:

1. The mass is at rest.
2. The mass moves with uniform velocity.
3. The mass accelerates.

Each situation may be viewed from two different frames of reference:

(a) Stationary origin O. Observer at rest.
(b) Moving origin O. Observer travelling with the mass.

For situation 1(a), in the rest frame (0) let the mass be m_0. Suppose that the mass is given a velocity v. For situation 2(a), based on our electromagnetic analogue, a stationary observer in frame (0) would expect the presence of a curled **h**-field around the mass m_0 as it passes by. The **h**-field for a point mass moving with speed **v** is shown in Figure 9.1.

At a point R in space a distance r from the mass, the gravitomagnetic field

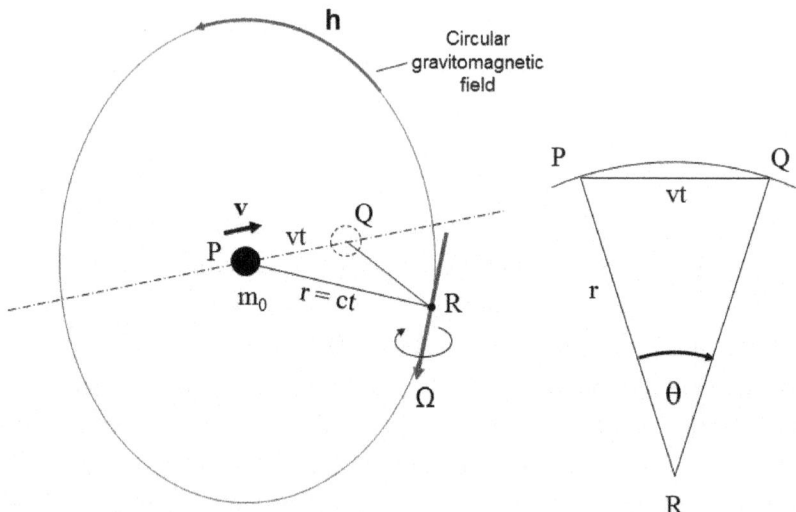

Fig. 9.1 Gravitomagnetic field for a moving point mass.

strength h is given by equation 8.12. The gravitomagnetic energy density is given by equation 8.13. Where a mass has gravitomagnetic energy due to its own gravitomagnetic field we must halve the value. Since $b = \eta h$ the gravitomagnetic self-energy of a moving point mass is

$$E = \tfrac{1}{2}\{-\tfrac{1}{2}bh\}V \tag{9.1}$$

In Figure 9.1 suppose that the mass m_0 is at point P. In time t information about the position of the mass can travel a spherical distance $r = ct$ and, in particular, to a point R in space. In the meantime the mass has moved forward a distance vt, to point Q. Now the triangle PQR is rather like the sector of a circle, with arc PQ. So, in approximate terms we have

$$PQ = vt = r\theta \tag{9.2}$$

We can think of θ as a twist in space due to the motion of the mass. Now the rate of twist, or angular velocity Ω, is given by

$$\Omega = \frac{\theta}{t} = \frac{v}{r} \tag{9.3}$$

Therefore, from equation 8.19 the induced gravitomagnetic field is

$$b = -2\Omega = -2\frac{v}{r} \tag{9.4}$$

So, from equations 8.12, 9.1 and 9.4, the gravitomagnetic self-energy, or kinetic energy, of the moving point mass is

$$E = \frac{1}{2}\left\{-\frac{1}{2}\left(-\frac{2v}{r}\right)\left(\frac{m_0\,r\,v}{V}\right)\right\}V = \tfrac{1}{2}m_0 v^2 \tag{9.5}$$

With the approximation used,

$$\eta = -2\frac{V}{m_0 r^2} \tag{9.6}$$

As the volume V shrinks to zero for the point mass, so the approximation for η also tends to zero, rather than tending to the near-zero value given by equation 8.15. Fortunately, however, in evaluating the kinetic energy of the point mass the term V cancels out, so it's not a problem.

To determine the internal energy of a stationary point mass m_0 we must allow for virtual vibrations of the mass in 3-D, thereby generating virtual twists θ_x, θ_y and θ_z. Combining equation 2.8, modified by a half, with equation 2.13 and using equations 8.14 and 9.6 the total energy is

$$E = E_x + E_y + E_z = 3\left\{\tfrac{1}{2}m_0\left[\frac{-m_0}{4\pi\gamma r}\right]\right\} = \frac{m_0}{\gamma\eta} = m_0 c^2 \qquad (9.7)$$

In the past, scientists investigating the relationship between mass and energy have not used the electromagnetic analogue of gravity, but have simply relied on Newtonian mechanics. Historically, this is how the link between mass and velocity was developed.

For situation 1(a), the mass at rest in the stationary frame (0) is m_0. From equation 8.8 or 9.7, the total energy of the rest mass in the frame (0) is m_0c^2. Let us assume that the mass is given a velocity v measured in frame (0). Then for situation 1(b) the total energy of the mass in frame (0) is its rest energy plus its kinetic energy:

$$E_0 = m_0c^2 + \tfrac{1}{2}m_0v^2 \qquad (9.8)$$

But for situation 2(b), the mass is at rest in frame (1) with rest energy:

$$E_1 = m_1c^2 \qquad (9.9)$$

Assuming that energy is conserved, then

$$m_1c^2 = m_0c^2 + \tfrac{1}{2}m_0v^2 \qquad (9.10)$$

So that

$$m_1 = m_0\left(1 + \tfrac{1}{2}\frac{v^2}{c^2}\right) \qquad (9.11)$$

As observed in frame (1) the mass seems to have increased in magnitude. This dependence of mass on velocity was first pointed out by Oliver Heaviside towards the end of the 19th century. Joseph J. Thomson, experimenting with electrons, concluded that, providing an electron's charge remained constant (conservation of electric charge), then an electron's mass depended on its velocity.

From Einstein's theory of special relativity the relationship between mass and velocity is shown to be

$$m_1 = \frac{m_0}{\sqrt{1 - \frac{v^2}{c^2}}} \quad (9.12)$$

After a great deal of experimentation with moving charged particles by German physicists at the beginning of the 20th century, initially with conflicting results, it was finally agreed that Einstein's formula was correct. Expanding the expression under the square root sign (√) in equation 9.12 as a series gives

$$m_1 = m_0 \left(1 + \frac{1}{2}\frac{v^2}{c^2} + \frac{3}{8}\frac{v^4}{c^4} + \frac{5}{16}\frac{v^6}{c^6} + \cdots \right) \quad (9.13)$$

So, equation 9.11 is seen to be an approximation to the theoretically predicted change of mass with velocity.

However, although the mass-velocity relationship has theoretical backing (eqn 9.12) and has been confirmed by experiment, it is my view that the relationship has been misinterpreted since it makes no allowance for the existence of the gravitomagnetic field. In accepting equation 9.11 (or equation 9.12) at face value it means that the principle of conservation of mass has been abandoned. If we consider the case of electromagnetism, there the principle of conservation of electric charge is retained and moving charges lead to the generation of magnetic fields. In forsaking conservation of mass we have rejected the pattern based on the electromagnetic analogue and automatically excluded the existence of the gravitomagnetic field.

In my view it is better to retain the principle of conservation of mass and keep the particle mass m_0 fixed. For situation 1(b) the change in energy due to the moving mass is now located in the gravitomagnetic field. For situation 2(b), since there is no relative motion between the mass and the observer there is no gravitomagnetic field present, so $m_0 = m_1$. Only by changing reference frames from (1) back to (0) can the energy difference be realised. In reality, bringing the moving mass to rest in frame (0) involves deceleration, which is covered by situation 3(a). It is the collapsing gravitomagnetic field and the accompanying induced gravitational field (eqn 9.14) which gives rise to the illusion of an increase in mass. I believe that this misunderstanding has, very probably, stymied progress in understanding how gravity may be controlled.

For condition 3(a), the observer should, theoretically, 'see' a curled induced gravitomagnetic $\partial \mathbf{b}/\partial t$-field around the point mass as it passes by. From the gravitational analogue of Faraday's law of induction,

$$\frac{\partial \mathbf{b}}{\partial t} = \nabla \times \mathbf{g}. \tag{9.14}$$

So the observer would conclude that the accelerating point mass experiences an induced gravity field in opposition to the acceleration (Fig. 9.2).

The force law for a gravitational force field (eqns 2.9 and 2.10) is

Force $\mathbf{F}$ = (mass m).(gravitational field $\mathbf{g}$) (9.15)

For condition 3(a), Newton wrote his second law as

Force = (mass m).(acceleration $\mathbf{a}$) (9.16)

But acceleration is not a force field. Newton's second law only exists because of the equivalence between $\mathbf{g}$ and $\mathbf{a}$. It would be better to work with the force law

$\mathbf{F} = m\mathbf{g}$ (9.17)

Here $\mathbf{g}$ is the induced gravitational field incurred by $\partial \mathbf{b}/\partial t$ due to the acceleration of a real mass. Furthermore, this explains why gravitational mass is the same as inertial mass.

Thus, we have the general force law for a source in a force field:

Force = (source).(field strength) (9.18)

This law applies equally for gravitation, electricity, magnetism and ideal fluid flow. Fundamentally, the patterns are all the same.

According to theory, the observer in condition 3(a) is bathed in radiation

FRAMES OF REFERENCE

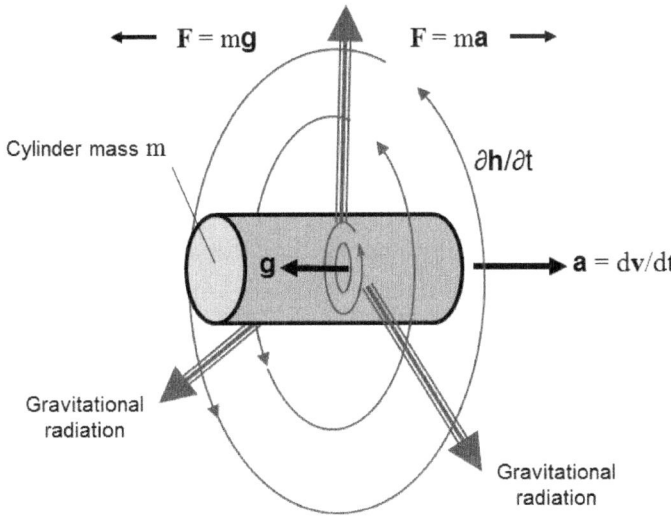

Fig. 9.2 Accelerating mass viewed from stationary reference frame.

apparently emanating from the accelerating mass. Why the radiation arises and where it comes from is discussed in Chapter 20.

Suppose an observer in condition 3(a) watches a free falling body, accelerate by, in the Earth's gravity field **g**. In the observer's frame of reference it can be argued that the formation of the $\partial\mathbf{b}/\partial t$-field around the body leads, in turn, to the generation of an induced gravity field **g** in opposition to Earth's **g**. The two gravity fields exactly cancel, so that the stationary observer would conclude that the falling body was not subject to any forces. Gravitational radiation does not arise in the case of an accelerating mass free falling in a gravitational field.

A free falling observer is in condition 3(b) and yet, without any external cues, he/she would assume that they were floating in a gravity-free environment, not subject to any force. And they would not be subject to gravitational radiation. So, there is a reciprocity between conditions 3(a) and 3(b).

When we accelerate forwards we are in condition 3(b). As all car drivers know, we experience being pressed backwards into our seats by the force of inertia. Corresponding with the view of observer 3(a), it feels as though a gravity field has arisen in the opposite direction to the acceleration. (Remember, the effect of acceleration must be communicated by contact.)

But can the observer 3(b) detect a time-changing induced gravitomagnetic field $\partial\mathbf{b}/\partial t$ in his, or her, frame of reference, which might give rise to an induced

g-field? Reciprocity between the reference frames 3(a) and 3(b) suggests that they can and that the induced **g**-field is generated, somehow, by the ether. But an ether containing nothing will not do! We need an ether containing energy, whereby the creation of energy gradients gives rise to forces.

To control inertia means being able to manipulate the gravitomagnetic field around a body or being able to manipulate the local conditions of space around a body. Either approach would provide the basis for a new means of propulsion. We must investigate the classical ether (Chapter 14) and the quantum vacuum ether (Chapter 18) to see how this might be achieved (Chapter 20).

10

SENSING ROTATION

If angular velocity Ω and induced gravitomagnetism **b** are equivalent then rotation sensors can be used to search for gravitomagnetism, but whether they can detect anything depends on their sensitivity. Note that both Ω and **b** are vectors, each having a magnitude and a direction associated with them.

Suppose we fix the frame of a gimbaled three-axis gyroscope to a turntable with the spinning flywheel's axle horizontal. As the turntable is rotated and the gyroscope is forcibly turned, the axle of the flywheel will tend to line up with the turntable's axis of rotation. This is vector addition at work. The change of direction of the flywheel's axle means that the gyroscope has detected an angular velocity Ω.

However, if we hang the gyroscope above the turntable, again with the spinning flywheel's axle horizontal, then as the turntable rotates below, the flywheel seems to be completely unaffected, with its axle remaining pointing in the same direction in space. For external angular velocity to have an obvious effect on a rotating body, then it seems that direct contact with the body is required.

The freely suspended gyroscope in a gimbaled three-axis frame, with its spinning flywheel of rotating mass, is the gravitational analogue of the electromagnetic search coil, with its ring of rotating electrons. According to our extended Newtonian gravitational theory, a rotating flywheel should interact with (sense, or detect) a gravitomagnetic force field just like a search coil interacts with (senses, or detects) a magnetic field. We know that unlike virtual magnetic poles attract and like virtual magnetic poles repel. On the contrary, although we can't be certain, we expect (eqn 8.11) like virtual gravitomagnetic poles to attract and unlike virtual gravitomagnetic poles to repel. However, due to the weakness of the gravitomagnetic field it is unlikely that any interaction will be detectable.

The freely suspended gyroscope can only detect strong gravitomagnetic fields. Although angular velocity -2Ω may be equivalent to an induced gravitomagnetic field **b**, angular velocity Ω has no force field of its own. For angular velocity Ω to affect a gyroscope, the rotation must be physically imposed.

Suppose we take a three-axis frame gyroscope to the North Pole and freely suspend it there, vertically, and start it rotating with its axle horizontal. Over one day the axle position will rotate horizontally by 360° relative to the surface of the Earth, whereas the axle's direction has, in fact, remained fixed in space (Fig. 10.1(a)). Now let's take our gyroscope to the Equator and suspend it vertically there and set it running with its axle in a west-east direction. After six hours the gyro axle would appear to us to have rotated to a vertical position, with the east end uppermost (Fig. 10.1(b)). After twelve hours, the axle would appear to be horizontal again, but what had been the east end would now be the west end. After eighteen hours, the axle is vertical again, with its original east end at the bottom. Finally, after twenty-four hours things are back as they were at the start. Again, we know that the apparent rotation of the flywheel's axle is due to the Earth's rotation, but that the axle's direction actually remains fixed in space.

In 1852, in Paris, Léon Foucault was the first scientist to demonstrate the rotation of the Earth using a suspended three-axis gyroscope.

A year earlier, Foucault had hung a pendulum bob from a long string attached to the high ceiling of the Panthéon, in Paris, and used it to demonstrate that the Earth was turning, since, as time progressed, the plane of oscillation

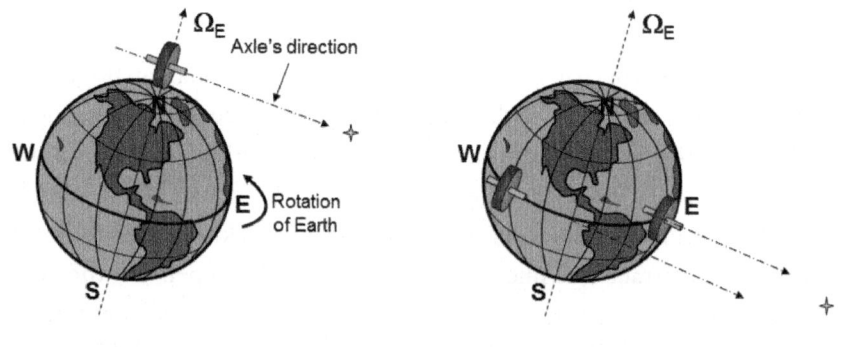

(a) Gyro at North pole.   (a) Gyro at fixed position on Equator.

Fig. 10.1 The axle of a free gyro is fixed in space.

rotated. The demonstration created a lot of interest for the Parisians and for many scientists, too.

As we saw earlier (Fig. 5.1), the formation of a hurricane is due to a depression, causing a radial in-rush of air to spiral around the Earth's vertical component of angular velocity $\Omega_E{'}$ at the centre of the depression. We noted (eqn 5.1) that the spiraling motion was caused by the Coriolis force $\mathbf{F} = m(\mathbf{v} \times 2\Omega_E{'})$, where m was a mass of air, or a raindrop, and $\mathbf{v}$ its velocity.

The long string of Foucault's pendulum meant that the bob, of mass m, roughly moved in a horizontal plane. As the mass m swung to and fro, with velocity $\pm \mathbf{v}$, it was subject to the Coriolis force, which caused the vertical plane in which the oscillation took place to rotate. The pendulum bob was forced around the vertical, just like the spiraling air and raindrops in a hurricane.

Using a gyroscope and a simple pendulum, Foucault was the first scientist to demonstrate that the Earth was rotating. Or did he demonstrate that the rest of the matter in the Universe was rotating around the Earth? We can't say. But, the model is easier if we assume a rotating Earth.

Fifty years later, in 1913, Georges Sagnac, another French scientist, had a slightly different idea to detect and measure the Earth's local component of angular velocity $\Omega_E{'}$. His technique involved interference fringes in light.

A wave pattern is a repeating cycle of up and down oscillations about a mean level where the maximum up or down undulation is called the wave amplitude. Consider the sine wave (Fig. 10.2).

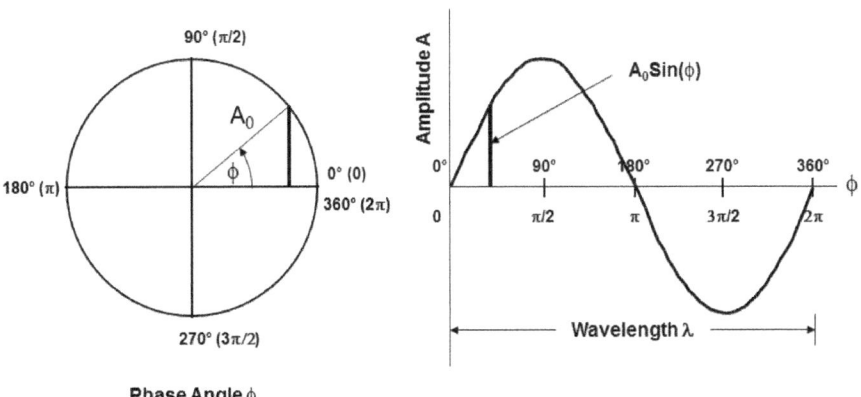

Fig. 10.2 The sine wave.

We call the length of one complete cycle the wavelength λ. The phase angle ϕ gives the position in the cycle and over one wavelength it rotates through 360°, or 2π radians. When the phase angle is just a function of distance we get a stationary wave, like that shown in Figure 10.2. If we allow the wave shape to move, like the ripples on the surface of a pond after a stone has been thrown in, then at any fixed position we will see the wave go through its up and down cycle, like a bobbing cork, as the wave shape moves by. Such a wave is called a travelling wave. A light wave is also a travelling wave. The phase angle ϕ of a travelling wave is a function of distance and time. The rate at which a complete up and down cycle is repeated at any fixed position is called the wave frequency f. Frequency used to be measured in units of cycles per second (cps) but this unit has been replaced by an SI unit called the Hertz (Hz). If the wave speed is c then frequency f and wavelength λ are related by the equation

$$c = f\lambda \qquad (10.1)$$

Consider two identical wave motions and lay one on top of the other (Fig. 10.3). If they match exactly, then we say that the waves are 'exactly in phase'. If all the maximum amplitudes of one wave coincide with all the minimum

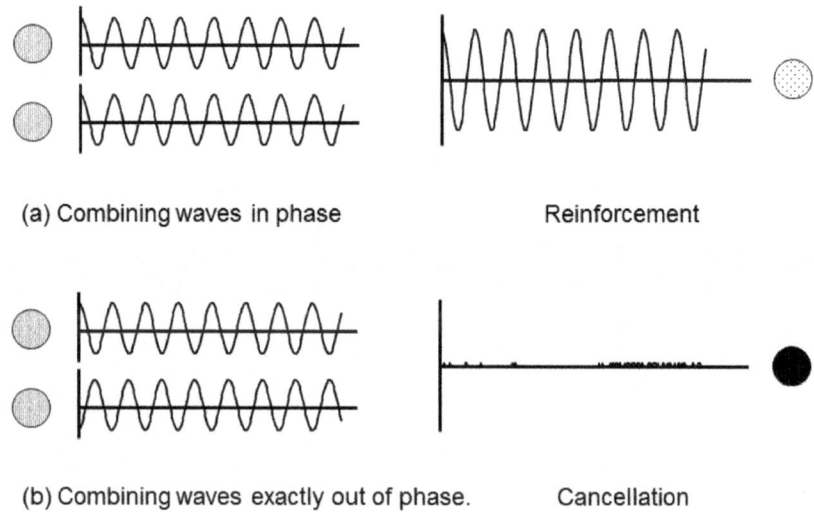

(a) Combining waves in phase   Reinforcement

(b) Combining waves exactly out of phase.   Cancellation

Fig. 10.3 Combining waves.

amplitudes of the other, then we say that the waves are 'exactly out of phase'. For any other relationship, we just say that the waves are out of phase.

When the two waves overlap we say that they interfere with each other. If both waves are exactly in phase, then the combined wave has twice the amplitude and its intensity is brighter. But, if the two waves are exactly out of phase then they cancel each other out, so that the combined wave has zero amplitude and zero intensity, giving blackness. Intermediate differences in phase angle between the waves will result in an intensity ranging between maximum brightness and a black region.

Suppose that we have a monochromatic (single frequency) source of light which is shone onto a pair of closely spaced parallel narrow slits (Fig. 10.4). The incident light is composed of a sequence of waves with random phase fronts but the two sequences of waves emerging from the slits have the same phase and are said to be coherent. These two waves spread out to form a combined image on the background screen. This was the experiment that the English scientist Thomas Young did in 1801. For a short period Young was the Professor of Natural Philosophy at the Royal Institution in London while his colleague, Humphry Davy, was the Professor of Chemistry. This was ten years before the arrival of the famous laboratory technician Michael Faraday. Young was also a linguist and is remembered

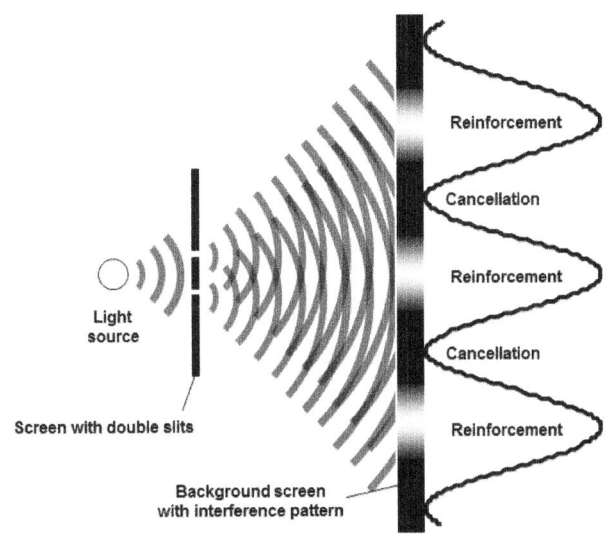

Fig. 10.4 Young's interference fringes.

for his part, along with the French polyglot Jean-François Champollion, in the decoding of the Egyptian hieroglyphics on the Rosetta Stone, now in the British Museum. What Young got in his light experiment was a band containing regions of dark and light images of the slits. At any point on the screen the light intensity was due to two rays of light, from different slits, each of which had travelled along slightly different path-lengths, so that they arrived at the screen at a different part in their phase cycles. Where they arrived exactly in phase there was a high intensity bright light, but where they arrived exactly out of phase, there was a low intensity light, or dark patch. These dark and light bands are called interference fringes. In doing this experiment, Young established the wave theory of light and thought that he had killed off Newton's corpuscular (particles of light) theory.

Just over 100 years later, Einstein turned the tables for a while, introducing light particles (photons) again. Finally, during the 1920s, it was realised that both particle and wave forms for light are legitimate; it really depends on the situation under investigation as to what is the best modelling approach to use.

Now we are ready to consider Sagnac's experiment, which involved a horizontal turntable, four plane mirrors placed at the four corners of a square and a coherent, monochromatic source of light (Fig. 10.5). The first mirror was half silvered so that the light beam split up and two light beams bounced around the square in opposite directions. Because the two beams of light started out as

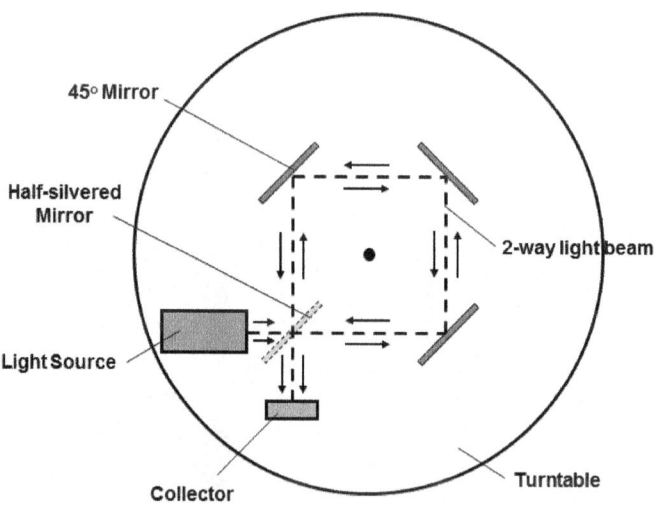

Fig. 10.5 Sagnac's experiment.

one beam, on being split the wave-forms of the two beams were 'exactly in phase'. After travelling right round the stationary square, the two beams of light arrived back and were combined at a collector.

Sagnac assumed that the rotational axis of one of the beams of light going around the square would point in the same direction as the vertical component of the Earth's rotation $\Omega_E{}'$, while the other would point in the opposite direction. So, he mistakenly expected that one beam of light would be slightly speeded up, while the other would be slightly slowed down. On being recombined at the collector, Sagnac looked for any sign of a change in intensity, indicating a phase difference between the two beams. However, he couldn't detect any change; the two beams appeared to be exactly in phase.

To check that rotation would really cause an effect, he rotated the square at just two revolutions per second and, sure enough, the intensity level changed noticeably.

We now know that Sagnac had had a bright idea, but that his apparatus was not sensitive enough to detect the Earth's vertical component of rotation at Paris, which was far too small for Sagnac's crude apparatus to detect. But he had shown a way of detecting rotation in general.

However, it wasn't until the advent of the laser, in the 1960s, that scientists could really make use of Sagnac's clever idea and that's when the development of a new type of gyroscope began. The active ring laser gyro (RLG) is a device somewhat similar to Sagnac's original square apparatus. The impetus for RLG development was the need to improve aircraft navigation.

Before describing how the RLG works we need to know about standing waves. Standing waves are a special form of interference between two waves travelling in opposite directions. Usually standing waves are formed when one wave interferes with its reflection, say from a plane surface, or a metal sheet in the case of electromagnetic waves.

As the waves merge we find that, under certain conditions, the envelope of the combined waves has fixed points where the two waves cancel out, leaving no vibration, and other fixed points where the two waves reinforce each other, giving maximum amplitude of vibration. The fixed points of the envelope where the two waves cancel are called nodes and the fixed points where the two waves reinforce each other are called anti-nodes. The distance between two nodes, or two anti-nodes, is $\lambda/2$, where λ is the wavelength of the waves.

Suppose that we have a cavity of length ℓ, filled with gas (for example, helium-neon) with mirrors at both ends. If we create a laser beam in the cavity then the beam will bounce backwards and forwards between the mirrors. Now, we can alter the length ℓ until standing waves, with nodes and anti-nodes, are formed in the cavity. When this happens, we have a resonant cavity. Suppose we now take our cavity, containing standing waves, and make it into a square-shaped cavity, using 45° mirrors at the four corners. The cavity surrounds an area A and has a perimeter of length P. In fact, it doesn't matter what shape the cavity is, providing that it encloses a planar area A, so it's easier to assume that it's a circular cavity (Fig. 10.6).

When the cavity is rotated the standing wave pattern remains motionless. This is like rotating a frictionless bucket of water, where the bucket turns and the water remains stationary.

Suppose the cavity has a peephole which allows one to look inside. As an observer rotates, with angular velocity Ω, round past the motionless standing waves a series of dark nodes and bright anti-nodes will be observed passing by the peephole, like a pattern of interference fringes. The rate at which the nodes, or anti-nodes, pass by the peephole is called the Sagnac beat frequency for the ring laser gyro. Theory shows that the beat frequency f_B is

$$f_B = 4\frac{(A \cdot \Omega)}{\lambda P} \qquad (10.2)$$

The beat frequency f_B is the apparent change in laser frequency caused by

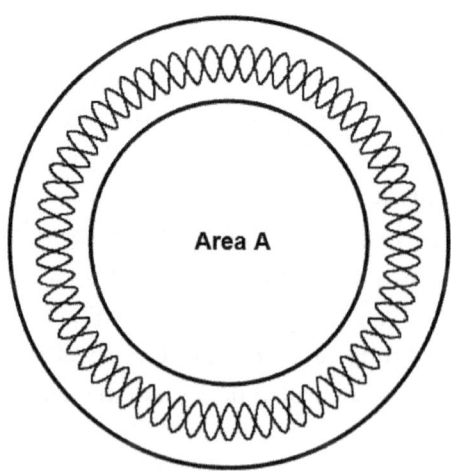

Fig. 10.6 Standing waves in circular cavity.

rotation Ω. In practice, a photo-detector in the cavity wall of the RLG actually detects the passage past the nodes and anti-nodes of the stationary waves.

The Earth's rotation can be detected with an RLG, providing it is sensitive enough. Generally, this means increasing the size of the enclosed area A.

Ultra Gyroscope-1 (UG-1) was the world's biggest active RLG when it was built, in 1999, by an international team of scientists based at the University of Canterbury, in Christchurch, New Zealand. The team was led by Professor Geoffrey Stedman. UG-1 is rigidly fixed to the Earth's surface with the perpendicular to its enclosed area A pointing vertically upwards. The area A = 21.0 m × 17.5 m = 367.5 m² and the perimeter P = 77 m. The laser frequency f = 474 THz = 474 × 10^{12} Hz, so the wavelength λ = 633 nm = 633 × 10^{-9} m. The number of nodes and anti-nodes in the cavity = P/λ = 121.6 × 10^6. That's more than 120 million nodes and anti-nodes!

Christchurch has a latitude λ = 44° South, so the vertical component of Earth's rotation Ω_E' relevant to the RLG is just over 10°/hr. The beat frequency for UG-1 is 1.513 kHz. This would be constant if the Earth's rotation was smooth and if UG-1 didn't suffer any disturbances, but it's not and there are! The most important disturbance is due to the tides around Christchurch tilting the land mass near the shore ever so slightly. The slight tilt in the land results in the perpendicular to the face of UG-1 moving slightly out of alignment with the local vertical. Corrections have to be made for this and any similar effect caused by any local seismic activity, to which this region is rather prone.

Once these corrections are made, the scientists start to see a more regular pattern for the beat frequency, but it is not uniform. The Earth's angular velocity vector Ω_E varies both in direction and magnitude. The Earth's axis of spin wobbles very slightly, like a child's spinning top. This is called the Chandler wobble, which has a period of about 432 days. In this time, the tip of the angular velocity vector Ω_E at the Poles traces out a circle with a diameter of between 4 and 6 m. Small indeed, but it is detectable by clever processing of data from UG-1.

The Prussian German philosopher Immanuel Kant predicted, circa 1755, that due to frictional effects associated with the Earth's tides, the magnitude of the Earth's rotation Ω_E must be reducing and, consequently, the length of a day must be increasing. Over 200 years later, Kant's prediction has been confirmed by data from UG-1, which shows that the length of each year is increasing by about 20 microseconds. Kant, and his fellow countryman Johann Lambert (after whom the unit of brightness is named), also predicted that the Universe contained other galaxies (island universes) apart from our own galaxy, the Milky Way. This was confirmed by the American astronomer Edwin Hubble during the 1920s. Hubble also noted that the Universe is expanding, a result now linked with the 'Big Bang' at the moment of creation. Kant also predicted the unification of the forces of electricity, magnetism and gravity. Gravity remains outstanding and the search for the link with electromagnetism is ongoing.

The rapid increase of fibre optic technology during the 1970s led to the development of the fibre optic gyroscope (FOG), where the cavity of the RLG was replaced by a fibre optic ring. The FOG is much cheaper, smaller and lighter than the RLG.

With the FOG two counter-rotating light beams, initially exactly in-phase, are introduced into the fibre optic ring at the same point. If the axis of the ring undergoes a rotation with angular velocity Ω, then on combining the two light beams at some point, a change in phase angle between them will have been introduced. From theory,

$$\text{change in phase angle} = 8\pi f \frac{(A \cdot \Omega)}{c^2}. \qquad (10.3)$$

The frequency of the light beams is f, the speed of light is c and the area enclosed by the ring is A.

The change in phase angle is detected electronically. In practice, the two counter-rotating beams, after passing around the fibre optic ring, are coupled together and shone onto a photodiode, which responds to light intensity. In this way, the number of interference fringes appearing over a time period can be detected, with each fringe (from bright through dark back to light) indicating a phase change of 360°. Knowing the phase shift means that the angular velocity Ω of the rotation can be determined.

Rather than just one fibre optic ring, FOGs are wound with several hundred fibre optic loops, to make them more sensitive. This is equivalent to increasing the area A.

Smaller solid-state devices, called micro-electro-mechanical systems (MEMS), are now being used to measure rotation. These devices are relatively cheap, small, rugged and very reliable. They will soon be fitted to vehicles, cameras and anywhere where rotational movement occurs, which needs to be controlled. Indeed, technology is moving at such a rate in this area, that this process has already started. The iPhone has a miniaturised silicon MEMS three-axis gyroscope and an accelerometer allowing it to sense motion and orientation.

In Foucault's pendulum, an oscillation was used to provide a velocity $\pm\mathbf{v}$, and the rotation of the plane of motion induced by the Coriolis force was used to detect Earth's vertical component of angular velocity Ω_E'. This technique can be miniaturised and used in MEMS devices to detect rotation.

In one device, called a vibrating structure gyro (VSG), instead of an oscillating pendulum a vibrating silicon ring is used. The vibrating ring has standing waves with nodes and anti-nodes which, due to the Coriolis force, try to rotate when an external angular velocity is imposed. An electronic system exactly counters the movement of the ring nodes and anti-nodes and the effort needed to do so is output, via signal processing, in terms of the angular velocity Ω causing the attempted rotation.

Thus, some of the sensing methods that we have for detecting rotation in the form of angular velocity Ω and as induced gravitomagnetism **b** are:

- a spinning flywheel
- Foucault's pendulum
- the RLG
- the FOG
- the VSG

However, we know that the gravitomagnetic permeability of space is exceedingly small ($\eta = 0.74 \times 10^{-27}$ m/kg), so inducing a gravitomagnetic field in a sensor across space will result in an exceedingly small rotational effect, so sensor sensitivity will be a very important factor in our quest to detect gravitomagnetism.

We have reached the frontier of research for this form of rotation sensing. To increase the sensitivity of devices any further means moving into the realm of quantum mechanics.

11

SEARCHING FOR GRAVITOMAGNETISM

While Newton was thinking about gravity a curious idea occurred to him. He suspended a wooden bucket full of water from a rope and twisted it around and around until the rope was really stiff. He waited until the water in the bucket was stationary and then let go. As the rope began to unwind, and the bucket rotated, the water gradually started to rotate due to viscous effects. But, in rotating, the water also started to rise up the inside of the bucket wall, an effect which we happily dismiss as being due to centrifugal force.

However, Newton wondered, if the bucket was rotated in empty space would the water still rise up the inside of the bucket? After all, if space is empty, how does the bucket of water know that it is rotating? What is it rotating relative to? Several centuries later, the Austrian physicist Ernst Mach suggested that rotation is relative to the background of the fixed stars (for Mach's principle see Chapter 21). But Newton decided that it was empty space, or the ether, that reacted to radial acceleration. By extension, is seems reasonable to suppose that empty space reacts, somehow, to linear acceleration, too.

Newton then posed the question. If space itself could be made to rotate, would it have a frictional drag effect on a stationary bucket of water, making it start to rotate and causing the water to rise up the side?

For the Earth modelled as a stationary spherical mass the gravitational strength on the surface is g (eqn 2.7), being approximately 9.81 m/s². However, since the Earth rotates with angular velocity Ω_E the surface gravity g is modified due to radial acceleration, depending on the latitude λ.

$$g = g - R_E \Omega_E^2 \cos^2\lambda \qquad (11.1)$$

The greatest effect is at the Equator, where $\lambda = 0$, and Earth's gravity field is the lowest. This gives rise to Earth's equatorial bulge.

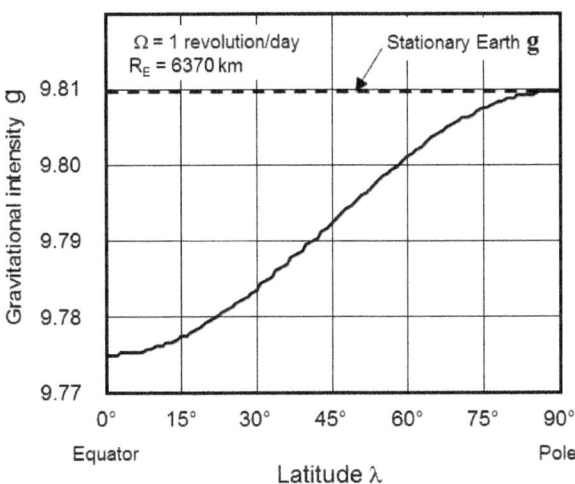

Fig. 11.1 Gravitational intensity at the surface of the rotating Earth.

Since Newton's time, others have wondered what would happen if the Earth was stationary and the whole of the Universe, including all the matter in it, rotated around the Earth with an angular velocity equal, but opposite, to the Earth's. Would the rotation create a gravitational modification of Earth's gravity field, equivalent to radial acceleration, giving the Earth an equatorial bulge? Based on reciprocity it would seem to be reasonable to assume that it would.

We have supposed that a moving body is surrounded by a gravitomagnetic field. Therefore, since the Earth is made of a myriad of bodies all rotating about a central axis, the summation of all the gravitomagnetic fields should generate a dipolar gravitomagnetic field (Fig. 11.2) which is coincident with the Earth's dipolar magnetic field.

In 1918 the Austrian scientists Josef Lense and Hans Thirring were the first to consider the possible existence of a gravitomagnetic field of a very large rotating mass. From Einstein's theory of general relativity they deduced that the rotation of the Sun would have an effect on the motion of the planets. They referred to the effect as 'frame-dragging' rather than gravitomagnetism. It was as though the vacuum of space is slightly viscous so that on rotating a large mass in it, locally space is dragged around.

How might we detect the Earth's gravitomagnetic field? Since we have assumed that induced gravitomagnetism is equivalent to angular velocity Ω, it is natural to consider using a rotation sensor to try to detect an induced gravitomagnetic field **b**. However, due the smallness of the gravitomagnetic permeability of free space (eqn 8.4) the induced gravitomagnetic field **b**, of the Earth, will be tiny.

In the magnetic case, a compass needle (magnetic dipole) experiences a torque in a magnetic field, causing it to twist as it tries to line up north-south. In the gravitomagnetic analogue, the gyro axle acts as a gravitomagnetic dipole. Like the compass, we expect the gyro axle in a gravitomagnetic field to experience a torque, causing it to twist as it tries to align with the field. Measuring such a torque will indicate the strength of the gravitomagnetic field present.

Given the idea of reciprocity discussed earlier, to detect the Earth's induced gravitomagnetic field we have two options of where to site our test gyro:

(a) Fixed in space, with the Earth rotating below.
(b) Fixed on the Earth, with space rotating about the Earth.

In 1960 Professor Leonard Schiff, at Stanford University, calculated that with an extremely sensitive gyroscope in a fixed-plane low Earth orbit it should be possible to measure two effects predicted by Einstein's theory of general relativity, namely:

1. the Earth's gravitomagnetic field or the effect of frame-dragging, predicted by Lense and Thirring and by Einstein's theory
2. the geodetic effect or the warping of space-time by the mass of the Earth, predicted by Einstein's theory.

Gravitomagnetism is predicted by the gravitational analogue with electromagnetism and can be viewed as a special relativity effect. The geodetic effect, on the other hand, is only predicted by general relativity. Both effects are expected to influence the direction of axial rotation of a spinning gyro.

Schiff discussed his idea with other physicists and engineers at Stanford University and gradually the idea for an experiment was conceived, with funding initially being obtained from NASA in 1964. So began the Stanford gyro experiment, now generally known as the NASA Gravity Probe-B (GP-B) satellite experiment.

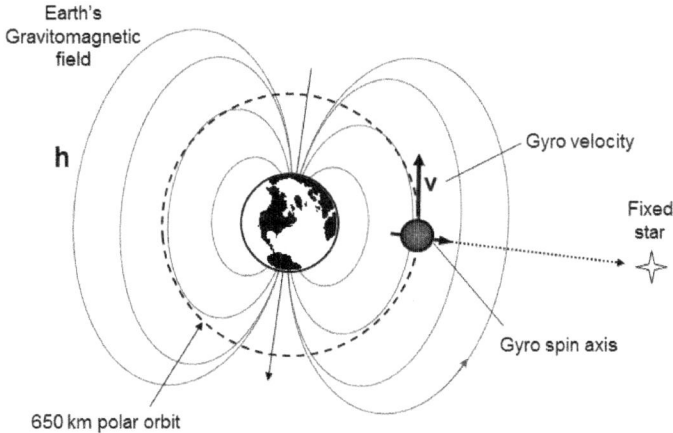

Fig. 11.2 NASA GP-B experiment.

Gravity Probe-B is one of the most expensive physics experiments in the history of mankind (although funding modern particle accelerators is much more expensive). Already, it has cost the USA more than $700 million, showing the importance that American scientists attach to detecting space-time warping and gravitomagnetism as a check on Einstein's theory.

In simple terms, the GP-B experiment involves placing a satellite containing a spinning ball-shaped flywheel, or gyro, in a polar orbit at an altitude of 650 km, above the rotating Earth (Fig. 11.2). The centre of the orbit corresponds with the Earth's centre but, as the Earth orbits the Sun, the orbital plane of the satellite remains fixed with respect to the background stars of the Universe. At the start of the experiment, the gyro's axis of rotation is set pointing at a particular fixed star.

It was predicted that the GP-B gyro would experience two separate torques, or twists, due to the Earth's induced gravitomagnetic field **b** and due to space-time warping, resulting in the direction of the gyro's axis moving and no longer pointing at the fixed star. These twisting motions are called precession.

For improved accuracy, the GP-B satellite actually carries four gyros. At the heart of the GP-B module is a cylindrical chamber, 2.3 m long and 250 mm in diameter, which contains the four gyros. These are quartz spheres, each 38 mm in diameter, which is about the size of a table-tennis ball. Another quartz sphere, of the same dimensions, acts as a test mass. At the end of the module is a 356 mm-long quartz-block tracking telescope. The four spheres are coated with a thin layer of niobium and kept at a temperature of 1.8 K, where they become superconducting. The four spheres are spun up to 170 Hz (~ 10,000 rpm) in a vacuum chamber, with all their spin axes aligned with the module centre-line, two rotating one way and two the other. To start with, the axes of the spherical gyros all lie in the orbital plane and all point at a fixed star, as seen through the short onboard optical tracking telescope. The star eventually chosen as the target was HR8703 in the Pegasus constellation. Helium thrusters on the module lock the telescopic sight line on to the star.

When a superconductor rotates it creates a current in its surface which generates a magnetic dipole field, the strength of which is called the London moment (after the German scientist Fritz London who first discovered it). The dipole axis always coincides exactly with the spin axis. The superconducting GP-B spherical gyros have a loop of wire around their circumference. Any change in the axial direction of spin is accompanied by a change in the direction of the magnetic dipole, resulting in a change of magnetic flux through the loop which is detected by a SQUID magnetometer. The movement of the gyro rotation axes relative to the telescopic sightline can be measured to an accuracy of 0.0001 seconds of angular movement (often referred to as arcseconds, to distinguish it from seconds of time). As a NASA engineer commented, the change in angle across a human hair seen ten miles away is about 0.001 arcseconds, so that gives some idea of the accuracy of the device.

The gyros are shielded from external magnetic fields (particularly the Earth's) by a lead bag which, itself, is surrounded by a mu-metal shield. Note that just as masses cannot be shielded from gravitational fields, it is assumed that rotating masses cannot be shielded from gravitomagnetic fields and the effects of space-time warping.

The disturbing force on the gyro (eqn 8.11) due to the gravitomagnetic field is

$$\mathbf{F} = m(\mathbf{v} \times \mathbf{b}). \tag{11.2}$$

This is the gravitational analogue (eqn 5.1) of the Coriolis force $\mathbf{F} = m(\mathbf{v} \times 2\mathbf{\Omega})$. For this reason (also see eqns 8.19 and 8.23), the induced gravitomagnetic field is often treated as an angular velocity Ω_M, where

$$\mathbf{b} = -2\Omega_M. \qquad (11.3)$$

For gravitomagnetism, the predicted angle of twist of the gyro axle in its orbit, averaged over one year, in radians per second is

$$[\Omega_M]_{AVERAGE} = 0.63 \times 10^{-14} \text{ rads/s}. \qquad (11.4)$$

If we convert this to arcseconds per year, we get

$$[\Omega_M]_{AVERAGE} = 0.042 \text{ arcsec/year}. \qquad (11.5)$$

That's an incredibly small angle to try to measure!

Our model, based on the analogue with electromagnetism, does not include space-time curvature effects so we cannot predict the GP-B gyro precession caused by space-time warping. NASA experts predicted this general relativity effect to be 6.6 arcsec/year. So space-time warping, although also very small, is predicted to be several orders of magnitude greater than the gravitomagnetic effect.

Lockheed Martin had overall responsibility for launching the GP-B experiment module, which took place at the Vandenberg Airbase, California, on 20 April 2004. Experimental data was accumulated over a year. In April 2007 NASA revealed that analysis of the GP-B experimental data had confirmed that the presence of the Earth did warp space-time. The results showed that the predicted geodetic effect agreed with the experimental measurement to within 1%. This is a most important result as it confirms that gravitation is a non-linear phenomenon. However, we should remember that Newton's model of gravity is a linear theory, which can only deal with two bodies at a time. Trying to model the real situation, of more than two bodies, results in a non-linear problem which is analytically unsolvable. Even so, Newton's model laid bare the underlying causes behind the motion of the individual planets in the solar system. In the same way, the linearised version of Einstein's gravitational theory, being simpler, may uncover hidden connections with gravity and other phenomena which the non-linear theory is too complex to reveal.

Trying to extract the effect of frame-dragging due to the Earth's rotation from the GP-B experimental data proved to be much more difficult than extracting the effect of space-time warping. It wasn't until April 2011, after much data processing, that scientists at Stanford University were finally able to claim that the GP-B experiment had detected the Earth's gravitomagnetic field and that its magnitude agreed with the theoretical prediction given by NASA.

We can now look at the other situation – that of trying to detect gravitomagnetism at a fixed place on Earth, due to space rotating around it. During the Cold War, in 1988, the scientists Vladimir Braginsky and Aleksander Polnarev, from the Soviet Union, and Kip Thorne, from the United States of America, put forward a joint proposal to use a Foucault pendulum, sited at the South Pole, to detect gravitomagnetism. A telescope's line-of-sight was to be locked onto a suitable star with the arc of the pendulum swing intersecting the line-of-sight in two points. Detection of the gravitomagnetic field was to be made by detecting any rotation of the plane of swing out of alignment with the telescope's sight-line. However, competition with the NASA GP-B experiment and perceived technical difficulties meant that the experiment never got any further than a paper study.

In 1967 the first person to suggest using a ring laser gyro to measure the gravitomagnetic field on Earth was the German scientist Heinz Dehnen, now Professor of Gravitational Physics at Constance University. Dr Ulrich Schreiber, a member of the German contingent of the international team of scientists experimenting with large RLGs at the University of Canterbury, in Christchurch, New Zealand, suggested that UG-1 might just be sensitive enough to detect Earth's gravitomagnetic field.

As described in Chapter 10, UG-1 was the world's largest RLG when it was first successfully operated in August 2000. Averaging the results over several hours, UG-1's sensitivity is estimated to be about 10^{-13} rads/s. The vertical component of the induced gravitomagnetic field at Christchurch, New Zealand, is predicted to be

$$[\Omega_{MV\text{-}NZ}] = 0.35 \times 10^{-14} \text{rads/s} = 0.023 \text{ arcsec/year}. \qquad (11.6)$$

So the sensitivity of UG-1 was very close to that needed in order to detect the gravitomagnetic field on Earth, albeit with beat frequency measurements averaged over several hours, or days.

In early 2004 a small team of academics from Lancaster University, including Professor Robin Tucker, Drs David and Anne Burton and Alan Noble, spent three months at the University of Canterbury, exploring the possibility of using the UG-1 RLG to detect and measure the gravitomagnetic field. This study was partly funded by Project Greenglow. Their approach was to mathematically model the beat frequency signal of UG-1 due to a rotating Earth and then to extract the theoretically predicted signal from the real beat frequency output to see whether it left a tell-tale signature for gravitomagnetism. But the technique was not successful. The Earth's gravitomagnetic field remained hidden within the RLG's beat frequency signal. Gravitomagnetism, like gravitational waves, is a very elusive quarry.

UG-1 has now been extended to form UG-2, an RLG with an area A = 833.7 m² and a periphery P = 121.4 m. Some consideration is now being given to building a gigantic RLG, possibly to coincide with the centenary, in 2025, of Michelson's large interferometer built in the USA. With a gigantic RLG it should, at last, be possible to detect the gravitomagnetic field due to Earth's rotation.

Before we dismiss gravitomagnetism as being too small ever to be of any use in the quest for gravity control, we should consider the parallel with magnetism. The Earth's natural magnetic field is quite weak, being of order 20×10^{-6} T, but we can generate very strong magnetic fields, of order 20T, using superconductors. That's an increase of 1,000,000 in induced magnetic field strength. So, perhaps something similar might be possible for gravitomagnetism!

Throughout the above chapter we have tacitly assumed that the gravitomagnetic field that we have been trying to detect is, to all intents and purposes, uniform. Coupled with the smallness of the gravitomagnetic permeability η, this seems to rule out laboratory experiments using small rotating bodies to detect any interaction due to gravitomagnetism. But, suppose that the angular velocity is not uniform; say that the spin increases, or the axis of spin changes direction. Then, according to equation 8.17 and our notion of equivalence, the body should experience an induced gravity field which curls around the changing angular velocity vector. This is Lenz's law in action. The permeability factor γ for gravity is quite large (eqn 2.3), so the effect of the induced gravity field may cause a reaction in a nearby body.

In the late 1980s, Eric Laithwaite, the Professor of Heavy Electrical Engineering at Imperial College London, started to put together an experiment to bring one spinning flywheel, moving in an arc, closely across the face of another spinning flywheel. The idea behind the experiment was that a change in angular velocity between the two flywheels might cause a reaction between them. The work was never completed.

Harvey Morgan, a retired US electrical engineer, had his own idea for a similar experiment to Laithwaite's with two adjacent flywheels axially aligned, one undergoing rotational acceleration, the other initially stationary but free to rotate. Morgan reported the results of his experiment in an article in the IEEE *Aerospace and Electronic Systems Magazine* (pages 5 – 10, January 1998). Quoting from Morgan's article, "A 2 pound lead flywheel was mounted on the shaft of a small, very high speed (26,500 rpm advertised) electric motor. Another flywheel was mounted on a ball-bearing shaft aligned with the motor shaft. The two flywheel's [sic] parallel faces were separated by about 1/16 inch. … When the motor was energized, it accelerated the lead flywheel towards it's [sic] top rated speed. The other flywheel, in response to the changing angular velocity and momentum of the lead flywheel, started turning briskly – in the opposite direction! The changing momentum field of the lead flywheel induced a torque in the other flywheel across an airgap. Newtonian mechanics does not predict that reaction."

This is an intriguing result which, as far as I know, has never been independently tested, so Morgan's observation remains unsubstantiated. Now the extended Newtonian model of gravity predicts that as the flywheel undergoes an increase in angular velocity $\partial\Omega/\partial t$, an induced gravitational field is generated internally, given by $g_i = -r\partial\Omega/\partial t$, where r is any radius less than the outer radius of the flywheel. This satisfies equation 8.18, where it is assumed that the induced internal gravitomagnetic field is $b = -2\Omega$. However, due to the smallness of η (eqn 8.15) it seems that the induced gravity field outside the flywheel must be nearly zero. But let us consider the case in more depth. Instead of a flywheel, think of a long cylinder of mass. As the cylinder rotates about its axis, the rotating mass creates a gravitomagnetic dipole. The solenoid is the electromagnetic analogue, where the rotating electric current creates a magnetic dipole. In the case of the solenoid, the external magnetic field is mostly confined to end-effects, since outside the solenoid the magnetic field along its length is negligible. Any change in electric current results in a change in the magnetic field at the solenoid ends, which gives rise to a curled induced electric field (eqn

7.4). Perhaps something similar happens with a rotating cylinder, and an external curled gravitational field is generated close to the end faces, when the rotating mass current changes. The rotational direction of such an induced gravity field would cause an adjacent axially aligned free cylinder to counter-rotate, as seen by Morgan.

Unfortunately, a rough calculation based on converting the formula for the magnetic dipole into a formula for the gravitomagnetic dipole with an increasing dipole strength indicates that the induced external gravity field is exceedingly small, which gives no support to Morgan's result. But maybe I've missed something. It seems to me that, since induced gravity fields play such a central role in the theoretical concept of gravity control, an experiment which suggests the existence of an induced external gravity field ought to be repeated. An idea for a more sensitive version of Harvey Morgan's experiment is described in Chapter 13. It opens up a new path of experimental study in the search for a way to control gravity.

12

WAVES, PARTICLES AND ATOM INTERFEROMETRY

Newton's suggestion, in the late 17th century, that a light beam could be considered as a stream of particles, called corpuscles, was overturned by Thomas Young, at the Royal Institution in London, during the very early 19th century. Young demonstrated that overlapping light beams created interference fringes, which could only be explained if light possessed wave properties. But, in the early 20th century, Einstein's explanation of the photoelectric effect was based on the concept of light particles, called photons. Gradually, scientists realised that circumstances dictated whether a light beam acted like a stream of particles, or as a wave motion.

During the First World War a young French aristocrat, Prince Louis-Victor de Broglie, with a degree in history, was stationed on top of the Eiffel Tower as a radio engineer in the French Army's communications branch. De Broglie's war-time radio work led him to develop a deep interest in electromagnetic waves and after the war was over he continued studying the subject at the University of Paris. De Broglie's elder brother, Maurice, was already a scientist who was heavily involved with organising the prestigious Solvay Conferences on Physics. (Ernest Solvay was a very successful Belgian industrial chemist and statesman who funded a series of international meetings in Brussels for invited world-class physicists.) The two brothers were thus able to discuss leading scientific developments of the time. As a scientist it is very useful to be able to talk with someone knowledgeable about a particular research topic, as you try to formulate your own ideas for making advances. Often the other person only acts as a 'sounding board' as you try to explain what your ideas are, gaining a clearer understanding of your own ideas in the process. At this stage trust is very important. Thus, Maurice was an ideal confidant for Louis-Victor. Later, Maurice was involved with organising the 1927 Solvay Conference on Quantum Mechanics, which his brother Louis attended, the reason being as explained below.

WAVES, PARTICLES AND ATOM INTERFEROMETRY

In 1923 Louis de Broglie submitted his PhD thesis for examination. In it he hypothesised that a beam of electrons might possess wave properties. A beam is analogous to a stream which implies a flow, so the idea of waves was a natural extension. In fact, de Broglie surmised, any stream of particles of matter might possess wave properties, with a wavelength given by

$$\lambda_D = \frac{h}{mv} \qquad (12.1)$$

where h is Planck's constant and p = mv is a particle's average momentum.

When two travelling waves of the same frequency combine they can interfere (Fig. 10.3). When two travelling waves with slightly different frequencies combine the interference pattern contains regions of amplitude swelling and fading (Fig. 12.1). The outline of the pattern is called the wave envelope. When two nearby notes on a piano are struck together you can hear 'beats' as the combined sound vibrations swell and fade in a continuous cycle. Each beat is associated with a wave envelope. With more waves (Fig. 12.1) of slightly different frequency each envelope becomes more pronounced and tighter. The wave envelopes move at a velocity v_g, called the group velocity, while the individual waves have their own speed, called the phase velocity v_p. The group velocity is always less than any of the phase velocities. In de Broglie's theory the relationship between the group velocity and the phase velocity is

$$v_p = f\lambda_D \qquad (12.2a)$$

$$v_p = \frac{c^2}{v_g} \qquad (12.2b)$$

In the case of electromagnetic waves in free space the group and phase velocities are the same and equal to c, the speed of light (eqn 10.1). So de Broglie's matter-wave idea was not an exact analogue with that of an electromagnetic wave. What de Broglie proposed was that the length of his matter wave λ_D was the width of the envelope of a group of waves which represented a particle. The particle's velocity v then corresponded to the group velocity v_g of the wave envelope.

What gave added support to de Broglie's idea of matter waves was an experiment done by the US physicist Arthur Compton in 1923, the same year that de Broglie submitted his PhD thesis for examination. Compton's experiment involved shining X-rays onto a block of paraffin wax and examining the scattered waves. Unexpectedly, Compton discovered that some of the reflected X-rays had, somehow, developed longer wavelengths. There was nothing in the wave theory of light to explain this effect. However, Compton realised that if he replaced his X-ray waves with quantum particles, which he called photons, then he could envisage elastic collisions between X-ray photons impacting with free electrons in the paraffin wax, resulting in an exchange of kinetic energy. Since the scattered photons still moved at the speed of light, any loss of energy by a photon due to a collision must result in a decrease in its frequency f and an increase in its wavelength λ. This result, now called the Compton effect, was further confirmation that electromagnetic waves could exhibit a particle nature. So de Broglie's reverse analogy that particles of matter might exhibit wave-like properties didn't seem so unreasonable.

The members of the Paris University panel for reviewing doctoral theses in physics were rather stunned by de Broglie's seemingly crazy hypothesis, so a copy of his thesis was sent to Einstein and his opinion sought. Einstein's view was that de Broglie's analogue with light waves was certainly a novel idea, but perfectly acceptable, and de Broglie was duly awarded his PhD. In fact, Einstein recognised the importance of de Broglie's speculation and passed the details of it to the quantum mechanics research group at Göttingen University, led by Professor Max Born.

Between 1922 and 1926, in the USA, Clinton J. Davisson and Charles Kunsman investigated the diffraction pattern of an electron beam (beam of particles) from the face of a crystal. They knew that their results were very similar to the diffraction pattern obtained when a beam of X-rays (beam of electromagnetic waves) was directed onto the crystal, but they hadn't drawn any conclusions from the similarity. The Göttingen group learnt of the US work and drew Davisson's attention to de Broglie's speculation about the dual particle and wave nature of an electron beam. With this knowledge, Davisson and Lester H. Germer carried out a new set of experiments. At last, the significance of their results became clear. Their electron beam behaved just as though it had a wavelength given by $\lambda_D = h/mv$. So, de Broglie's hypothesis was confirmed, for electrons, at least.

At the same time, in the UK, George P. Thomson carried out a slightly different experiment with an electron beam striking and passing through a thin metal plate (10^{-8}m thick), which also confirmed de Broglie's matter-wave idea. The emerging electrons created a broad spot and a series of concentric diffraction fringes on a photographic plate, showing that electron beams had wave-like properties.

Louis-Victor de Broglie won the Nobel Prize in Physics in 1929 for his matter-wave analogue. (Alfred Nobel was a very successful Swedish industrial chemist and the inventor of 'Dynamite'. On his death in 1896 he left a large part of his fortune to fund yearly prizes for outstanding achievements in Chemistry, Physics, Physiology or Medicine, Literature and Peace. The prizes were first awarded in 1901. The Nobel Prize for Economic Science was created in 1968.) Clinton Davisson and George Thomson had to wait until 1937, when they were awarded the Nobel Prize for their confirmatory detection of matter waves in electron beams. In an extraordinary turn of events, it was George Thomson's father, John J. Thomson, who discovered the electron in 1897. It was the first elementary particle of matter to be isolated and for this and his work on the conduction of electricity by gases he was awarded the Nobel Prize in Physics in 1906.

One of Max Born's assistants, at Göttingen University, was Werner Heisenberg (another was Wolfgang Pauli). In 1925 Heisenberg spent a late spring-time holiday in Heligoland, a small island in the North Sea. The island had been annexed from Denmark by Britain in 1807, during the Napoleonic War, and then passed to Germany in 1890, in exchange for their colony of Zanzibar, off the east African coast in the Indian Ocean. While relaxing in Heligoland, Heisenberg invented a mathematical model of the atom using matrix algebra. It was a very abstract model and presented no picture to help the imagination, unlike Bohr's mini-solar system model (see Chapter 16) of 1912. Half a century earlier, Michael Faraday had pleaded with theoreticians (Maxwell, in particular) to turn their mathematical hieroglyphics into something understandable to experimenters. It is most important for theoretical and experimental scientists to have a shared picture to work with.

In the same year, 1925, Erwin Schrödinger, an Austrian professor of mathematical physics at Zurich University, was also investigating how to improve Bohr's model of the atom. The atom has some elasticity associated with deforming the electron orbits (think of Maxwell's displacement current), so one way of formulating a mathematical model might be to combine elastic waves with matter waves. There is no evidence that Schrödinger used this approach, but it gives a good 'feel' about the internal atomic vibrations. Indeed, the US physicist Richard Feynman has since observed that it is not possible to follow Schrödinger's original thinking, as part of the model came from pure guesswork by Schrödinger. But with this approach the wave equation can be used to model the elastic vibration of the orbital electrons within the atom. And the motion of an electron particle is replaced with an electron wave. For a solid bar internal vibrations in 1-D (say the x-direction) are represented by the wave equation

$$\frac{\partial^2 \Psi}{\partial x^2} = \frac{1}{c^2} \frac{\partial^2 \Psi}{\partial t^2} \qquad (12.3)$$

During the wave vibration ψ is the displacement of the solid (treated as an electron displacement) in the x-direction at time t and c is the wave speed. In this case, the waves are longitudinal and c is the speed of sound. The standard way of solving the wave equation is called the method of separation of variables. The solution is the product of a function just of distance x, written X(x), and a function just of time t, written T(t). If we choose $X(x) = \psi(x)$ and choose T(t) as a cosine wave with a de Broglie wavelength λ_D, then we get the displacement wave equation

$$\frac{\partial^2 \Psi}{\partial x^2} + \left(\frac{\omega^2}{c^2}\right)\Psi = 0 \qquad (12.4)$$

Where $\omega = 2\pi f = \dfrac{2\pi v_p}{\lambda_D}$, $c = v_p$ and $\lambda_D = \dfrac{h}{m_e v}$

On substituting for ω^2/c^2 we get the equation

$$\frac{\partial^2 \Psi}{\partial x^2} + \left\{\frac{4\pi^2 m_e^2 v^2}{h^2}\right\}\Psi = 0 \qquad (12.5)$$

This is Schrödinger's famous linear differential equation for the 1-D displacement ψ of a confined vibrating electron. The result can be extended to 3-D (x, y and z directions) to model the movement, or vibration, of an electron confined within its orbit in an atom. Applying the boundary conditions satisfied by ψ within the atom leads to the requirement that only discrete electron energy levels are permitted. This agrees with the requirements stemming from the much simpler Bohr model of the atom (Chapter 16). The solution of the Schrödinger equation gives rise to stationary (or standing) waves for the orbital electrons within the atom. Schrödinger's theory was called *wave mechanics*, to distinguish it from Heisenberg's approach which was called *matrix mechanics*. The symbol ψ was given a new name and called the *wave function*.

In the derivation of Schrödinger's equation we have cross-patched elastic and light wave models, mixing longitudinal and transverse waves, which is a highly dubious analogue and, really, no better than guessing. But Schrödinger's model does give a useful picture of what is happening inside the atom, unlike Heisenberg's abstract model. Max Born was initially dissatisfied with the derivation of the Schrödinger equation, which he knew was partially based on guesswork. However, when he realised that the square of the wave function, ψ^2, could be interpreted in terms of the probability of finding a particle at a point inside a wave packet, Born was happier with it. But this is just an interpretation which has to be accepted, it is not something that can be proved. Nevertheless, usage has confirmed the applicability of the idea. One of the most important properties associated with the spread of the wave function ψ is that of quantum tunnelling, whereby the influence of a quantum entity (e.g. an electron) can penetrate, or tunnel, across a barrier.

For a matter-wave representation, as more waves of different frequency are grouped together to model a moving particle, the narrower, or tighter, becomes the width λ_D of the envelope of the wave group (Fig. 12.1). As λ_D becomes shorter, so the possible position x of the particle within the envelope becomes less spread. But since $p = h/\lambda_D$, as λ_D decreases so the particle's momentum p increases.

When an infinite number of waves of different frequency are used to represent a particle, we get a pulse where λ_D is just about zero, so the position x of the particle is almost exactly defined, but the particle's momentum p approaches infinity. On the other hand, when the momentum p is exactly known, as it is with a wave group containing just one wave of known frequency, then the length of the wave envelope λ_D is infinitely long. Consequently, the position x of the particle within the infinitely long wave length λ_D is unknown. As an example, suppose we consider a particle of light. The photon has a known electromagnetic

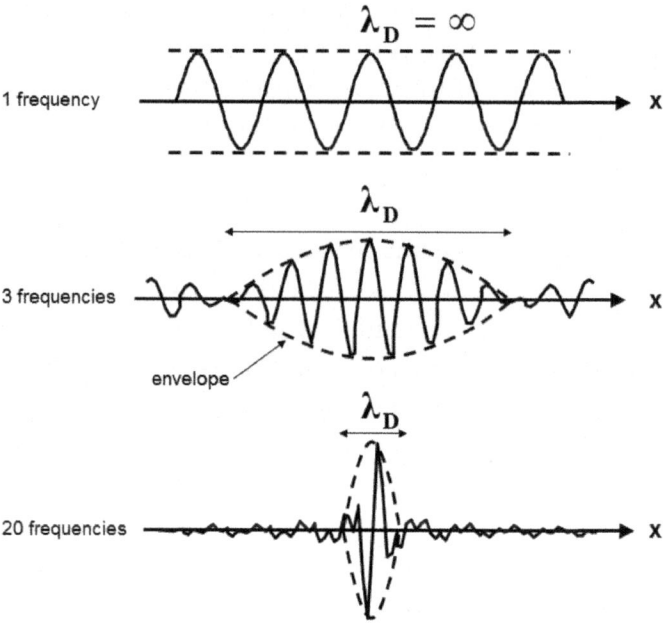

Fig. 12.1 Particle representation using de Broglie wave.

wavelength λ and a known momentum $p = h/\lambda$. But for a single frequency representation of a particle the de Broglie wavelength λ_D is infinite. Therefore the position of the photon within the de Broglie wavelength is unknown. In other words, where a particular photon resides in an electromagnetic wave is unknown.

In everyday life, we are not aware that streams of particles possess wave properties and we can easily see why this is so. Consider a stream of golf balls flying along with an average velocity of 20 m/s (45 mph). Assuming that a golf ball has a mass of 0.06 kg (~ 2 oz.), then the de Broglie wavelength of the stream is $\lambda_D = 6.626 \times 10^{-34}/(0.06 \times 20) = 5.52 \times 10^{-34}$ m, which is so incredibly small that there is no chance of measuring it. But the electron beam experiments of Davisson and Germer and of Thomson showed that the de Broglie wavelength of an electron beam could be measured. λ_D is about 10^{-12} m.

Visible light is composed of a band of electromagnetic waves with wavelengths λ all of order 10^{-6} m. If we use a microscope to look at something, we cannot resolve any details smaller than 10^{-6} m across. If we could view the same thing with an electron beam, say with a wavelength $\lambda_D \sim 10^{-12}$ m, we could see much more detail than is possible with light. Very quickly scientists began to speculate about the possibility of making a lens to focus an electron beam, analogous to the convex lens for a light beam. The phenomenon they needed to mimic was refraction. This they succeeded in doing fairly quickly and, amazingly, in 1931, less than a decade after de Broglie's original speculation about the existence of matter waves, the first electron beam microscope was built by Max Knoll and Ernst Ruska, in Germany. Within a few years, electron microscopes were being manufactured for commercial use. More than half a century later, in 1986, Ernst Ruska was finally awarded the Nobel Prize in Physics for his development of the electron microscope. Nobel Prizes are only awarded to living scientists so he only just made it, because he died in 1988.

However, back in 1925, Werner Heisenberg was not impressed with Schrödinger's model of the atom and wondered whether there might be a flaw in the matter-wave approach. This led him to examine the idea of matter waves in more detail. In 1927 Heisenberg stated his 'uncertainty principle' regarding all matter waves. If the error in measuring a particle's momentum is Δp and the error in measuring a particle's position within the wave group is Δx, then the following inequality holds:

$$\Delta p . \Delta x \geq \hbar \qquad (12.6)$$

The symbol $\hbar = h/2\pi$ and is called h-bar. The symbol h is Planck's constant.

Albert Einstein had supported de Broglie's idea of matter waves but he was not happy with Heisenberg's uncertainty principle. His famous quote on the subject was, "God does not play dice". He wondered whether there were hidden, or missing, variables which might invalidate the principle. However, experimental results connected with the uncertainty principle have supported it. The Nobel Prize Committee was quickly convinced of the correctness of the idea and Heisenberg was awarded the Nobel Prize in Physics in 1932. Nowadays, the wave-like nature of matter has been established for sub-atomic, as well as atomic, particles.

It was the British theoretical physicist Paul Dirac, later the Lucasian Professor of Mathematics at Cambridge University, who showed that Heisenberg's *matrix mechanics* could be transformed into Schrödinger's *wave mechanics* and vice versa. Thus Heisenberg's initial dislike of Schrödinger's matter-wave approach was shown to be misplaced. Moreover, Dirac extended Schrödinger's model so that it was consistent with Einstein's theory of special relativity. In doing so, he discovered that the electron possessed intrinsic angular momentum, or spin (see Chapter 16). Schrödinger and Dirac shared the Nobel Prize in Physics in 1933, for their contributions to the quantum theory of the atom.

Following the success of the RLG and the FOG rotation sensors, the attention of some researchers has turned to the idea of replacing the circulating light particles (photons) with circulating particles of matter, in the never-ending challenge to increase sensitivity.

For gyros employing counter-rotating light particles, using interferometry to determine the phase shift due to rotation, we make use of equation 10.2:

$$\text{phase shift} = 8\pi f \frac{(A \cdot \Omega)}{c^2} = 8\pi \frac{(A \cdot \Omega)}{\lambda c} \qquad (12.7)$$

where we have used $c = f\lambda$.

The analogous result for a matter gyro, using interferometry to determine the phase shift, is

$$\text{phase shift} = 8\pi \frac{(A \cdot \Omega)}{\lambda_D v} \qquad (12.8)$$

The average velocity of the particles is v.

Comparing the gyro phase shift for light particles with the gyro phase shift for matter particles gives

$$\text{Matter phase shift} = \left(\frac{mc^2}{hf}\right) . \text{light phase shift} \qquad (12.9)$$

Using counter-rotating beams of electrons in close proximity in matter gyros can cause difficulties, because electrons are charged particles. Using uncharged atoms avoids such problems and has the added bonus that a much greater increase in sensitivity can be achieved. The typical mass of an atom is m ~ 10^{-26} kg and for light, f ~ 10^{15}Hz, so the ratio (mc^2/hf) indicates an increase in gyro sensitivity of order 10^9, if atom beams are used rather than light beams. That's a huge increase in sensitivity.

Atom interferometers need an *atom laser*, so called because it's roughly analogous to the laser. Instead of generating a coherent (in phase) beam of photons all with the same frequency, the atom laser must generate a beam of coherent

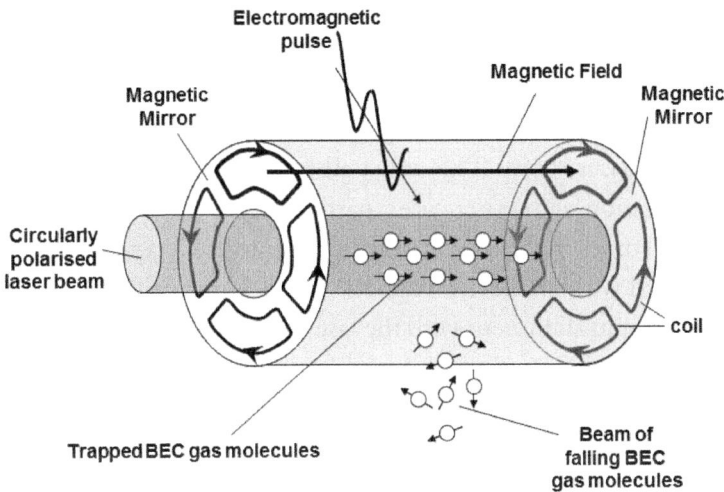

Fig. 12.2 The atomic laser.

particles of matter all with the same wavelength λ_D. More generally, we start with a cloud of pure gas molecules, all having the same mass. In the gas state the molecules whizz about in all directions with different velocities. Since $\lambda_D = h/mv$, we see that for all the molecules of the gas cloud to have roughly the same wavelength means that all the molecules must have roughly the same velocity. To achieve this it is necessary to cool the gas down. As the temperature is lowered, the velocities of the gas molecules become smaller and nearer in value. Consequently, the molecules begin to collect together as the de Broglie wavelengths λ_D become longer. As absolute zero (0 K) is approached, the molecules slowly condense into a single state all with the same average wavelength. However, the molecules are not frozen into a solid (rather like water vapour condensing into water droplets on a cold surface, but not forming ice). This special state is called a Bose-Einstein condensate (BEC) and it only works for some gases.

The possible existence of this quantum state of matter was predicted by Einstein and the Indian physicist, Satyendra Nath Bose, in the 1920s, but it wasn't until 1995 that the formation of a BEC was first demonstrated, in the USA, by a research group at Boulder and by another research group at the Massachusetts Institute of Technology (MIT). Since then, other research groups in the US, and groups in other countries too, have also made BECs.

An atom laser needs a cavity to contain the BEC of gas molecules and an aperture to let the ultra-cold molecules out in a controlled beam. The gas molecules have quantum spin of $\pm h/2\pi$ and, consequently, a tiny magnetic dipole field (see Chapter 16). Using a circularly polarised laser beam the spin axes of the gas molecules are lined up and trapped in a magnetic field created between two magnetic mirrors (Fig. 12.2). This magnetic trap forms the analogue of the laser cavity. This must all be done in a vacuum chamber, to avoid contamination with other gases, particularly air.

To release some of the BEC molecules from the trap, the spin axes are briefly disturbed with an electromagnetic pulse. Most BEC molecules return to their original spin alignment with the laser beam after the pulse has gone and these remain in the magnetic trap, but a few escape and fall under gravity. By applying several pulses a beam of falling BEC gas molecules is formed.

In 1997 the research group at MIT, led by Professor Wolfgang Ketterle, reported that they had managed to split a falling BEC beam and photograph the formation of an interference pattern as the two beams spread out and overlapped. This was the first crude demonstration of interferometry using an atomic laser.

The race is now on between research teams to control the movement of BEC molecules by building de Broglie wave guides, using micro-fabrication techniques similar to those used in MEMS devices. Such techniques will open up the way for all forms of sensors based on atom interferometry, as well as devices for quantum computing. Already devices have been built to demonstrate atomic gyroscopes, gravimeters and gravity gradiometers, with greatly increased precision over earlier devices, so we are beyond the 'proof of principle' stage. Nevertheless, it will probably be another decade before we see commercialisation of some of these devices.

The European Space Agency (ESA) is currently planning its 'HYPER' mission, scheduled to go into orbit some time around 2015. It is intended that one of the sensors on board the experiment module will be a cold atom interferometer, with which it is hoped to measure the Earth's gravitomagnetic field in real time. In the GP-B experiment, data had to be collected over a year to obtain an averaged measurement of Earth's gravitomagnetic field, which just shows how rotation sensor technology has advanced over the past forty years.

13

TWINNED FIELDS AND SOME EXPERIMENTAL IDEAS

The gravitomagnetic **h**-field was predicted to exist (under various names) by many scientists over the last century, but it is only fairly recently (April 2011) that the NASA Gravity Probe-B space experiment has confirmed its existence. However, in Earth-based experiments the gravitomagnetic **h**-field has remained hidden, due to the incredibly small value of the gravitomagnetic permeability of space η (eqn 8.15) and the minimising effect that it has on the induced field **b** in a rotation sensor. Until the sensitivity of rotation measurement is vastly increased, the situation won't change. Unless, that is, a case is discovered where the magnitude of the gravitomagnetic permeability η is high.

An electron has a negative point charge e, given by

$$e = -1.602 \times 10^{-19} \text{ C} \tag{13.1}$$

And it has a radial electric field **E**. But an electron also has a point mass m_e, given by

$$m_e = 9.109 \times 10^{-31} \text{ kg} \tag{13.2}$$

Thus, an electron must have a radial gravitational field **g**, too. **E** and **g** can be thought of as twinned fields. One is then led to wonder whether a magnetic field **H**, created by a moving electric charge, is twinned with a gravitomagnetic field **h**, created by a moving electron mass, as one feels it ought to be.

If the **H** and **h** fields are, indeed, twinned, so that the induced fields **B** and **b** are connected, then the assumption that **b** is equivalent to angular velocity Ω means that **B** ought to be linked with rotation, too.

In fact, at the magnetic molecular level there is a direct link between angular velocity Ω and induced magnetism **B**. The molecules responsible for ferromagnetism form nuclear magnetic dipoles which precess, or rotate, when they are placed in an external magnetic field. This phenomenon was first investigated by Sir Joseph Larmor, an Irish mathematical physicist and one-time Lucasian Professor of Mathematics at Cambridge University, the position recently held by Stephen Hawking. An analogue of this magnetic phenomenon is the gyroscope in a gravitational field. When one end of the gyroscope's axle is mounted on a model Eiffel Tower, the gyroscope then precesses around the tower. In 1908 Professor Owen Richardson, an English physicist working at the Cavendish Laboratory at Cambridge University, extended Larmor's original work. At the microscopic level, when a molecular magnetic dipole is rotated about its north-south axis with an angular velocity Ω, a small increase in the induced magnetic field **B** occurs given by

$$\mathbf{B} = \frac{m_e}{e}(-2\Omega) = 5.69 \times 10^{-12}(-2\Omega) \tag{13.3}$$

At the macroscopic level, or visible scale, we must multiply the effect by the number of molecules involved. The idea of molecules, where some substances only exist as pairs, or combinations, of atoms was proposed by Count Amadeo Avogadro, who was the Professor of Physics at the University of Turin at the beginning of the 19th century. Avogadro is better known for his hypothesis, given in 1811, that "Equal volumes of gases (at the same temperature and pressure) contain the same number of molecules". At the time few scientists paid any attention to Avogadro's idea. Fifty years later, sadly after his death, further research confirmed the brilliance of Avogadro's hypothesis. His original idea for gases was extended to all substances and led to the introduction of Avogadro's number N_A, which defines the number of molecules in the mole of a substance, where the mole is the SI unit of the amount of matter and is associated with atomic weight. But what is the value of N_A? In his PhD thesis of 1905, Albert Einstein was one of the first scientists to empirically determine N_A. Unfortunately, he made an algebraic error in his thesis (famous scientists can make mistakes) but when this was corrected a few year later, by a student, the predicted value for N_A was found to be in good agreement with today's accepted value.

$$N_A = 6.02 \times 10^{23} \text{ molecules/mole} \tag{13.4}$$

So, if N is the number of molecules per unit volume of a substance and ρ is its density then

$$N = \frac{(\rho \times N_A)}{\text{Atomic Weight}} \tag{13.5}$$

Let us assume that the magnetic molecule is made of iron. The atomic weight of iron is 55.85 kg/mole and its density is 7.9×10^3 kg/m³, so that its molecular density N is

$$N = 8.5 \times 10^{28} \text{ molecules/m}^3 \tag{13.6}$$

However, magnetic saturation sets a limit to the size of Richardson's predicted effect. Experimental work (Fig. 13.1(a)) done in 1915 by the American physicist Professor Samuel Barnett confirmed Richardson's theory and equation 13.3 is now known as the Barnett effect.

Also in 1915, in the midst of the First World War, Albert Einstein and the Dutch physicist Wander de Haas showed that the effect works in reverse, too. Suppose we place an iron bar axially inside a solenoid, or long coil. The magnetic permeability μ of iron is much greater than the free space value so when the current in the coil is switched on, a powerful magnetic field is induced in the iron bar and an electromagnet is created.

What Einstein and de Haas did was to suspend an iron bar by a thread, attached to one end, so that the bar dangled inside the coil. When the coil current was switched off, the induced magnetic field in the bar, generated by the coil, vanished but the bar rotated very slightly (Fig. 13.1(b)). As the inverse of equation 13.3 shows, the magnetism lost by the bar appears as rotation. In fact, this result demonstrates the conservation of quantum angular momentum at the macroscopic level.

In 1911 Arthur Schuster, Professor of Physics at Manchester University, suggested that terrestrial magnetism might be caused by the Earth's rotation. At the time there was no real evidence to support the idea, but later astronomical data of the magnetic fields and the angular momentums of the planets and the Sun did seem to support it. Consequently, in 1947 Professor Patrick Blackett, also at the University of Manchester, reconsidered the idea. But, experiments with neutrally charged rotating metal spheres, of 1m diameter, did not reveal

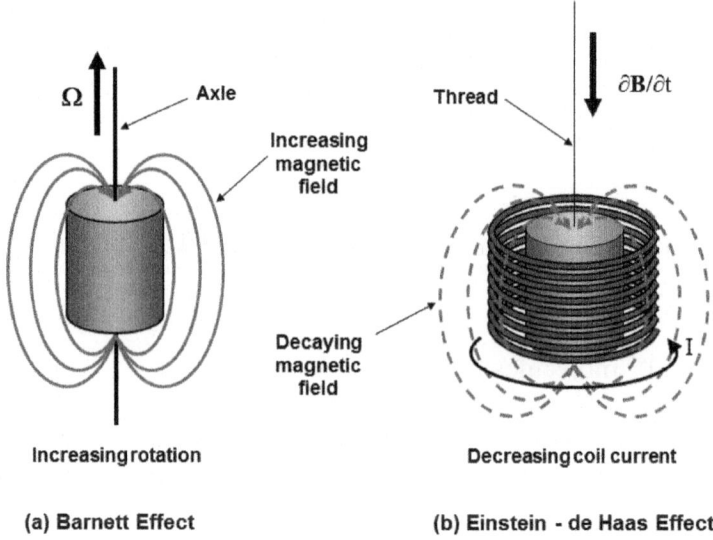

Fig. 13.1 Iron bar and rotation.

the creation of any magnetic fields, which led Blackett to drop the idea. In today's terminology we might say that it was proposed that the Earth's gravitomagnetic field **h** was responsible for creating the Earth's magnetic field **H**. But, in the gravitational case, where the **h**-field is predicted to accompany a moving mass or to arise due to a change in **g**-field, we assume it does so without the presence of an **H**-field. The twinning between **h** and **H**, therefore, is not symmetrical, nor is it between **g** and **E**. In recent years a much more complex model, called geodynamics, has been developed to explain the existence of the Earth's magnetic field. It is assumed that the Earth's outer core contains molten iron alloys while the inner core is a hot solid. In simple terms, the heat convection from the inner core causes spiral, or solenoidal, electric currents to form in the outer core, which axially align with the axis of rotation. It is the combined effects of these currents which generate the Earth's magnetic field.

In the reverse situation, when an **H**-field is created, say by a changing **E**-field, we expect that an **h**-field is created, too. So is it possible to separate **h** from **H** in the electromagnetic case? We know that it is possible to screen against electric and magnetic fields. But in the absence of negative mass it is not possible to screen against a gravitational field, so we assume, likewise, that it is not possible to screen against a gravitomagnetic field. If this is so, then by generating

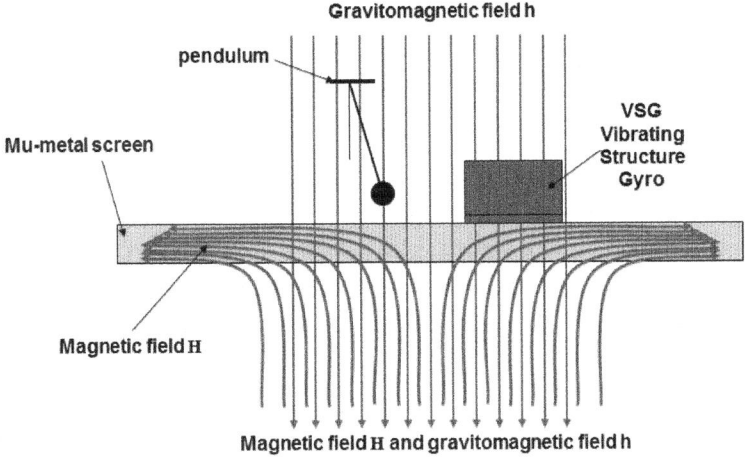

Fig. 13.2 Separation of gravitomagnetism from magnetism.

a strong magnetic field, assumed to be harbouring a gravitomagnetic field, we may be able to use a mu-metal screen to separate the two fields (Fig. 13.2).

We now need to use our imagination to think of some experiments which might be done to detect the gravitomagnetic field. We might try using a vibrating structure gyro (VSG), a rotation sensor briefly described in Chapter 10, placing it above the mu-metal screen (Fig. 13.2). However, any slight penetration of the magnetic field through the screen might spoil the operation of the VSG.

From the gravitational analogue of the Lorentz force, one component is

$$\mathbf{F} = \eta m(\mathbf{v} \times \mathbf{h}) = m(\mathbf{v} \times \mathbf{b}) \qquad (13.7)$$

The predicted sideways force **F** is that experienced by a mass m as it moves with velocity **v** across a gravitomagnetic field **h**. The η-factor means that the force **F** is likely to be very small. In Chapter 11 we looked at the US-Soviet Russian proposal to use a Foucault pendulum to detect the Earth's gravitomagnetic field based on equation 13.7.

In our bench-top experiment, the detection of the gravitomagnetic field above the mu-metal screen might be achieved by noting any change of periodicity of an oscillating pendulum (Fig. 13.2), with a non-magnetic bob. The path taken by the bob would be slightly curved by the gravitomagnetic field. This simple detection method would not be affected by any stray magnetic field. However, it would be much less sensitive than using a VSG and would only work for a fairly strong gravitomagnetic field. A more sophisticated experiment might involve directing a horizontal beam of neutral particles just above and across the mu-metal surface onto a target screen. The presence of the gravitomagnetic field would cause the position of the target point, where the beam strikes the screen, to be moved sideways.

The phenomenon of superconductivity was discovered in 1908 by Kammerlingh Onnes, the Professor of Experimental Physics at Leyden University in the Netherlands. He found that at extremely low temperatures certain metals had no resistance to electric current. Such metals, in this state, may be used to generate very strong magnetic fields. Thus, if the notion of separating twinned fields has any validity, it seems likely to be with superconductors that we might detect gravitomagnetism. Another phenomenon associated with superconductivity is a form of anti-gravity. When a superconductor in a gravity field is placed above a magnet it floats, something referred to as Meissner levitation, named after the German physicist Walther Meissner, who discovered the effect in 1933.

In 1992 Dr Evgeny Podkletnov, a Russian scientist working in Finland, reported in a peer-reviewed paper that test masses suspended above a levitated rotating superconductor appeared to get very slightly lighter. He claimed that the spinning superconductor shielded a test mass from the Earth's gravity field. It seems that most scientists either missed the paper, or disregarded it. Later, in 1996, Podkletnov submitted a further paper for publication, backing up his previous findings. This time, two science journalists in the UK, Dr Robert Matthews and Dr Ian Sample, sensationalised the claim in a newspaper article with the eye-catching title "Breakthrough as scientists beat gravity".

Dr Podkletnov suddenly found himself in the spotlight on the stage of world science being subjected to a barrage of questions by investigative science journalists. Theoretically, because we only have one sort of mass, gravitational shielding is not supposed to be possible. Severe questioning of a scientist who claims to have made a breakthrough is only natural, as other scientists want to know the apparent secret so that they can check to see whether it is true. Dr Podkletnov was, therefore, hard pressed to explain his amazing claim. It turned out that he was a materials scientist, whose expertise was in working with a new form of superconducting material called YBCO (pronounced as ibco, standing for yttrium barium copper oxide). Dr Podkletnov was not a gravitational physicist at all, but seemed to have come across the anti-gravity effect by chance. The clamour was very intense and Dr Podkletnov felt obliged to withdraw his paper.

Gradually the uproar subsided and most experts in general relativity and superconductivity ignored Dr Podkletnov's claim. Just a few people wondered if this was the first evidence that gravity could be altered, or created, artificially. A few months later I received a copy of Dr Podkletnov's withdrawn paper, sent to me by a scientist at the MoD. I was asked what I made of it. Not a lot! But it certainly aroused my curiosity and I felt that I needed to know more about superconductors. After all, if the twinned field idea has any merit then superconductivity is likely to be a key factor. Moreover, since superconductivity is a quantum mechanical effect, it could be that a quantum gravitational effect holds the key to gravity control.

Superconductors, in their superconducting state, prevent the penetration of external magnetic fields. Resistance-free persistent currents run in the surface of a superconducting body and inside it the electric field **E** is zero, as is the magnetic field **H**. From our twin field notion it is clear that the persistent currents i of a superconductor are also persistent mass currents I, although they are extremely small. Even so, it means that superconductors are linked with gravitomagnetism. From equations 13.1 and 13.2 we see that the ratio of the mass current I to the electrical current i is

$$\frac{I}{i} = \frac{m_e}{e} \qquad (13.8)$$

And the ratio between the induced fields **b** and **B** created by the mass and electric currents is

$$\mathbf{b} = \left(\frac{\eta}{\mu}\right)\left(\frac{m_e}{e}\right)\mathbf{B} \qquad (13.9)$$

Since 1989 the US physicists Ning Li and Doug Torr, both at the University of Alabama, have been interested in the idea that it might be possible to detect an induced gravitomagnetic field **b** associated with a superconductor. This is dependent on the value of the gravitomagnetic permeability η associated with superconductors. However, several US theoretical physicists have argued that the value of η assumed by Li and Torr in their superconducting study is too large. In their view, although a gravitomagnetic field may exist in theory, it would be far too small to detect in practice.

Prior to Podkletnov's superconductivity experiments, Li and Torr also investigated the idea that the gravitational analogue of Maxwell's equations (eqns 8.16 and 8.17) might hold sway within a superconductor. If so, then by changing the gravitomagnetic field $\partial \mathbf{b}/\partial t$ it might be possible (eqn 8.18) to use a superconductor to generate a detectable gravitational field.

We can deduce the relationship that Li and Torr used in their analysis as follows. Because there are no free magnetic poles (eqn 4.10) and since $\mathbf{B} = \mu \mathbf{H}$ we have

$$\nabla \bullet \mathbf{B} = 0 \tag{13.10}$$

On taking the divergence of a curl the result is always zero. This is a particular property of this double vector operation. This means that equation 13.10 is satisfied if we let

$$\mathbf{B} = \nabla \times \mathbf{A} \tag{13.11}$$

A is called the vector magnetic potential. From our twinned field notion it follows that

$$\mathbf{b} = \nabla \times \mathsf{A} \tag{13.12}$$

A (note Arial font) is called the vector gravitomagnetic potential. But in the gravitomagnetic case we have also concluded that the induced gravitomagnetic field **b** is equivalent to an angular velocity (eqn 8.20). Therefore we can write

$$\mathbf{b} = -2\mathbf{\Omega} \tag{13.13}$$

Furthermore, we have also shown (eqn 8.21) that a velocity **v** exists (in this case linked with the persistent currents) such that

$$\nabla \times \mathbf{v} = 2\mathbf{\Omega} \tag{13.14}$$

So, we may assume that

$$\mathbf{b} = \nabla \times (-\mathbf{v}) = \nabla \times \mathbf{A} \qquad (13.15)$$

Substituting in equation 13.9 gives

$$\nabla \times (-\mathbf{v}) = \left(\frac{\eta}{\mu}\right)\left(\frac{m_e}{e}\right)(\nabla \times \mathbf{A}) \qquad (13.16)$$

Neglecting any constants of spatial integration, we see that

$$\mathbf{v} = -\left(\frac{\eta}{\mu}\right)\left(\frac{m_e}{e}\right)\mathbf{A} \qquad (13.17)$$

Altering the persistent current means altering the gravitomagnetic field **b** and altering the vector magnetic potential **A**. Taking the time differential of the above equation gives

$$\frac{\partial \mathbf{v}}{\partial t} = -\mathbf{g}_i = -\frac{\partial \mathbf{A}}{\partial t} = -\left(\frac{\eta}{\mu}\right)\left(\frac{m_e}{e}\right)\frac{\partial \mathbf{A}}{\partial t} \qquad (13.18)$$

This shows that changing the vector magnetic potential **A** of the superconductor with respect to time leads to the generation of an induced gravity field $\mathbf{g}_i$. This is just another version of the gravitational analogue of Faraday's law of induction (eqn 8.17). But, can $\mathbf{g}_i$ be detected?

Dr Podkletnov's focus of research was on the investigation of spinning type II YBCO superconducting discs for emergency energy storage purposes. Type II superconductors are different to ordinary superconductors in that in their superconducting state they allow external magnetic fields to pass right through them in a regular array of magnetic lines called magnetic vortices. Within the magnetic vortices the material is in a normal (non-superconducting) state, while throughout the rest of the body it is superconducting. We should note, in passing, that the magnetic permeability μ within the vortices is greater than its free space value. It seems reasonable, therefore, to assume that the same is true for the gravitomagnetic permeability η, too.

Details from Podkletnov's withdrawn paper reveal that he used a large superconducting YBCO annular disc in his experiment, having a diameter of 275 mm. The top of the annulus was covered in metal foil and the lower surface was of a non-superconducting material. The superconducting disc was levitated through the Meissner effect using three pairs of radially aligned alternating current solenoids (Fig. 13.3), creating radial alternating magnetic vortices within the disc in the process. The three main solenoids partly enclosing the YBCO disc were then activated. The internal magnetic fields combined to form an asymmetric magnetic field which caused the disc to spin.

During his original experiment Podkletnov had noticed what he took to be a small gravitational anomaly. It was a chance observation, but being a good researcher, he explored the quirky result further. In his withdrawn paper Podkletnov stated that when the YBCO disc was spun to about 4000 rpm, with the three main solenoids operating at between 2 and 3 MHz, then a test mass hung above, the disc lost about 0.2% of its weight. Even when the disc was stationary (but levitated) a tiny weight loss of about 0.06% was noted. Although the weight losses were very small, if they were really due to gravity modification then a very significant effect had been discovered. But was it a real effect or simply an experimental glitch?

Professor Ning Li, together with a group of scientists and engineers at the NASA Marshall Space Flight Center, including Ron Koczor, David Noever and Glen Robertson, formed the Delta-G Team. During the period 1995 to 2002

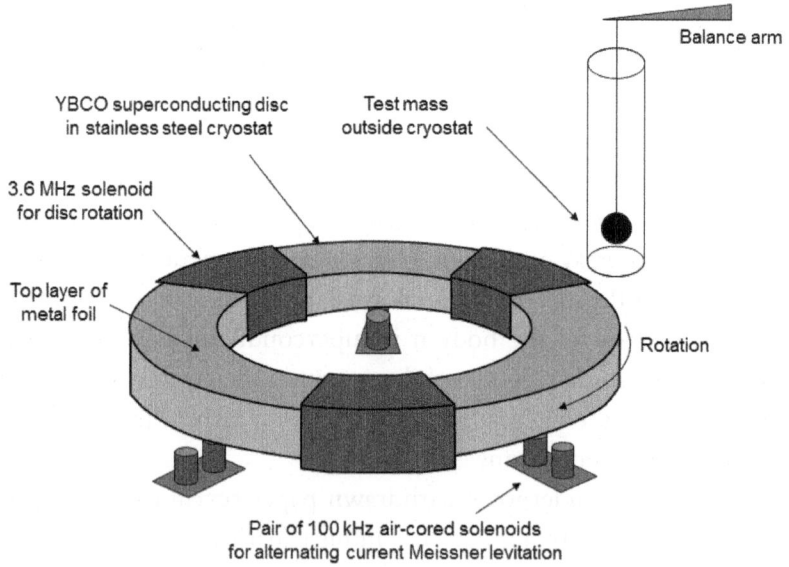

Fig. 13.3 Podkletnov's gravity-shielding experiment.

the team attempted to replicate Podkletnov's experimental observation. According to press reports at the time, Boeing researchers, at the top-secret Phantom Works in Seattle, also took a close interest in the NASA study and instituted Project GRASP (Gravity Research for Advanced Space Propulsion). Through my contact with the British inventor Tony Cuthbert, I learnt that the Boeing engineer running the GRASP programme was Jamie Childress. I was intrigued, because David Hatcher Childress is the author of the book, *The Antigravity Handbook*. I wondered whether Jamie and David were related as it's so useful to be able to discuss a controversial subject in private.

However, it seems that the Delta-G Team was not able to confirm Podkletnov's result and, subsequently, the NASA and Boeing programmes of research were terminated.

In the meantime, during 2001, BAE Systems funded a small study at Sheffield University to investigate some aspects of Podkletnov's experiment. This study, which was part of Project Greenglow, was led by Dr Clive Woods. Dr Podkletnov visited Sheffield University and described the details of the manufacture of his YBCO disc, as this was thought to be an important factor. The experimental work at Sheffield University was carried out by Dr Steve Cooke. Limited funding meant that Podkletnov's method of using Meissner levitation and interacting radio frequency (rf) fields to generate rotation of the YBCO disc was not attempted. Instead, a small YBCO disc, of 50 mm diameter, formed the lid of a cryostat mounted on the vertical axle of a motor. While the cryostat rotated and the YBCO disc was in its superconducting state the weight of a test mass suspended just above the disc was observed. No weight loss was detected. In a modified experiment, a superconducting YBCO disc was excited with a 13.56 MHz signal but, again, no weight loss of the test mass suspended overhead was noticed. So, the limited Sheffield experimental results provided no support for Dr Podkletnov's claim of local gravity modification.

However, this still leaves the possibility that it was Podkletnov's method of electromagnetic excitation of his superconducting YBCO annular disc that gave rise to a gravitational effect. The three pairs of air-cored solenoids used for alternating-current (ac) Meissner levitation of the YBCO annulus in its superconducting state create alternating magnetic vortex arrays along the edge of the annulus, with radial vortex cores. The orthogonal magnetic H-fields associated with the main solenoids, used to rotate the YBCO annulus, may interact with the magnetic vortex arrays around the edge. The YBCO rotation may also stretch the alternating magnetic vortex cores in a circular direction within the annulus. In the analogue with fluids, stretching a vortex core increases its spin. From our twinned field hypothesis we expect an alternating magnetic field to be accompanied by an alternating gravitomagnetic h-field. Due to the time-varying change in the H and h fields we expect them to be enclosed by alternating E and g fields in the vertical plane (eqns 7.4 and 8.17). Since the YBCO annulus and solenoids are contained in an electrically shielded container, only the effect of the g-field can percolate outside. At any point in the alternating induced g-field (or near field low frequency gravitational wave) the average g-amplitude probably averages to zero, although the YBCO annulus rotation may introduce some g-field asymmetry. Even so, a test mass in the external oscillating induced gravitational field, suspended from the arm of a chemical balance, might well indicate a weight change as the balance response time is likely to be poor, resulting in an averaged weight measurement. Note that the three solenoids (Fig. 13.3), used by Dr Podkletnov to rotate his YBCO disc, can be thought of as a toroid with gaps which links his experiment with an idea suggested by Dr Forward, of Hughes Aircraft Company, for generating a gravitational effect (see section 21.1).

In 2006 Dr Martin Tajmar and Dr Clovis de Matos of ESA announced that by accelerating, or decelerating, a spinning ring of niobium type II superconductor they had changed the gravitomagnetic field h of the ring, resulting in the induction (eqn 8.17) of a tiny, but detectable, gravitational field g. Interestingly, they used gas jets to spin their superconducting ring, not the solenoid method used by Dr Podkletnov. Their claim was quickly investigated by a research team in New Zealand working with UG-2, the world's largest ring laser gyro (see Chapter 11). According to their calculations, any change in the gravitomagnetic field of a spinning superconducting lead disc placed alongside one arm of UG-2 should result in a disturbance of UG-2's beat frequency. But nothing was detected, throwing the Tajmar and de Matos result into doubt. And so the experimental research goes on.

Earlier, in 2001, Dr. Podkletnov had made a further claim that, while working at a restricted security laboratory in Moscow, he had developed an apparatus which could generate a gravitational impulse. Like a scene from a spy novel, Dr Woods and I had a clandestine meeting with Dr Podkletnov, in the foyer of the

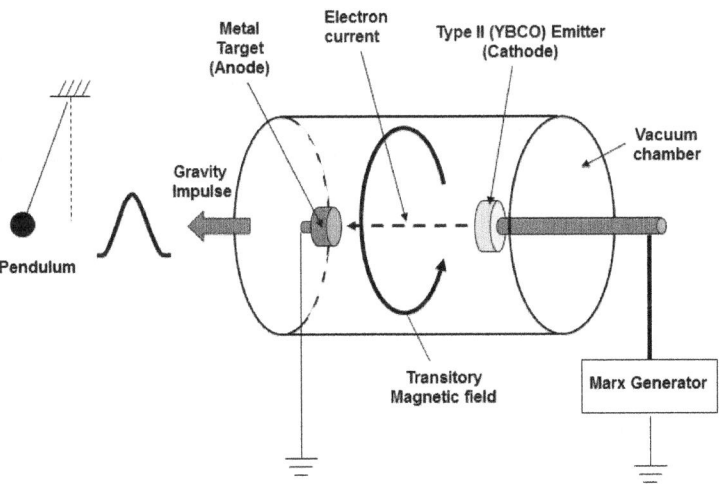

Fig. 13.4 The Podkletnov gravity impulse generator.

Swiss Cottage Hotel in central London, during the August Bank Holiday of 2001. Dr Podkletnov described his apparatus and I produced a PowerPoint diagram of it which, later, he confirmed as being an accurate representation. That diagram (Fig. 13.4), now attributed to *Jane's Defence Weekly*, has since appeared in several places. In simple terms, the apparatus consisted of a cylindrical quartz glass vacuum chamber with a diameter of 1m and a horizontal axial length of 1.5m. At one end of the cylinder was a type II (YBCO) superconducting cathode, or emitter, and at the other end was an anode, or target collector. The vacuum chamber was housed in a Faraday cage, with additional screening material to absorb any ultra-high frequency electromagnetic signals. A Marx generator was used to create an electric field between the electrodes and when the potential reached about 2 MV a massive discharge occurred, forming a current of about 10^4A which lasted a few milliseconds. According to Dr Podkletnov, the electromagnetic effects were confined within the shielded apparatus, but a narrow collimated gravitational impulse emerged from the anode end of the chamber, which was detected with a simple pendulum.

With our twinned field notion, we can see that accompanying the momentary massive electrical discharge current i is a small fleeting mass current I, given by

$$I = \left(\frac{m_e}{e}\right)i \qquad (13.19)$$

The transient currents i and I give rise to transient **H** and **h** fields. Consequently, as $\partial \mathbf{H}/\partial t$ changes then so does $\partial \mathbf{h}/\partial t$. This implies that a changing magnetic field results in an induced gravitational field, given in equation 8.17 as

$$\nabla \times \mathbf{g} = \eta \frac{\partial \mathbf{h}}{\partial t} \qquad (13.20)$$

With this thought in mind, one simplistic explanation for the generation of the gravity impulse is as follows. Accompanying the confined huge transitory magnetic field surrounding the discharge is a transitory gravitomagnetic field which, from equation 13.20, generates a transitory gravitational field, or dipole, within the chamber. Because gravitational fields cannot be screened, the induced gravitational field extends out of the chamber. However, one would expect the induced **g**-field to be exceedingly tiny, unless within the vacuum tube, during the bright discharge, η has a value much greater than its free-space value.

The general opinion of most of those scientists who have considered Podkletnov's claim is that the so-called gravity impulse is due to some other effect and is not related to gravitation at all. An acoustic shock is felt to be a strong possibility. Another possibility is that the electrons striking the anode have enough energy to create extreme X-rays. The frequency f of such emissions is given by

$$f = \frac{Ve}{h} \approx 5 \times 10^{20} \text{ Hz} \qquad (13.21)$$

V is the breakdown voltage of 2 MV, e is the electric charge of an electron and h is Planck's constant. In our discussion with Dr Podkletnov he mentioned the possibility of X-rays and said that no precautions had been taken to safeguard against them. In my mind's eye I could see a collimated beam of X-rays emanating from the anode forming a brief photonic mass (eqn 13.23) current, accompanied by a gravitomagnetic field and the generation of a transient gravitational beam (eqn 13.20) passing out of the quartz chamber towards the pendulum bob. But, again there was the η term. Was I being too fanciful (see section 21.3)?

The possible explanations given above are all based on conventional physics, but perhaps some other cause was responsible for the apparent gravitational effect. However, as no foreign scientists were permitted to view Podkletnov's apparatus in Moscow, that appeared to be the end of the matter. Then I learnt that in 2003 a team of scientists at Hathaway Consulting Services in Toronto, Canada, had attempted to replicate the Moscow experiment, with the full support of Dr Podkletnov. A large evacuated tube, containing a type II (YBCO) superconducting cathode at one end and an anode at the other end, were positioned between the charged spheres of two Van de Graaff machines. The spheres were charged with opposite polarity to create a potential difference of approximately 1 MV and, on electrical breakdown, the discharge was directed along the tube, from emitter to collector. Gravitational force detectors were placed just outside, at the end of the tube, but no gravity impulse was detected during the discharge. Thus, the prevailing view remains that Dr Podkletnov is mistaken in his claims about gravity beams.

At the moment the theory linking superconductivity with gravity remains experimentally unconfirmed, so most scientists just keep an open mind and watch out for developments. Perhaps, therefore, we should start at a more basic level and build up our knowledge. For example, we might investigate whether a suspended disc of type II (YBCO) superconductor in changing from its superconducting state, with its magnetic vortex array, to its normal state following a rise in temperature, undergoes rotation as one might expect from the Einstein-de Haas effect. In this case, though, any effect must be associated with the decay of the array of magnetic vortex tubes percolating the disc, as spinning ferric molecules play no part.

In terms of looking for an inertial analogue of the Einstein-de Haas effect, we might consider suspending a fine torsion balance just above the surface of a fixed type II superconducting disc which is allowed to warm so that the magnetic vortex array decays. From our notion of twinned fields, theory suggests (eqn 13.20) that as the magnetic vortex field decays, an induced **g**-field will curl around the direction of changing gravitomagnetic **h**-field. Although any rotation of the torsion balance may seem highly unlikely, given the smallness of the inertial permeability η of free space, the experiment may be worth a try in the search for a breakthrough.

As an alternative to the proposed superconducting inertial version of the Einstein-de Haas experiment we might try using a torsion balance, or torsion ring, placed inside a conducting coil and look for any signs of a transient rotation when the current is ramped up, or down (Fig. 13.5). This proposed experiment is also based on the idea that as $\partial \mathbf{H}/\partial t$ changes, so does $\partial \mathbf{h}/\partial t$, thereby creating (eqn 13.20) a curled gravitational field. Again, due to the smallness of the inertial permeability η of free space, any rotation may be inhibited, but in the search

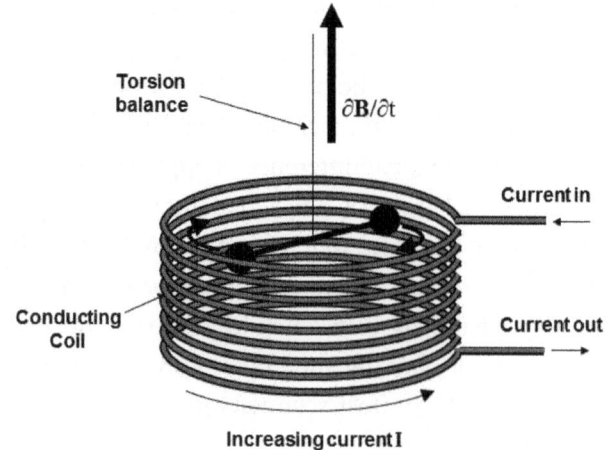

Fig. 13.5 Torsion balance suspended in coil with changing current.

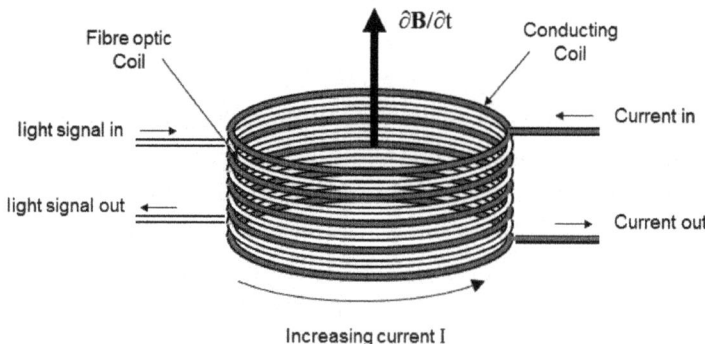

Fig. 13.6 Coincident conducting and fibre optic coils.

for an advance as many ideas as possible need to be explored.

For increased sensitivity, we can replace the torsion balance with a fibre optic coil coincident with the conducting coil. A single frequency light signal is fed into the fibre optic and a shift in frequency is looked for while the current in the conducting coil is ramped up, or down. The basis for this idea is Einstein's gravitational red-shift (eqn 19.9).

Perhaps, as a first go, we might try placing a fibre optic gyro (FOG) in a conducting coil, with their axes coincident. Switching the coil electric current on and off is one form of ramping. Although the detection of an effect may seem highly unlikely, again, the idea may be worth a try. The experiment is reminiscent of Faraday's electromagnetic induction experiment of 1831, where his observation of just a flicker of the galvanometer needle led on to the development of the colossal electrical industry.

In the late 19th century Maxwell, although unaware of the existence of electrons, built an apparatus to try to determine whether a conducting copper coil possessed a gyroscopic property. However, his experiment was not successful. Nevertheless, since electrons have a rest mass m_e, then a conducting coil containing an electron current must be subject to a gyroscopic effect. But the effect is exceedingly small and Maxwell's apparatus was not sensitive enough to detect it.

So, as Maxwell's attempt to detect the gyroscopic property of a conducting coil reminds us, even if the idea is right, the success, or failure, of an experiment hinges on the level of detection sensitivity. This applies to the experiments described above.

From quantum mechanics (Chapter 15) we know that a photon possesses energy E, given by

$$E = hf \qquad (13.22)$$

where h is Planck's constant and f is the photon's frequency. Combining this with Einstein's relationship between mass and energy (eqn 8.8) we see that a photon has effective, or photonic, mass given by

$$m = \frac{hf}{c^2} \qquad (13.23)$$

This is another aspect of the idea of twinned fields. Consequently, just as a copper coil containing an electron current must be subject to a gyroscopic effect so, too, must be a fibre optic coil with a rotating light beam. So in theory, at least, a fibre optic coil could be used as a gyroscope to detect rotation. However, a visible light photon has a photonic mass of about 10^{-35}kg, which is considerably less than the mass of an electron, which is about 10^{-31}kg, so it seems exceedingly unlikely that the gyroscopic property of a fibre optic coil with a circulating light beam would be detectable directly. Nevertheless, the drift velocity of the current of electrons is only about 1m/hr, whereas photons in glass move at two-thirds the speed of light in a vacuum, of order 2×10^8m/s. So detection may be possible, after all. Only an experiment will tell.

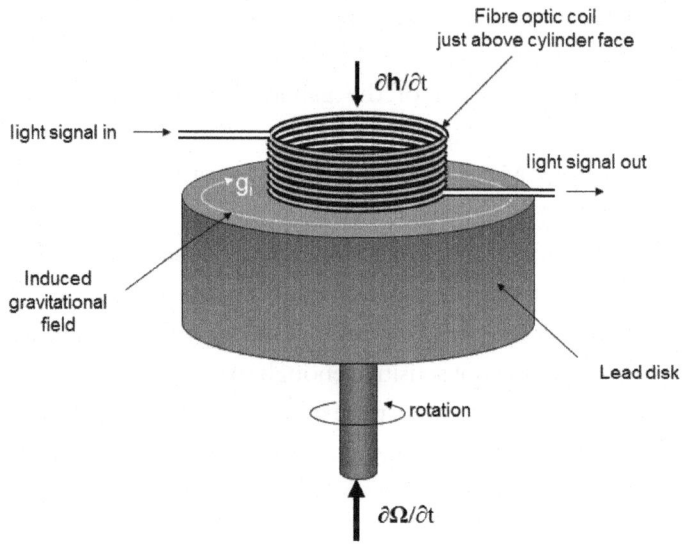

Fig. 13.7 A more sensitive version of Harvey Morgan's experiment.

Actually, the possible gyroscopic property of a fibre optic coil, generated by a photonic mass current, is not the method used by a FOG to detect rotation. Achieving the sensitivity needed to detect very small rotations with a FOG relies on a different physical phenomenon, namely the Sagnac effect and the effect that acceleration has on the frequency of light (eqn 19.11). In particular, it is the photon's change of frequency in response to the change in tangential acceleration during a change in angular velocity Ω that is exploited. To increase the sensitivity, counter-rotating light beams are used.

At this stage it is interesting to combine the last two experiments, dispense with the conducting coil, keep the fibre optic coil for detection purposes and rotationally spin up the torsion balance so that the two end masses accelerate. Then, according to equation 13.20, the two masses should each generate an induced gravitational field. For a greater effect we could replace the two end masses of the torsion balance with a suspended torsion ring. In fact this is a non-superconducting version of the Tajmar-de Matos ESA experiment. If we replace the torsion ring with a disc, as shown in Fig. 13.7, then we have a more sensitive version of Harvey Morgan's experiment, mentioned in Chapter 11.

Although angular velocity Ω does not possess an influence field, a changing angular velocity $\partial\Omega/\partial t$ may result in a change in gravitomagnetic field $\partial \mathbf{h}/\partial t$,

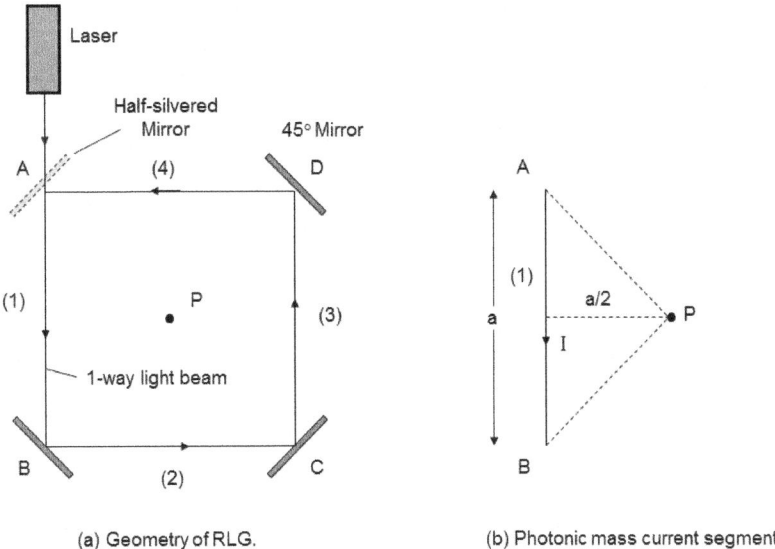

Fig. 13.8 Mallett's proposed RLG experiment.

thereby creating an induced gravity field $\mathbf{g}_i$, which does possess an influence field. If this field extends outside the disc then, while the angular velocity is changing, a frequency shift should occur between the input and output signals to the fibre optic cable. From Newton's third law, an estimate of the induced gravitational field is $r\partial\Omega/\partial t$, where r is an internal radius of the disc. The frequency shift Δf is given by Einstein's red-shift formula (eqn 19.9), where the path length Δs is $2\pi rN$, where N is the number of turns of the fibre optic cable.

The experiments described above were aimed at stimulating some new ideas with the hope that they might lead to an Earth-based laboratory experiment to detect the gravitomagnetic field, or its effect. The search for the missing link between electromagnetism and gravitation provided an added reason. Having written this chapter, it came as quite a surprise to learn of the gravitomagnetic experiment proposed by Professor Ronald Mallett, of the University of Connecticut, which has close links with some of my ideas. As an example of synchronicity, we had both considered the idea that since photons have effective mass then light beams form photonic mass currents which must generate gravitomagnetic fields. Since Professor Mallett is a theoretical gravitational physicist, this gave me a great fillip. However, I had already concluded that the external gravitomagnetic field, outside of the light beam, was too weak to be detected and had, consequently, dismissed this type of experiment. Clearly, I needed to re-examine my idea.

Mallett's proposed experiment is a variation of the GP-B experiment (Chapter 11), but in the form of a laboratory table-top experiment. He envisaged using a ring laser gyro (RLG), with just one circulating laser beam (Fig. 13.8a), to create a gravitomagnetic field and using a neutron, with intrinsic spin (Chapter 16), as an atomic gyro. On placing the neutron at the centre of the RLG, any neutron spin precession would be tantamount to detecting the existence of a gravitomagnetic field.

Using a linearised (weak) gravity field approximation Professor Mallett derived the strength of the gravitomagnetic field at the centre of the RLG. However, on looking at Mallett's theory, in the respected journal *Physics Letters* (A 269, 8 May 2000), I was confronted with several pages of esoteric equations which could only be interpreted by an expert in the subject. I turned to Professor Robin Tucker, our adviser for Project Greenglow, for help and was told that, although he had some criticisms, the theory was basically valid.

This reinforced my view that the complexity of the theoretical work involved in determining the gravitomagnetic field strength obscured the underlying idea and was an impediment to opening up the subject and enlisting help from talented experimentalists from other areas of science and engineering. Assuming the validity of the electromagnetic analogue with gravity, using this method offers a much simpler and clearer approach. The Biot-Savart model is used to determine the magnetic field around a straight wire carrying an electric current, so we can use the analogue to derive the gravitomagnetic field around a straight photonic mass current. Each side of the square RLG is of length a. Due to the finite length of the photonic mass current I, the result is slightly different to that given in equation 6.3. It turns out that the gravitomagnetic field h_1 at a perpendicular distance a/2 from the centre of the laser beam of length a (Fig. 13.8b) is

$$h_1 = \frac{I}{\pi a \sqrt{2}} \qquad (13.24)$$

From equations 2.5, 8.6 and 13.24 the induced gravitomagnetic field **b** at the centre of the RLG, made up of four photonic mass current segments, is given approximately by

$$b = \eta_0 h = 4\left(\frac{4\pi G}{c^2}\right)h_1 = \left(\frac{G 8\sqrt{2}}{a c^2}\right)I \qquad (13.25)$$

Now the intensity **I** of laser beam is given by

$$I = \frac{\text{Power}}{\text{Cross-sectional area of beam}} = Ic^2 \qquad (13.26)$$

So the induced gravitomagnetic field **b** at the centre of the RLG is

$$b = \left(\frac{G\,8\sqrt{2}}{a\,c^4}\right) I \qquad (13.27)$$

The precession of a gyro in a gravitomagnetic field is the analogue of a search coil precessing in a magnetic field, where the rate of precession is called the Larmor frequency. In either case, the gyro and the search coil experience a turning moment, or couple, which causes the precession. Theory shows that the magnitude of the rate of precession Ω is given (eqn 11.3) by

$$\Omega = \tfrac{1}{2} b \qquad (13.28)$$

So, finally, we get Mallett's result for his proposed RLG experiment:

$$\Omega = \left(\frac{G\,4\sqrt{2}}{a\,c^4}\right) I \qquad (13.29)$$

To increase the strength of the gravitomagnetic field Mallett proposed using a stack of RLGs. If we replace Mallett's RLG stack with a fibre optic coil of height ℓ, containing N turns, we can think of this as the analogue of a solenoid. Reading across from magnetic theory (and assuming that the speed of light in glass is roughly the same as that in the vacuum) we see that the induced gravitomagnetic field along the coil centre-line is

$$b = \left(\frac{4\pi G}{c^2}\right)\left(\frac{N}{\ell}\right) I = \left(\frac{4\pi G N}{\ell\,c^4}\right) I \qquad (13.30)$$

To detect the presence of the gravitomagnetic field Professor Mallett proposes using the Faraday effect, discovered by Faraday in 1845. When a linearly polarised ray of light is shone along a magnetic field line the angle of polarisation is rotated. The angle turned is dependent on the magnetic field strength, the path length of the light ray and a constant related to the material in which the magnetic field is established. Mallett assumes that the same will be true for the gravitomagnetic field and hopes that by shining a polarised light ray along the RLG stack centre-line of length L to detect the rotation θ caused by the presence of the induced gravitomagnetic field **b**

$$\theta = v\mathbf{b}L \tag{13.31}$$

The parameter v is the Verdet constant, which has only been determined for magnetic cases but not, as yet, for gravitomagnetic cases.

However, we can be fairly confident that the Faraday rotation effect will apply to gravitomagnetic cases for two reasons. Firstly, it follows from our twinned field notion. Secondly, it is known that light is dragged around by a rotating medium. That is when the plane of polarisation of linearly polarised light shone along the axis of rotation of a transparent medium is turned, albeit slightly. Since angular velocity Ω is equivalent to induced gravitomagnetism **b** this supports the view that the Faraday rotation effect will occur in the gravitomagnetic case.

The motivation which led Professor Mallett to develop his experimental idea is his deep interest in the possibility of time travel. One way of achieving time travel is with the rotation of the space-time metric linked with a strong gravitomagnetic field. However, the c^4 parameter in the denominator of equation 13.30 means that the gravitomagnetic field generated by an RLG stack will be extremely weak. To strengthen the gravitomagnetic field Mallett has considered the idea of using *slow-light*. That is light, or rather the group velocity of a light beam, moving at a few metres per second. However, several scientists have challenged the underlying principle behind the use of slow-light in the way anticipated by Mallett. Currently slow-light is a hot topic for research. DARPA, the US Defense Advanced Research Project Agency, has spent millions of dollars investigating applications of slow-light in optical fibres, although this is probably for optical computing purposes. So, whether Mallett's experiment to generate a detectable gravitomagnetic field is feasible remains to be seen.

All along, the difficulty in detecting a gravitomagnetic field outside a mass current has been due to the extreme smallness of η, the gravitomagnetic permeability of free space. Perhaps the gravitomagnetic field of a mass current might be detected within the current, where the internal magnitude of η is large (eqn 20.7). Michael Faraday glimpsed the possibility that this might be so in his own gravitational experiments. He noted (10140: 30 August 1849) that magnetism has a strong effect on an iron bar inside a current carrying coil but has no effect on an iron tube outside. "Which shews the great difference within and without. Something of a like nature may occur with gravitation effects. So go on experimenting…" However, he was not able to capitalise on his insight.

In April 2002 Professor Raymond Chiao, of the University of California, claimed that it ought to be possible to use a superconducting sheet as a special antenna, or transducer, to convert incident microwave radiation to gravitational radiation and vice versa. Getting the plane of the incident electromagnetic wave polarisation correctly positioned to stimulate the magnetic vortex arrays in the superconductor surface might be an important factor. If Chiao's idea is right, then gravitational radio communication ought to be possible, through matter as well as across space. However, Chiao's experimental investigation of the proposed technique was not successful, leaving his idea hanging, but not dismissed.

In May 2003 Dr Robert Baker co-chaired, with Dr Paul Murad of the US DoD, an international conference on methods of generating and receiving high frequency gravitational waves, to try to throw some light on the subject. This was attended by Dr Walter Johnston, from BAE Systems, Warton, representing Project Greenglow. NASA personnel and engineers from Boeing also attended, along with many academics from the USA, Russia, China and Europe. But, no breakthroughs were revealed.

For many years there were rumours that scientists in the USSR had developed new technologies based on having discovered the link between electromagnetism and gravitation. One particular fear was that the Soviet Union had developed a communications system which allowed their military forces to transmit messages through the Earth to underground bunkers and through the sea to nuclear submarines. There was a concern that NATO, being unable to eavesdrop, would be blind to Soviet intentions. There was even speculation that Soviet scientists had developed an imaging system which allowed their security services to 'see' inside buildings. With hindsight, it seems that the rumours probably had some basis, being linked with Soviet scientific research on neutrino beams (see Chapter 16).

If the Russians have discovered a link between gravity and electromagnetism then they have the key to gravitational propulsion technology, too. However, when the Cold War ended during the 1990s no details of any such secret technologies leaked out and it was assumed that they didn't exist. If they do, then NATO forces may have no counter-measures to defend against them, leaving them extremely vulnerable.

14

THE LUMINIFEROUS ETHER

It might be thought that carrying out experiments to investigate how to make a vacuum is fairly pointless. After all, what is the point of having a chamber containing nothing? But, it was Otto von Guericke's original development of the air pump to create a vacuum chamber that led others to think about using condensing steam to achieve the same result. This led to the steam pump which, in turn, led on to the Age of Steam Power, which ushered in the first Industrial Revolution.

In 1654 von Guericke, the Mayor of Magdeburg, in Germany, demonstrated the power of the vacuum to his highness Ferdinand III, the Holy Roman Emperor. He made a globe from two copper hemispheres and using his air pump he evacuated the region inside. A team of eight horses attached to each hemisphere failed to pull the globe apart, showing the considerable force of air pressure on the outer surface of the globe that we are, mostly, unaware of.

The Ancient Greeks had their five elements, namely earth, water, air, fire and the ether. These are equivalent to our five states of matter, being solid, liquid, gas, plasma and the vacuum. The Greeks were sure about their first four elements and so are we about our first four states of matter. But they weren't sure whether the ether was a plenum (completely full of something) or a void (containing nothing at all). Over 2000 years later, we are still not sure whether a vacuum is filled with a very fine medium of something or is truly empty.

In Maxwell's treatise on the *Dynamical Theory of the Electromagnetic Field*, published in 1864, he said that based on the transmission of heat and light from the Sun he believed that a luminiferous ether must exist which filled the vacuum of space and that it was so fine that it permeated through all bodies. Maxwell assumed that electric, magnetic and gravitational potential energies could be stored in the ether, by stressing the medium. However, since it is generally accepted that energy is a positive quantity, Maxwell was puzzled by the fact that gravitational potential energy was negative. In order to keep the energy positive overall the ether would have to contain an infinite store of energy, which didn't seem very likely. In our everyday life we are not aware that the ether has any energy, never mind that it might contain an infinite amount. But, then, who was really aware of air pressure until von Guericke carried out his famous demonstration?

Lord Kelvin wondered whether atoms could be made out of knots made of ether. He mused over the idea that the simple atom might be formed by a vortex ring of ether, while more complex atoms might be formed by vortex rings which contained several intertwined loops, or knots.

Hertz's discovery of electromagnetic waves, in 1888, seemed to confirm Maxwell's view of the existence of the ether. Surely, wave motion needed a medium? Many scientists were persuaded by this argument. So, it was generally accepted that the ether was like an extremely fine fluid which was stationary with respect to the Sun and that the Earth moved through it. The ether was thought to be so fine that it passed right through the Earth without being disturbed.

Based on the analogue with a fluid, it was assumed that the speed of light measured on Earth would be affected depending on whether the light waves were moving upstream or downstream in the ether flow, or moving perpendicularly across it. In the same way, when viewed from the bank, the speed of a motorboat in a river depends on the direction it is headed and the direction of the current.

In 1887 the US scientists Albert Michelson and Edward Morley used a large interferometer to try to detect the motion of the Earth through the ether. However, they could not detect any change in the interference fringes, outside the limits set by experimental error. The speed of light was unaffected. Was the ether stationary with respect to the Earth? Accepting this proposition was rather like saying that the Earth was the centre of the Universe and everything else moved around it. This idea had already been dismissed, so it seemed that the Michelson-Morley experiment implied that the ether didn't exist. Michelson was the first American scientist to win the Nobel Prize in Physics, for this work, in 1907.

Nowadays, astronomers tell us that our Sun is moving through our galaxy towards the star Vega at a speed of more than 40,000 mph, taking the Earth with it. So, why didn't the Michelson-Morley experiment detect this motion?

In 1893, just when it seemed that a conclusion about the non-existence of the ether had been reached, the Irish mathematician, George Fitzgerald, suggested that it might still be there. Perhaps, he suggested, the ether had a strange effect on distances measured in the direction of motion, causing them to be compressed, or contracted.

The idea of Fitzgerald's contraction appealed to the Dutch theoretical physicist, Hendrik Antoon Lorentz. He developed the idea further and deduced a set of 3-D linear transformations between inertial frames of reference, where length contractions were built in for distances in the direction of motion. Amazingly, with these transformations, now known as the Lorentz-Fitzgerald transforms, it was possible to transform Maxwell's equations for electromagnetism between inertial frames of reference moving with different velocities.

It would be nice if the Lorentz-Fitzgerald transforms could, also, be used between motions rotating at different speeds. But unfortunately they can't, because rotating frames of reference are not inertial frames, since radial accelerations are present.

A stationary electric charge has its own special frame of reference called the 'rest frame' in which there is no magnetic field. Using the Lorentz-Fitzgerald transforms we can impose a uniform velocity on the charge. Now, when viewed head-on, the charge is found to be surrounded by a circular magnetic field. Thus, magnetism is seen to be a relativity effect (see Chapter 9). The reverse process applies, too, and any magnetic field surrounding a moving charge can theoretically be transformed away by moving to the rest frame of the charge.

So, what about the existence of the ether? Well, finally, or so it seemed, Albert Einstein provided the answer in his famous paper on special relativity, published in 1905. He stated that the speed of light is constant and independent of the inertial frame of reference in which measurements of it are made. It doesn't matter in which inertial frame of reference a light beam is emitted; in any other inertial frame of reference, moving at a different speed, the observed speed of light will still be the same. He also stated that the laws of physics must be the same in all inertial frames of reference. All that was needed to change frames of reference were the Lorentz-Fitzgerald transformations.

With the introduction of Einstein's theory of special relativity, incorporating the Lorentz-Fitzgerald transforms, the existence of the luminiferous ether was superfluous to requirements. Consequently, whether the ether really existed, or not, was a philosophical issue which could be avoided. So although during the 19th century most scientists had been convinced that the ether existed, following Einstein's paper on special relativity, many scientists during much of the 20th century concluded that the ether probably didn't exist. By implication, the vacuum was empty, completely devoid of anything.

The crux of the special theory of relativity is that for any medium the speed of light remains the same in all inertial frames of reference moving at different uniform velocities. What is seldom mentioned is that for this to be true the same light viewed in different frames of reference must have a different frequency and this is linked with 'time dilation'. Photonic mass is unlike ordinary mass in that it cannot have a rest mass m_0. Nor can photonic mass be accelerated in a uniform medium. Photons always travel through such a medium at a constant speed c with a finite apparent mass. Any attempt to accelerate, or decelerate, photons travelling in a uniform medium will result in them changing frequency.

The concept of the photon, also introduced by Einstein in 1905, offers another explanation for why the ether was not detected. A photon has effective mass and, therefore, is surrounded by a gravitomagnetic field, the strength of which is dependent on the frame of reference from which the source of light is viewed. When these effects are allowed for, the resulting frequency change turns out to be a first order approximation to the special relativity form of the Doppler shift, obtained from the Lorentz-Fitzgerald transforms. So, although the speed of light remains constant, as observed by Michelson and Morley, its frequency, and its photonic mass, change.

As a final point, it's interesting to consider a common sense view about the existence of a medium when we move from one inertial frame of reference to another. Suppose a stationary observer on a windless day holds a toy windmill. The observer is in the rest frame of the windmill. If he moves forward with the windmill through the air, the windmill will begin to rotate. So, a change in reference frame for the windmill causes an effect, due to the motion through the medium. Look at it in reverse and keep the windmill still, but let the observer walk towards it. The observer is now in a frame of reference moving with respect to the windmill, but there is no effect on the windmill! To get the equivalent reverse effect the observer must take the air with him/her, which can be done by creating a wind of the same speed at which he/she walks. Now the windmill turns. No medium, no effect! Special relativity may have rendered the ether superfluous but it didn't say that it doesn't exist. Perhaps the vacuum of space isn't empty after all.

15

THE QUANTUM OF ENERGY

As a solid body is heated more and more we imagine that its atoms, being held in a lattice structure, vibrate more and more, as though rattling the cages of their confinement. Internally, there is a whole mixture of vibrating frequencies, some being more prominent than others. On reaching the surface, heat energy is radiated away in the form of electromagnetic waves covering a range of frequencies. However, our eyes only respond to the narrow band of frequencies in the visible light spectrum (in the 10^{14}Hz band). When a body is first perceived to be hot we see its surface take on a dim red colour. As the temperature increases some higher frequency orange appears with the red, the colours merging gradually to give yellow; then even higher frequency blue appears (yellow and green make blue, so we don't see green separately); and finally, as the body starts to vaporise, it becomes white hot, as all the colours of the visible light spectrum merge together.

Heat radiation experiments were carried out in laboratories during the 19th century using an oven with a fine temperature control. The interior of the oven was totally shielded from outside influence and, prior to heating, was in total darkness. Consequently, the oven interior was referred to as a black body cavity.

On being heated to a particular temperature the oven was left to settle until a state of internal temperature equilibrium was reached. In this state it was assumed that the oven walls were all at the same temperature and that the same amount of heat being radiated by the walls was also being absorbed by the walls. A sample of the heat energy radiating around inside the oven was then allowed to escape from a small hole in the door and this was analysed to determine the various energy contributions from the electromagnetic waves involved. At this time, in accordance with Maxwell's premise, it was assumed that electromagnetic radiation needed a medium to travel through, namely the ether.

Experimentation showed that the heat radiation emitted from the hole in the oven door was only dependent on its temperature. It didn't matter what the walls of the oven were made of. In 1879 Josef Stefan, an Austrian physicist, deduced from experimental data that the total emitted power of heat radiation across the whole spectrum from the hole in the oven door was proportional to the fourth power of the oven's temperature T. Stefan's research student was Ludwig Boltzmann. Later, Boltzmann developed the theory of statistical mechanics, a theory based on the idea that a heated body may be treated as a vibrating body made of many interlinked particles, the overall properties of which can be defined in statistical terms of probabilities and averages. Boltzmann introduced the idea that the energy E associated with each degree of freedom, or independent mode of vibration, of a body was proportional to its temperature T.

$$E = \tfrac{1}{2}kT \tag{15.1}$$

The constant k is Boltzmann's constant.

$$k = 1.3807 \times 10^{-23} \text{ J/K} \tag{15.2}$$

Boltzmann set out to prove Stefan's empirical law theoretically, using classical thermodynamics theory. This he achieved in 1884.

Near the end of the 19th century, black body cavity experiments over a range of temperatures had enabled German and Austrian scientists to determine a parameter called the emissive power $E(\lambda,T)$ for a particular temperature T, such that $E(\lambda,T).d\lambda$ was the heat energy being radiated per second per square metre within the narrow wavelength band from λ to $\lambda+d\lambda$. The units for $E(\lambda,T)$ are $W.m^{-2}/m$. They were then able to plot out the emissive power $E(\lambda,T)$ against wavelength λ for a range of temperatures T (Fig. 15.1). For each temperature T, the total power radiated per square metre for the whole spectrum is equal to the area under that T-curve.

In 1893 Wilhelm Wien, a German scientist, derived his 'displacement law', showing that as the temperature T of a body increased, the maximum heating effect was displaced to shorter wavelengths λ_{max}. The result, supported by experimental evidence (Fig. 15.1), is expressed as

$$\lambda_{max}T = \text{a constant} \tag{15.3}$$

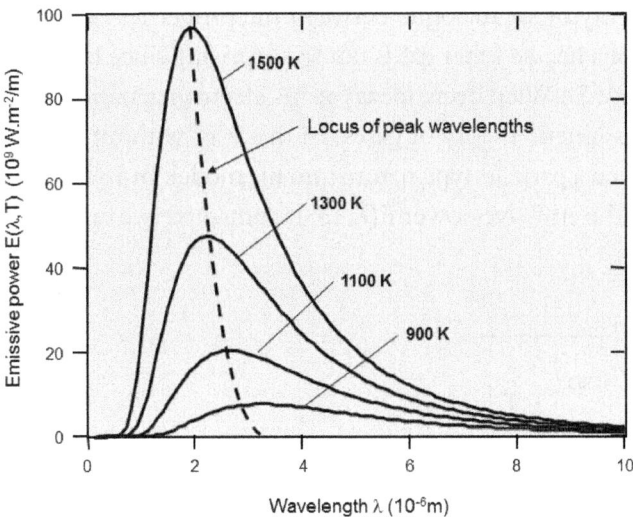

Fig. 15.1 The black body spectrum.

That is, as the temperature T increases so the peak wavelength λ_{max} decreases. From experiment the constant was found to be 2.93×10^{-3} m.K.

Wien extended his displacement law and was able to show that the maximum value of the emissive power $E(\lambda,T)$ was proportional to the fifth power of temperature T. Wien then thought about developing a model for the radiation emanating from a black body cavity.

Wien had noticed that Maxwell's velocity distribution curves for heated gas molecules looked similar to the black body radiation curves. Perhaps heat radiation was analogous to the movement of gas molecules. The kinetic theory of gases has a long history, stemming from ideas originally proposed by Titus Lucretius during the 1st century BC in ancient Rome. In the modernised theory, it is assumed that gases are composed of microscopic spherical particles, or molecules, which move about with different velocities v. Several scientists before Maxwell had developed the theory, including Robert Boyle in the 17th century and Daniel Bernoulli in the 18th century. Using Newton's laws of motion coupled with statistical averaging, Maxwell was able to model the motions of gas molecules. Introducing n as the mean number of molecules per unit volume he derived velocity distribution curves for n against the change in velocity from v to v+dv, where dv signifies a small increase in velocity, for a range of temperatures T.

The possibility of an analogue between the properties of a heated gas and the properties of a heated ether gas is not so surprising since both are associated with energy flow. So Wien drew ideas for his electromagnetic radiation theory from Maxwell's kinetic theory of gases. In this way, without realising it, Wien almost introduced a particle-type nature into his model. In 1896 Wien published his formula for the emissive power $E(\lambda,T)$. It contained two empirical constants, c_1 and c_2.

$$E(\lambda,T) = \left(\frac{c_1}{\lambda^5}\right)\frac{1}{\exp\left\{\frac{c_2}{\lambda T}\right\}} \qquad (15.4)$$

The term *exp* stands for the exponential function, often designated as e (and not to be confused with the charge of an electron). The constant $e \approx 2.718$ is a naturally occurring constant in nature, like the constant $\pi \approx 3.142$. The brackets { } indicate the power to which *exp* is raised.

With the appropriate choice of c_1 and c_2 to get the best fit to the experimental $E(\lambda,T)$ curves, Wien's formula seemed to work reasonably well. However, on closer investigation of the narrow bandwidth associated with light he found that, in getting a good match to the experimental results for the shorter wavelength ultra-violet part of the visible spectrum, the match covering the longer wavelength infra-red end of the spectrum was poor.

In Britain in 1900, Lord Rayleigh attempted to derive his own formula for $E(\lambda,T)$. Being an expert on sound vibration, Rayleigh's approach was to consider a black body cavity containing ether and electromagnetic waves as being analogous to a box full of air containing sound waves. The dimensions of the box supported various modes of oscillation, all of which vibrated independently. He assumed that each mode contained the same amount of energy. A correction to Rayleigh's model by James Jeans led to the following expression for the emissive power $E(\lambda,T)$ of a black body

$$E(\lambda,T) = \frac{2\pi c k T}{\lambda^4} \qquad (15.5)$$

The constant k is Boltzmann's constant (eqn 15.2) from statistical mechanics.

Their predicted results agreed well with the experimental emissive power

$E(\lambda, T)$ data for longer wavelengths, at the infra-red end of the spectrum, but for the shorter wavelengths, at the ultra-violet end of the spectrum, their predicted value for $E(\lambda, T)$ quickly became infinite. This was true even for low temperatures, which didn't make sense. Scientists abhor infinities; it means there's something wrong with the model. The failure of the Rayleigh-Jeans model at ultra-violet wavelengths has since become known as the 'ultra-violet catastrophe'.

In the meeting of the Berlin Physical Society, in October 1900, the German physicist Max Planck pointed out that merely by adding '-1' in the denominator of Wien's formula (eqn 15.4) the predictions would match the experimental radiated heat energy distribution curves exactly across the whole spectrum. But why was this so? Planck was then obliged to painstakingly work backwards to try to determine how to derive the modified formula theoretically. Although it appeared to be only a minor modification, the ramifications in explaining the reasoning behind it led to a revolution in physics that astounded the scientific world. As Planck later confessed, "This was the most strenuous period of academic work in my whole life."

Planck assumed that the inside wall of a black body cavity was lined with a number of point-like electromagnetic dipole oscillators. Today, we would think of these oscillators as vibrating atoms, but when Planck was formulating his model the concept of the atom had not been developed in any depth. He assumed that the heat energy E of each oscillator was dependent on its vibrational frequency f. For thermal equilibrium he assumed that each oscillator resonated in sympathy with the frequency of the heat energy radiated and absorbed. An oscillator with a resonant frequency f also resonated at the harmonic frequencies 2f, 3f and so on. Planck then introduced a parameter h, so that energy could only be exchanged in discrete quantities $E = hf$ by each oscillator. So his oscillator could exchange energies hf, 2hf, 3hf and so on. He then had to work out all the possible oscillatory modes for the N oscillators to get the energy for each mode, then sum the results to get the total energy E_N and finally divide by N to determine the averaged energy E per oscillator.

Planck established that the averaged energy $E = E_N/N$ for one oscillator was

$$E = \left(\frac{hc}{\lambda}\right) \frac{1}{\exp\left\{\frac{hc}{\lambda k T}\right\} - 1} \tag{15.6}$$

Having introduced a discrete energy exchange for each oscillator vibrating at a frequency f, Planck hoped that the combination of N and h in his model would smooth out the discreteness, allowing a continuous change in energy with continuous change in frequency. He thought that this could be achieved by making N, the number of oscillators, tend to infinity while making h tend to zero. But his result for E (eqn 15.6) wasn't dependent on N and if he let h tend to zero he would get E = kT, since $exp\{0\} = 1$. He was very perplexed. If every mode of oscillation emitted the same amount of energy then this would lead to the ultra-violet catastrophe encountered by Rayleigh and Jeans (eqn. 15.5). So, Planck had derived part of the formula that he was searching for, but he was stuck with his model where the radiated energy was exchanged in discrete amounts, rather than a range of continuous amounts that he had expected.

Next, Planck had to multiply the average energy E of one oscillator by the number of resonating modes at that frequency per unit volume of the ether. Finally, after some rather tortuous work, Planck completed the derivation of his formula for the emissive power distribution of a black body at a temperature T.

$$E(\lambda, T) = \left(\frac{2\pi h c^2}{\lambda^5}\right) \frac{1}{\exp\left\{\frac{hc}{\lambda kT}\right\} - 1} \qquad (15.7)$$

Planck had derived the formula that he wanted and determined the theoretical values of Wien's constants c_1 and c_2. But, in doing so he had used a method which involved adding discrete amounts of energy and had introduced a new constant h. The magnitude of h turned out to be extremely tiny.

$$h = 6.626 \times 10^{-34} \text{ J.s.} \qquad (15.8)$$

The dimensions of 'h' were the same as those of angular momentum, measured in SI units of Joule-seconds.

Planck was not aware of any natural phenomenon that h might represent and he was very apprehensive about his new constant. But, taking a bold stand, he declared that for an oscillator vibrating with frequency f, the smallest unit of energy that it could emit or absorb was E = hf. He called this quantity of energy a quantum. He added that energy could only be emitted or absorbed by an electromagnetic oscillator in integer quantum amounts of E, 2E, 3E and so on.

Energy could only be transferred in discrete packages, rather than in any amount as had, hitherto, been assumed. He then had to wait and see what the reaction was by his colleagues and other scientists to his bizarre idea.

With hindsight we can see that Planck had stumbled on one of nature's secrets; that radiation involves the emission of discrete amounts (quanta) of energy, not continuous amounts. At high frequencies (short wavelengths) blackbody oscillators are not able to generate large enough energy quanta for emission, so that high-frequency radiation is curtailed, thus explaining why there is no ultra-violet catastrophe in reality.

Einstein was fairly quick to see the implications of Planck's discovery. He used it in his 1905 paper to explain the photoelectric effect, where a beam of electromagnetic waves striking a particular metal surface can result in the emission of electrons with a range of velocities up to a certain maximum.

In 1902 the German physicist Philip Lenard showed that increasing the intensity of incident radiation, although resulting in an increase in the number of emitted electrons, made no difference to the maximum velocity that any of the emitted electrons could attain. However, keeping the intensity fixed but increasing the frequency of the incident radiation did cause an increase in the maximum velocity of some of the emitted electrons. This did not seem to make sense. Surely, increasing the intensity of radiation, so that more incident electromagnetic energy hit the metal surface, should increase the overall kinetic energy of the freed electrons and, therefore, the maximum velocity of some of them? As Professor Tucker once said to me, "It's like shining a brilliant (high intensity) red light laser on to a metal surface but getting no release of electrons. And then shining a faint (low intensity) torch light with a green filter onto the same metal surface and getting electron emission." What had changing the frequency got to do with it?

Einstein pictured the situation in a new way. Rather than the uniform spread of energy in the beam of incident radiation, he saw it as a stream of electromagnetic particles, or photons (a term introduced later by Arthur Crompton), each containing a quantum of energy. It was rather like a stream of grape-shot fired from a cannon gun. All the photons had the same amount of energy. When a photon hit an electron in the metal surface, if it had enough energy, it could free it, or knock it out of the surface. The energy of the photon became the kinetic energy of the emitted electron. (It's not quite as simple as that as some incident energy is retained by the metal, the amount depending on the work function.) Increasing the intensity of the radiation meant more photons striking the metal surface, but each one still had the same amount of incident energy. More photons meant more emitted electrons, but no change in the distribution of their kinetic energies. However, increasing the radiation frequency meant that the energy of each incident photon was increased. Now when the photons struck the surface electrons they stung more and, if freeing them, gave them a greater kinetic energy, increasing the maximum velocity of some of them. So, Einstein's use of Planck's quantum of energy explained the photoelectric effect.

Earlier, in 1887, Hertz had discovered that bathing the spark gap of his apparatus (Fig. 7.5) with ultra-violet light reduced the voltage at which a spark occurred, but he had no idea why. We now know that it was caused by the photoelectric effect, where ultra-violet photons striking the terminal ends freed surface electrons forming an electron cloud in the gap, thereby providing a lower resistance conducting path for an incipient spark.

Little did Planck realise, in 1900, that he was starting a whole new field of physics, now known as quantum mechanics. It took a long time, including a world war, for the scientific community to accept Planck's strange idea of energy quanta but, finally, he was awarded the Nobel Prize in Physics for this work in 1918. Einstein received the Nobel Prize in 1921 for his explanation of the photoelectric effect. In 1924 the Indian mathematical physicist Satyendra Bose sent Einstein a paper which derived Planck's law treating heat radiation as a quantum gas. This was much like Wien's earlier attempt, except that Bose's quantum gas particles (discrete particles now called photons) could overlap, while matter gas particles could not. Einstein recognised the significance of the different rules applying to quantum particles and matter particles. He pointed out that if matter gas particles could ever be made to behave like quantum gas particles, by overlapping, then they could be condensed into a form of matter with bizarre properties. Half a century later, Planck's and Einstein's groundbreaking work led to the invention of quantum devices like the laser and the atomic clock. Even Einstein's idea of matter gas with quantum properties has been realised with the formation of very low temperature Bose-Einstein condensates leading to the development of the atomic laser (Fig. 12.2).

Analysis of the Sun's radiation shows that it is a black body radiator with its maximum frequency peaking in the red-yellow part of the spectrum, indicating a temperature of about 5700 K. Other stars in the Universe are also black body radiators, some hotter and some cooler than the Sun. However, the biggest black

THE QUANTUM OF ENERGY

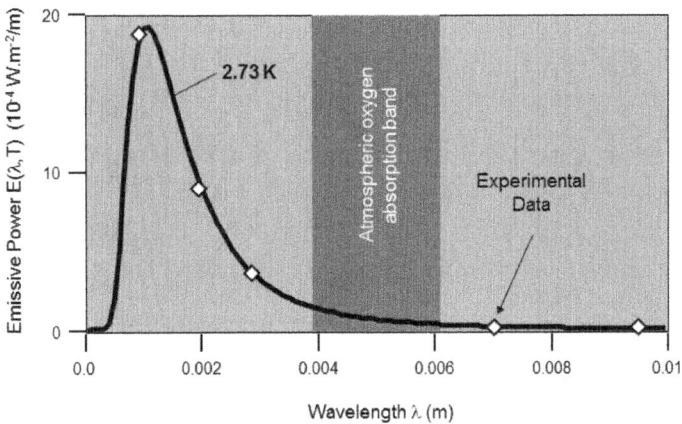

Fig. 15.2 Planck radiation curve for the remnant of the Big Bang.

body cavity, by far, is the Universe itself. In 1964 Arno Penzias and Robert Wilson, two research technicians at the Bell Research Laboratory in the USA, were investigating the background noise signals picked up by microwave satellite communication systems. They aimed the gigantic horn of their microwave receiver towards various points in deep space, but always got roughly the same noise results. Noise is the static that you hear on the radio when you are not tuned in to a station. You can also see the results of noise on the television screen when the TV station goes off air.

The two technicians plotted out the noise amplitude against wavelength very carefully and found that for the millimetre wavelength part of the spectrum, the results fitted a black body radiation curve for a temperature T = 2.73 K (Fig. 15.2). By chance, Penzias and Wilson had discovered evidence of a phenomenon that other scientists had been investigating theoretically for more than thirty years. The interior space of the Universe is like a slightly warm oven, 13.7 billion years after the 'Big Bang' fireball which created the Universe in the first place.

Planck's investigation of the electromagnetic radiation from hot black bodies assumed that the energy was transmitted via an all-pervading continuous medium called the ether. Having concluded that energy was transmitted through the ether in discrete quantities called photons, within a few years the ether was abandoned as being surplus to requirement. But do photons need a medium, or not? For at least 2000 years scientists and philosophers have argued about the nature of the vacuum of space. We are still trying to understand it!

16

ATOMIC PARTICLES, SPIN AND THE ATOMIC CLOCK

It was the Ancient Greeks who first proposed the idea that matter might be decomposed into basic building blocks. And it was Democritus, around about 450 BC, who introduced the word atom to describe these fundamental pieces of matter. Gradually over several thousand years, the elements, with their own identifiable properties and their unique atomic structures, were isolated, with progress slowly speeding up after the start of the European Renaissance.

In 1898 the French scientists Henri Becquerel and Pierre and Marie Curie discovered some active substances which naturally decomposed, radiating away some form of minute particles. All three were awarded the Nobel Prize in Physics in 1903 for their pioneering research into radioactivity. Earlier, in 1880, Pierre Curie, working with his brother Paul, had discovered the piezoelectric effect where a squeezed crystal of quartz generates positive and negative charges on opposite faces of the crystal. This effect has since been greatly exploited in the digital technology revolution. The Curies were an extraordinary family. Having shared the Nobel Prize in Physics in 1903, Marie Curie then won the Nobel Prize in Chemistry, in 1911, for her discovery of the elements radium and polonium, being the first person to win two Nobel Prizes. And their daughter Irène Joliot-Curie won the Nobel Prize in Chemistry in 1935, jointly with her husband, Frédéric Joliot, for their synthesis of radioactive elements.

In 1884 Joseph J. Thomson succeeded Lord Rayleigh as Head of the Cavendish Laboratory at Cambridge University. In 1897 Professor J.J. Thomson discovered the electron, the fundamental electric charge, for which he was awarded the Nobel Prize in Physics in 1906. Ernest Rutherford, a postgraduate student from New Zealand, started at the Cavendish Laboratory in 1895. Rutherford was inspired by Thomson's discovery of the electron and wanted to discover other fundamental particles of matter. He began his research work at the Cavendish Laboratory by investigating X-rays, newly discovered by the

ATOMIC PARTICLES

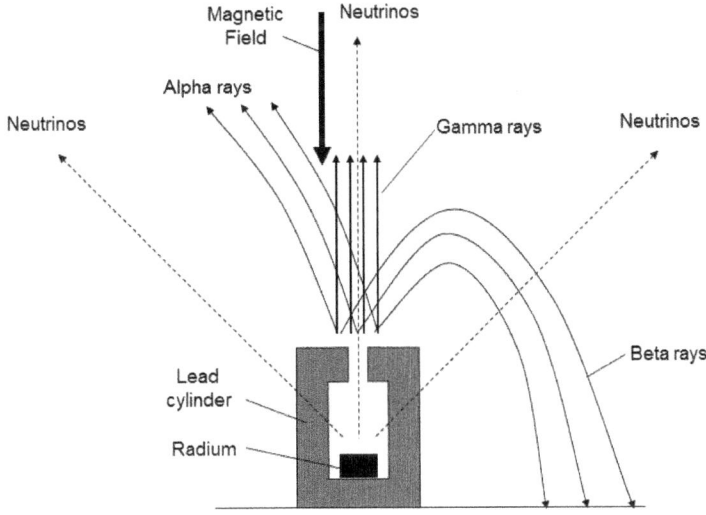

Fig. 16.1 Alpha, beta and gamma rays and the mysterious neutrinos.

German physicist Wilhelm Röntgen in 1895 (winning him the first Nobel Prize in Physics), and radioactive substances, discovered by Becquerel and the Curies in 1898. Placing a sample of radium at the bottom of a small lead cylinder causes the radiated particles to be spewed out at high speed in a vertical direction. Directing a magnetic field vertically down into the cylinder (Fig. 16.1) causes any charged moving particles to veer off-course (eqn 6.11). In this way Rutherford separated the radiation out into three separate groups which he labelled as alpha, beta and gamma radiations. We now know that alpha radiation is composed of positively charged helium atoms, beta radiation is composed of negatively charged electrons and gamma radiation is a form of electromagnetic radiation similar to X-rays, but with even shorter wavelengths. Rutherford won the Nobel Prize in Chemistry for this work in 1908.

A decade later, as scientists tried to put a theory together to explain the experimental results, they encountered a problem. Based on Einstein's formula $E = mc^2$, it was clear that some of the mass of the radium had been converted into radiated energy, but when the sums were done some of the energy associated with the beta radiation was missing. In 1931 the Austrian physicist Wolfgang Pauli suggested that another group of particles, with the energy apparently lost by the beta radiation, must have been present in the radiation, too. Pauli said they had no charge, no rest mass and moved at the speed of light, thus explaining why they were not detected. Recent experiments show that they actually have an extremely tiny mass and move at slightly less than the speed of light. The proposed particles were first called neutrinos and designated with the Greek letter ν, pronounced nu. However, following a later reformulation, they became anti-neutrinos represented by ν̄, or nu-bar.

In 1907 Rutherford moved to Manchester University and interested himself in the atom. An early model of the atom, suggested by J. J. Thomson, was in the form of a round plum, or Christmas, pudding studded with currants. The solid pudding was assumed to be positively charged, while the interspersed surface currants, representing the electrons, were negatively charged and the atom held together by electrically attractive forces. Rutherford asked his research assistants, Hans Geiger and Ernest Marsden, to experiment with firing high speed, positively charged tiny alpha particles at a sheet of metal foil. To detect where the particles emerged on the other side Geiger and Marsden used a fluorescent screen. When an alpha particle hit the screen it produced a flash of light called scintillation. The number of scintillations was then counted for each angle over a measured time period. In the experiment most of the alpha particles passed straight through the metal foil with minor deflections, showing that the apparent solidness of the atoms of which the foil was comprised was an illusion. However, a few of the alpha particles were deflected by more than a right angle, bouncing back towards the source. Rutherford concluded that the positively charged alpha particles must have been repelled by highly concentrated positively charged atomic centres, or nuclei, within the metal foil. In a like manner, a spinning bicycle wheel seems to form a solid body and you cannot poke your finger through it. But we know that the wheel is mostly empty, with the rim and the hub held in position by the spokes. If you throw a handful of gravel at the spinning wheel many of the stones will pass through, while those striking an occasional spoke will ricochet away, but those striking the hub will be reflected back. Rutherford's interpretation of the Geiger-Marsden experiment led him to propose a new model of the atom. He based his idea on the analogue of a miniature solar system, where the Sun represented the positively charged atomic centre, or nucleus, while the planets represented the negatively charged electrons orbiting the nucleus. As another example of synchronicity in science, Rutherford was unaware that in 1904 the Japanese physicist Hantaro Nagaoka had also proposed a solar system model of the atom. But the analogue is not exact, since the planets lie almost in a single orbital plane around the Sun, whereas electrons can lie in any orbital plane around the nucleus.

The typical diameter of an atom is 10^{-10}m. Suppose we scale up an atom to the size of a football pitch 100m long, so that the orbiting electrons are whizzing around the touchline. The nucleus of the atom is on the centre spot and at this scale it can be represented with a grain of sand, since its diameter is about 10^{-14}m. Thus the atom is almost entirely empty of matter.

Since an atom occupies a volume of about 10^{-30}m^3, then the 1mm tip of a blunt needle contains roughly 10^{21} atoms (see, also, eqn 13.6). The mind just boggles at such huge numbers.

Rutherford felt that the atomic nucleus probably contained a mixture of positively and neutrally charged particles. In 1932 James Chadwick, one of Rutherford's research team, discovered the neutron, confirming Rutherford's earlier hunch. Thus, the atomic nucleus could be decomposed into further pieces of matter, namely positively charged protons and uncharged neutrons. It is the number of protons in the nucleus that gives the elements their separate identity. Chadwick received the Nobel Prize in Physics in 1935 for his discovery of the neutron.

It was realised that a strong nuclear force must exist deep within the atom to hold the positively charged, and therefore repelling, protons together in the nucleus, along with the uncharged neutrons. Now the two forces that we know of are gravitation, with mass as its source, and electricity, with charge as its source. Since the uncharged neutrons and the positively charged protons both have mass it seems reasonable to guess that the attractive force holding the neutrons and protons together within the nucleus is an extremely strong variant form of gravity, of extremely limited range (since the orbiting electrons, of mass m_e, are not affected). But for the gravitational force to overcome the electrical force within the nucleus implies that the gravitational permeability γ there is somehow suppressed (eqn 2.2), making it much less than its free space value. The energy stored in the strong nuclear field is negative, as it is for a gravitational field.

We now know that the strong nuclear force is 100 times stronger than the electromagnetic force and that it only operates within the nucleus, with a range of 10^{-15}m. Applying Einstein's formula $E = mc^2$, the predicted energy density of a nucleus is found to be huge, being about 10^{34}J/m^3. But when, in 1933, Professor Rutherford was asked whether it would ever be possible to extract any energy from the nucleus of an atom he replied, "The idea is pure Moonshine."

Since neutrons are uncharged they were ideal particles to fire at the nuclei of various elements to try to break them up and scatter the pieces, since they were not repelled by the positively charged nucleus. Using this method, the German scientists Otto Hahn and Fritz Strassmann obtained some of Rutherford's moonshine in 1938, in the form of the binding energy released by splitting uranium nuclei. Hahn was awarded the Nobel Prize in Chemistry in 1944 for his discovery of what became known as nuclear fission. In 1942 the Italian physicist Enrico Fermi, by then working in the USA, built the first nuclear reactor. In the Second World War both sides funded atomic weapons research. At breakneck speed, with US money, the Allied scientists developed the atom bomb and the USA used it in 1945, to shorten the war against Japan. By 1956 atomic energy was being generated for peaceful purposes by the USA and the UK.

In 1913 the Danish scientist Niels Bohr briefly joined Rutherford's group at Manchester University. When Bohr applied Planck's idea of energy quanta to the solar atomic model for hydrogen he realised that if the orbiting electron could only have discrete amounts of angular momentum of $1(h/2\pi)$, $2(h/2\pi)$, $3(h/2\pi)$ etc., then the orbital radii could only take on certain values, now called Bohr orbits. For convenience, $h/2\pi$ is often replaced with $\hbar$ and referred to as h-bar.

The idea of quantum numbers was introduced to define a particular quantum state. The first number was related to an atomic electron's Bohr orbit and, hence, its quantised energy. The second number was connected with an atomic electron's orbital shape (circular, or elliptical) and, hence, its orbital angular momentum. The third number related to the orientation of an atomic electron's orbital plane when immersed in a strong external magnetic field. Now that scientists had a mathematical model of the atom, they could play with it and improve it. The planetary orbits of electrons were replaced with shells of orbiting electrons. Magic numbers 2, 10, 18, 36, 54 and 86 defined the maximum number of electrons in any shell, starting with 2 in the lowest. If an orbiting electron absorbed an incoming photon, with one quantum of energy, the increase in energy of the electron would cause it to move to a higher orbital shell. On the other hand, if an orbiting electron emitted a photon of energy, the loss of energy meant that the electron would have to drop down to a lower orbital shell. The change of energy ΔE of the atom, caused by a photon being emitted could be written as

$$\Delta E = E_j - E_k = hf_{jk} \tag{16.1}$$

The frequency f_{jk} was that of the emitted photon, or electromagnetic wave, and E_k and E_j were the atom's energies associated with the higher and lower electron orbital shells. Bohr won the Nobel Prize in Physics for his theory of the structure of the atom in 1922.

This all fitted in very nicely with the radiation emitted by hydrogen, observed using spectral analysis. Heating hydrogen gas energises the atomic electrons causing them to change orbit. The emitted radiation from the gas is focused into a beam, passed through a slit and the beam is then spread out using a prism. On focusing the resulting spectrum onto a screen, several images of the slit are formed, one corresponding to each frequency of radiation emitted by the hydrogen gas (Fig. 16.2). For example, if an electron drops from the fifth orbit to the second orbit the hydrogen atom will emit a photon with a frequency of 0.69×10^{15} Hz.

The model for atoms with more shells was more complex, but understanding was growing. However, we are deluding ourselves if we think that our model of the atom represents reality. It's just a shared model between scientists, which allows them to keep track of what is going on. An electron is

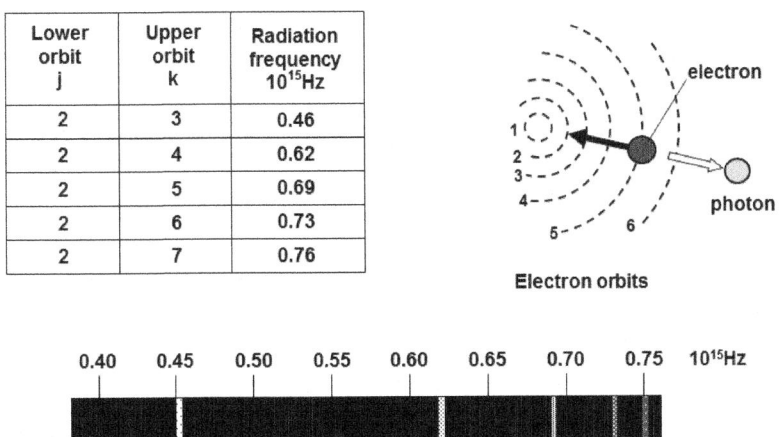

Fig. 16.2 Radiation spectrum for hydrogen.

not a little planet and a photon is not a tiny comet. But, for imagination purposes it may, sometimes, be helpful to think of them in this way. As long as the model predicts results that agree with experiment, then it is acceptable. We are only talking about a picture analogue, after all.

The electron shell model of the atom has been reused to model what goes on in the nucleus itself. In the nuclear shell model, interacting neutrons and protons orbit in separate concentric shells with another set of magic numbers defining the maximum number of neutrons and protons which can fill any shell. This model of the nucleus has been most successful in explaining why some atoms are more stable than others, winning Marie Goeppert-Mayer, who first spotted the analogue in 1949, a part share in the Nobel Prize in Physics in 1963.

Once scientists got used to the idea that radiated energy occurred in discrete lumps, they realised that the vibration in solids probably occurred in a similar way. As sound, or heat, passes along a bar we can think of energy quanta being passed along. These energy quanta are called phonons, to distinguish them from the electromagnetic energy quanta called photons.

In 1925 the Dutch physicists Samuel Goudsmit and George Uhlenbeck introduced the idea that an electron had spin, like the Earth spins on its axis as it orbits the Sun. However, there was a major difference with ordinary spin, because an electron could only have a discrete spin of $+½\hbar$ or $-½\hbar$, usually referred to as up and down half-integer spin states. This was a case where the analogue didn't read across exactly. But now, at least, it was clear that Planck's constant h was connected with the phenomenon of spin, where spin 1 means the particle has its own, or intrinsic, angular momentum of $\hbar$.

Note that for positive, or right-handed, spin the axis points upwards (see Fig. 5.2). An electron with up, or positive, spin has positive angular momentum $+½\hbar$. In conventional terms this means that when viewed from below (looking in the direction of the axis) the electron has clockwise rotation with a positive angular frequency ω, giving a positive frequency $f = \omega/2\pi$. An electron with down, or negative, spin has negative angular momentum $-½\hbar$ and viewed from below rotates anticlockwise, giving a negative frequency. In both cases, with positive spin up, or negative spin down, the spin energy of an electron is positive.

It was a long road which led to the discovery of particle spin. It began with Newton's experiment, in 1667, shining a beam of sunlight through a prism to create a rainbow, or spectrum, showing that white light was composed of many frequencies; then Joseph von Fraunhofer's introduction of a slit into the beam of sunlight, in 1814, which revealed dark spaces, or lines, in the spectrum showing that some frequencies were missing; via Michael Faraday's very last experiment, in March 1862, to see whether a magnetic field might move the positions of the dark lines in the spectrum; and, although Faraday noticed nothing, with Pieter Zeeman's successful repeat of Faraday's experiment, with more sensitive instrumentation, in 1896, which showed that the dark lines did move. There was just one problem: Zeeman's experiment threw up an anomaly. There were too many dark lines formed when the magnetic field was switched on.

Thus it was in 1925 that Goudsmit and Uhlenbeck provided the answer to Zeeman's anomaly. When an atomic electron changed its orbit it emitted, or absorbed, a photon. If the atom was immersed in a strong magnetic field during this process, the emission, or absorption, of the photon had two different results depending on the orbiting electron's spin state. Hence, there were extra dark lines in the spectrum.

Since a spinning charge was known to generate a magnetic field, it was realised that if the outer orbit of an atom contained unpaired orbiting electrons then the atom would have a magnetic dipole field. Where atoms combined to form molecules, then they too could possess a magnetic dipole field. Earlier, in 1922, before the concept of electron spin had been suggested, the German physicists Otto Stern and Walter Gerlach had used an oven to generate a cloud of silver molecules which they squeezed through a short narrow horizontal slit to create a molecular beam. The beam was then passed between the poles of a G-shaped magnet, where the upper pole was sharpened to a knife edge, parallel to the beam. Thus, the strength of the magnetic field between the poles increased in strength in the vertical direction towards the sharp edge. After passing between the poles of the magnet the emerging silver molecules hit a metal plate and condensed leaving two horizontal traces. To explain the result it was concluded that the silver molecules must be magnetic dipoles. As the dipoles passed between the magnet's poles, depending on their axial dipole orientation and on their position in the increasing magnetic field, the silver molecules were deflected along two different paths. This, further, supported the idea of electron spin.

It was soon discovered that neutrons and protons also had half-integer spin states. In turn, this gave support to the probable existence of the antineutrino.

When a neutron (spin ½) was extracted from the nucleus of an atom it decayed into a proton (spin ½) and an electron (spin ½). The decay was attributed to the presence of another nuclear force, shown to be 10^{12} times weaker than the force which holds the nucleus together. To distinguish between the two, one was called the weak nuclear force while the other became the strong nuclear force. In the decay process, while overall particle charge was conserved, particle energy and particle spin, or angular momentum, seemed not to be. However, if an antineutrino $\bar{\nu}$ with spin -½ (that is, angular momentum $-½\hbar$) accompanied the proton p and the electron e in the decay of the neutron n then the conservation laws were satisfied. In symbols we can write this as

$$n \rightarrow p + e + \bar{\nu} \qquad (16.2)$$

So, it seemed likely that the antineutrino did exist and was linked with the weak nuclear force. It wasn't until 1956 that Pauli's 1931 speculation about the existence of the antineutrino was finally confirmed. In an experiment carried out with a large tank of water placed next to a nuclear power station, the US physicists Frederick Reines and Clyde Cowan searched for antineutrinos escaping from the reactor via their interactions with water molecules. When an antineutrino hit a hydrogen nucleus in the water it dislodged the positive charge from a proton which became a neutron. The freed charge e^+ was the anti-particle of the electron, known as the positron. Almost instantly, the positron combined with an electron resulting in annihilation of the pair in a characteristic burst of energy, or scintillation. Photomultiplier tubes on the inside walls of the tank detected each flash of light. But, although the number of escaping antineutrinos was enormous, their extreme speed and exceedingly small mass meant that the chance of one hitting an atom was very remote. Consequently the count of light flashes was very small, being just a few per hour. However, switching the nuclear power station off and on made a difference to the results, indicating that Reines and Cowan had detected antineutrinos. In the same year, 1956, the Hungarian physicists Sándor Szalay and Gyula Csikai were the first to produce evidence of the existence of the neutrino in a cloud chamber photo where a helium nucleus was seen to abruptly change direction as it emitted an unseen neutrino. Scientists have now ascertained that the neutrino can only have right-handed spin (+½), rotating axially about its direction of motion, while its anti-particle pair, the antineutrino, can only have

Force	Source				Carrier		Range	Strength
	Matter	Spin ℏ			Boson	Spin ℏ		
CLASSICAL FIELD THEORY								
Gravity	+ mass		Attraction		graviton ?	2 ?	Universal	1
Electro-magnetism	± charge		Attraction		photon	1	Universal	10^{40}
ATOMIC THEORY								
	Nucleus		Atom		Quantum Vacuum		10^{-10} m	
	Electron (e)	½						
Strong Nuclear	Neutron (N)	½	Nucleus		gluon	1	< 10^{-15} m	10^{42}
	Proton (P)	½						
	Quarks U D C S T B Colour R+G+B = W	½	Quark confinement Neutron N Proton P					
Weak Nuclear	Radioactivity α, β, γ		Neutron decay		W⁺ W⁻ Z	1	< 10^{-17} m	10^{30}
	Neutrino (ν)	½			Higgs	0		

Fig. 16.3 The four forces of nature.

left-handed spin (-½). More types of neutrinos have now been discovered, so the original antineutrino is sometimes referred to as the electron-antineutrino.

So, to recap on the four forces of nature, if the gravitational force has strength 1, then the weak nuclear force will be of strength 10^{30}, the electromagnetic force will be of strength 10^{40} and the strong nuclear force will be of strength 10^{42}. But, because roughly equal numbers of positive and negative charges exist in the Universe the force of electromagnetism is neutralised. The strong and the weak nuclear forces are confined within the atom leaving gravity, the weakest force of all, to dominate the Universe.

Well, if protons, neutrons, electrons and neutrinos had spin, what about the other particles? Based on their spin states, scientists separated the particles out into fermions and bosons.

Fermions are particles of matter which have half-integer spins. These include the electron and the neutrino (and their anti-particles) called leptons, which don't respond to the strong force, and the proton and the neutron, called baryons (consisting of three quarks, see below), and the various quarks, which do respond to the strong force.

In 1925 Wolfgang Pauli introduced a rule which proposed that no two electrons in the same orbital shell around the nucleus of an atom could have the same spin state. If one electron had spin $+\frac{1}{2}$, conventionally called spin-up, then the other electron must have spin $-\frac{1}{2}$, or spin-down. Spin was the fourth quantum number used to characterise a quantum state. Pauli's original rule was quickly extended to form the Pauli exclusion principle which states that fermions occupying the same species cannot have identical quantum numbers. The fact that atoms, although mostly empty, combine to form solid matter, rather than overlapping or passing through each other, can be explained by Pauli's exclusion principle. It was this set of rules that led to the magic number sequence for electrons orbiting in atomic shells and which helped to make sense of Dmitry Mendeleyev's periodic table of the elements. Pauli was awarded the Nobel Prize in Physics in 1945 for his discovery of the exclusion principle.

Bosons are particles which carry force fields around and they have multiple integer (whole number) spins. The photon is the carrier particle for the electromagnetic force. When electric fields **E** interact we can treat the **E**-field adjustment, in particle terms, as being made by virtual, or short-lived, photons passing information around. The particle form of an electromagnetic wave, with its combined oscillating electric **E**-field and magnetic **H**-field, is given by a distribution of real photons. The spin axis of each photon points in the direction in which the wave travels and has right-handed spin 1.

Gravity is expected to follow the same pattern as that for electromagnetism. In wave terms, the linearised gravitational wave is formed (Fig. 8.7) by a sinusoidal gravity **g**-field and an orthogonal sinusoidal gravitomagnetic **h**-field. So far, no such waves have been detected. The gravitational analogue of the photon (eqn 13.22) is the graviton containing a quantum of energy hf and with right-handed spin 1 about the direction of wave motion. From wave-particle duality, we assume that the gravitational wave can be represented by a distribution of real gravitons. Whereas electric fields store positive energy, gravity fields store negative energy. During a gravity field disturbance we assume that positive energy gravitons redistribute the negative gravitational energy about. In Chapter 8 we agreed that for a linear theory of gravity effective mass cannot create its own gravitational field. In Einstein's theory of gravity, where effective mass does create its own gravitational field, the model is non-linear.

However, the graviton is still assumed to have a quantum of energy hf, but it has right-handed spin 2.

Describing the strong nuclear force and its boson, or carrier particle, called the gluon is much more complicated. In developing the standard model for particle physics, scientists have now postulated that the basic building blocks of matter are quarks (so far unobserved and maybe unobservable). The term quark was originally introduced by Murray Gell-Mann, who won the Nobel Prize in Physics in 1969 for his work on classifying the elementary particles of matter. According to the experts there are six types of quarks, called flavours, namely up, down, charm, strange, top and bottom. Only the up and down quarks are stable; the others rapidly decay. All quarks have spin ½ but have fractional electron charges as follows: up, charm and top have charge $+\frac{2}{3}$, while down, strange and bottom have charge $-\frac{1}{3}$. Their masses are all different. Theory also allows for anti-quarks. Particle physicists predict that the proton contains one down and two up quarks, with overall charge 1, while the neutron contains two down and one up quark, with overall charge 0.

The discovery of the presence of quarks inside neutrons and protons was made by accelerating a bunch of electrons to very high speed and making them hit a fixed target resulting in a burst of sub-atomic particles. Although quarks were not liberated in this process, examination of the particles in the debris, their tracks and their rates of decay agreed with the predictions made by theoreticians of the strong nuclear force. The work was conducted at the Stanford Linear Accelerator Center (SLAC) in the USA, between 1967 and 1973, by Jerome Friedman, Henry Kendall and Richard Taylor, winning them the Nobel Prize in Physics in 1990.

Now the quarks are fermions, subject to Pauli's exclusion principle, which means that quarks with the same quantum numbers of spin should not co-exist in the same species. So how is it possible for two up quarks to reside in a proton? To get round this the theoreticians introduced a new quantum number called the colour charge, which can take three values, namely red, green and blue (this has nothing to do with real colours). A mix of three colours gives white, which results in zero, or neutral, colour charge. Colour charge is based on the analogy with electric charge, of which the latter can only take two values, namely positive and negative charge. As in electromagnetism, opposite colour charges attract, while like colour charges repel. So, quarks can be colour charged red, green or blue, while anti-quarks can be charged minus-red, minus-green and minus-blue. Also, following the analogy of charge conservation in electromagnetism, colour charge is conserved.

In the proton the two up quarks have different colour charges, so their quantum numbers are different and the Pauli exclusion principle is satisfied. Furthermore, a stable object must contain three different coloured quarks, so that its overall colour is white and it is neutrally colour charged. Thus, the proton contains a red quark, a green quark and a blue quark.

In an electromagnetic interaction between charged bodies there is an exchange of photons, which are emitted and absorbed by the charged bodies. In the case of gravitation, analogy suggests that a gravitational interaction between masses is mediated by an exchange of gravitons. Extending the analogy further, into the realm of the strong nuclear force, it has been proposed that an interaction between the colour charged quarks, which make up the protons and neutrons, is facilitated by an exchange of gluons, which are emitted and absorbed by the quarks. In the process of gluon emission and absorption a quark must change colour in order to conserve colour charge. Photons, gravitons and gluons are all expected to posses spin, be electrically uncharged and have zero rest mass. However, in addition, gluons are also predicted to carry both a colour charge and an anti-colour charge. Thus, there are nine possible colour charge combinations for the gluons but, for symmetry reasons one combination is not allowed, leaving only eight types of gluons.

Fundamentally the nuclear strong force binds the quarks together in forming the protons and neutrons. The leakage of the nuclear strong force from the interiors of the protons and neutrons gives rise to a secondary effect, binding the protons and neutrons together in the nucleus. We might surmise that it is the leakage of colour charge, into the region between the protons and neutrons in the nucleus, which causes a reduction in the gravitational permeability γ there, making the attractive gravitational force between the proton and neutron masses so much stronger.

Within the neutron and the proton the effect of the colour charged quark interactions makes it seem as though the quarks are connected by rubber bands. When quarks are close the bands are loose and each quark's colour charge appears to be weak, so that each quark appears to be almost force free. But as quark separation increases the bands tighten and each quark's colour charge appears to increase, so that the inter-quark force becomes strong. The accepted explanation for the strength of inter-quark forces was developed by the US physicists David Gross, David Politzer and Frank Wilczek, in 1973, leading to them sharing the Nobel Prize in Physics for their work, in 2004. Their model assumes the existence of the quantum vacuum (see Chapter 18), where positive and negative particles of energy can briefly spring into life and quickly disappear again. Such ephemeral particles are called virtual particles. In the case of the strong nuclear force it is virtual gluons, with their combined colour and anti-colour charges, which fleetingly appear and disappear. Each quark is surrounded by a cloud of virtual colour charged gluons (Fig. 18.1(b)) which makes the colour charge of the quark effectively stronger the further away, in atomic terms, that you are from it. On the other hand, as you move nearer to the quark the cloud of virtual gluons thins and the quarks' effective colour charge weakens. Thus, at large distances the inter-quark force is stronger, while at short distances the inter-quark force reduces considerably, giving an appearance of being almost force free. This phenomenon is called asymptotic freedom and explains why quarks probably cannot be extracted from neutrons and protons. It is the small residue of the internal inter-quark colour forces that escapes from the confines of the protons and neutrons which acts as the glue to bind them together in the nucleus.

Based on the speculation that the nuclear strong force is gravitational in character, the part played by the colour charged virtual gluons is as an obscurant of the gravitational field. Asymptotic freedom then comes about as follows. As a pair of quarks move closer together the density of the virtual colour-charged gluons in the intervening region decreases. The same is true for the density of virtual photons appearing in a Casimir cavity as the width decreases (Chapter 18). The upshot of this is that there is an increase in the gravitational permeability γ in this region, resulting in a reduction of gravitational attraction between the quark masses. When a pair of quarks move further apart, the density of the virtual colour-charged gluons in the intervening region increases, so that the gravitational permeability γ decreases, resulting in the gravitational attraction between the quark masses increasing. If we knew how to create colour charge outside of an atom then we might be able to turn gravitons into gluons and, thereby, control gravity.

Finally, according to the experts in particle physics, the carrier particles for the weak nuclear force responsible for radioactivity, where particles with charge and mass are emitted, are the electrically charged W^- and W^+ bosons and the neutral Z boson, all of which have spin 1. As we have seen, synchronous breakthroughs in understanding in science occur quite often. In 1967 Abdus Salam, at Imperial College London, and Steven Weinberg, at Harvard, in the USA, independently proposed theories which unified the weak nuclear force with the electromagnetic force. In 1979 Salam and Weinberg, together with Sheldon Glashow of Harvard, who had first introduced the W and Z particle approach, were awarded the Nobel Prize in Physics, although it wasn't until 1983 that experiments at CERN (the European Centre for Nuclear Research) involving colliding proton and anti-proton beams confirmed the validity of the Salam-Weinberg theory. Examination of the tracks left by the remnants of the particle collisions revealed the tell-tale presence of W and Z type bosons via the effect they had on the paths taken by the scattered sub-nucleonic particles. Carlo Rubbia and Simon van der Meer won the Nobel Prize in Physics for this work in 1984.

The photon, or electromagnetic boson, is unrestricted and moves throughout space at the speed of light, giving it an effective mass although its rest mass is zero. Theory dictates that the rest masses of the W^-, W^+ and Z bosons are also zero. However, proton-smashing experiments indicate that while moving they develop very large masses. Thus, one explanation for why the weak nuclear force is confined to within the neutron is that while conveying the force about, the bosons develop huge inertias bringing them almost immediately to rest.

The source of mass developed by the moving weak nuclear force bosons is of great interest to particle physicists. Suppose we try to push a copper body between the poles of a horseshoe magnet. We would notice that the body encounters a strong resistance to movement. In air the body has little inertia, so the enormous increase in inertia of the body as it is pushed between the magnetic poles might be thought to be due to an increase in the mass of the body. But we know that the increase in inertia is actually due to the creation of eddy currents in the copper body, caused by the background presence of the magnetic field, which tries to prevent the motion in accordance with Newton's third law (Lenz's law).

In a similar manner, Professor Peter Higgs, of Edinburgh University, proposed the idea of another background field, called the Higgs field, formed by a sea of virtual particles called Higgs bosons, through which the weak nuclear field bosons moved. We have already introduced the idea of a quantum vacuum field filled with virtual photons to which we have added virtual gluons. So, to accommodate the Higgs field we must just extend the contents of the quantum vacuum to include Higgs bosons. But unlike the magnetic field, which is vector in character, the proposed Higgs field is a scalar field, meaning that at any point in space the field has a magnitude but no direction. The Higgs boson is the carrier particle for the Higgs field and it is the fundamental source of mass.

In terms of analogy, we can think of the Higgs vacuum field as being more like viscous liquid, say glycerine. Particles moving through a viscous liquid in any direction experience a drag, which causes them to decelerate. But from Newton's second law (eqn 2.15), force and acceleration are linked with mass, so retardation of the particles as they move through the liquid might be interpreted as their gaining inertial mass.

Photons and gluons, as well as weak nuclear field bosons, all pass through the quantum vacuum, but only the W^-, W^+ and Z bosons interact with the Higgs field. To explain why this is so the particle physics experts, particularly Steven Weinberg, introduced the idea of symmetry. We are all familiar with the idea of the rotational symmetry of a sphere, where viewing it from a different direction leaves its shape unchanged. A slender cylinder stood vertically in a gravitational field on a horizontal plane has circular symmetry. However, if the cylinder is knocked over the symmetry is lost, or broken. In a much more abstract way the four forces of nature all have links with forms of symmetry. Newton's laws of motion remain unchanged for different inertial frames of reference as do Maxwell's equations for electromagnetism. With the strong nuclear force there is symmetry in the way that the quarks interact with each other. When it comes to spin, although photons and gluons exhibit symmetry, having left and right forms, there is experimental evidence showing that the W^-, W^+ and Z bosons only have left-handed spin so that their spin symmetry is broken. It is postulated that this broken symmetry results in the W^-, W^+ and Z bosons absorbing Higgs bosons.

Using $\Delta E = (\Delta m)c^2$ with Einstein's variation (eqn 17.5) of Heisenberg's uncertainty principle (eqn 12.6), we see that the absorption of a large mass from the vacuum can only be done over a very short time, which means that the temporal existence of a weak field boson with mass is extremely short. Moreover, in absorbing a large mass a weak field boson's inertia increases causing it to rapidly decelerate, severely limiting its range of influence.

So, to summarise, in terms of the influence exhibited by the Higgs field, we can think of the Higgs boson as the carrier force particle. Note that the Higgs boson has no direction of spin (spin 0), which is consistent with the Higgs field being a scalar field. It is the attachment of a Higgs boson to the moving weak nuclear field bosons which gives them mass. It also gives mass to the quarks and the electrons, too. Consequently, the Higgs boson is deemed to be the root source of gravitation. However, the mass of protons and neutrons comes primarily from the vibrating energy of the constituent quarks and gluons, while atomic and molecular mass comes partially from binding energy, all via Einstein's $E = mc^2$.

Peter Higgs' idea was received with scepticism by some physicists when it was first proposed in 1964. In Werner Heisenberg's opinion the Higgs idea was rubbish. Stephen Hawking also expressed his doubts that the Higgs boson would be found. New ideas are often disturbing. Nevertheless, many particle physicists cautiously accepted the idea championed by Professor Higgs and supported experiments aimed at either verifying, or disproving, it. The search became somewhat of a mystical quest, like the search for the Holy Grail, with some excited scientists referring to the Higgs boson as the 'God particle'!

Searching for the existence of the Higgs boson was undertaken at the Fermilab accelerator in the USA and at the CERN collider in Europe. Each experiment involved the examination of the particle scattering tracks of many particle collisions. The predicted lifetime of a Higgs boson is only 10^{-22} s, after which it is predicted to decay into a pair of W and Z bosons, or a pair of photons, or a pair of quarks, so the scattering tracks have to be captured electronically and the results examined computationally. Scientists at CERN estimate that the results of over 300 trillion proton-proton collisions have been examined. This gargantuan effort paid off on 4 July 2012, when scientists at CERN claimed to have detected the influence of a particle with the predicted Higgs boson characteristics in a number of their experiments. Further analysis verified the claim and this led to the announcement on 8 October 2013 that Professor Peter Higgs had been awarded the Nobel Prize in Physics.

The worldwide research programme to investigate the standard model of particle physics has cost billions in whatever unit of currency is chosen. The size and complexity of the CERN proton-proton collider is awe-inspiring. The drive and commitment of the particle physicists must also be recognised. So the ultimate source of gravitation has now been pinpointed deep within the heart of the atom. Spin, or angular momentum, is also present. So it would seem reasonable for a non-particle physicist to conclude that gravity and gravitomagnetism must exist deep within the heart of the atom, too! But will this particle physics approach lead to the discovery of how to control gravitation in the way that we can control electromagnetism?

The Pauli exclusion principle does not apply to bosons. Thus, photons with identical properties can overlap and combine together to give rise to large macroscopic effects, such as those which occur in the laser. Free electrons, being matter particles, do not usually combine (like charges repel), but in some materials under certain circumstances, such as very low temperature, it seems that pairs of electrons (fermions with spin $\pm \frac{1}{2}$) can briefly combine to form Cooper pairs (bosons with integer spin). This is part of the explanation for the strange phenomenon of superconductivity (see Chapter 13), where matter particles act like quantum force particles.

Since atoms and molecules are composed of fermions some will possess non-zero intrinsic spin. In an ad hoc manner Pauli's exclusion principle can be applied to all particles of matter with odd integer times $\frac{1}{2}$ spin. Under special conditions of low density and low temperature, some gases with molecules of $\frac{1}{2}$ integer spin exhibit macroscopic effects related to the Pauli exclusion principle. Optical pumping is used to inject angular momentum into the gas molecules which leads to spin polarisation, in other words, molecules with the same spin condition, or spin alignment. These spinning gas molecules tend to avoid each other, leading to an increase in length of the mean free path between molecular collisions. This results in some dramatic changes in the properties of these special gases. Optical pumping is also used in the magnetic trap used in the atomic laser (Fig. 12.2).

In 1929 the clock maker Warren Marrison, in the state of New Jersey, USA, had a brilliant idea. He used the oscillating piezoelectric effect of a quartz crystal to control the speed of an electric motor which drove the hands of a clock. This groundbreaking invention heralded the arrival of the quartz crystal clock and, later, the quartz crystal watch. However, the frequency of quartz crystal oscillations tends to drift with time, making quartz crystal clocks prone to errors over long periods. For more accuracy a stable reference system is needed to keep the quartz crystal oscillation regular. This requirement eventually led to the development of the atomic clock.

The first experimental atomic clock was made by Harry Lyons at the National Bureau of Standards (NBS) in the USA in 1949. In fact, it used a beam of ammonia gas molecules, so was really a molecular clock. The first accurate atomic clock, called Caesium 1, was built in 1955 by Louis Essen and Jack Parry at the National Physical Laboratory (NPL) in the UK. It is now on display in the Science Museum in London.

The history of the atomic clock can be traced back to the work of Otto Stern, who moved to the USA in the 1930s, and his experiments with molecular beams, followed by the work of the US scientist Isador Rabi, who won the Nobel Prize in Physics in 1944 for his experiments on the resonance properties of atoms.

The resonant frequency f_{RES} for caesium atoms is exactly 9.192631770×10^9 Hz, which is in the microwave band. Now an atom of caesium has fifty-five orbital electrons and, as the magic numbers show, this means that there is just one electron in its outer orbit. The magnetic property of this element is dependent on the spin property of the outer orbital electrons of its atoms. Thus, an atomic beam of caesium atoms contains a mixture of atoms with spin-up and spin-down.

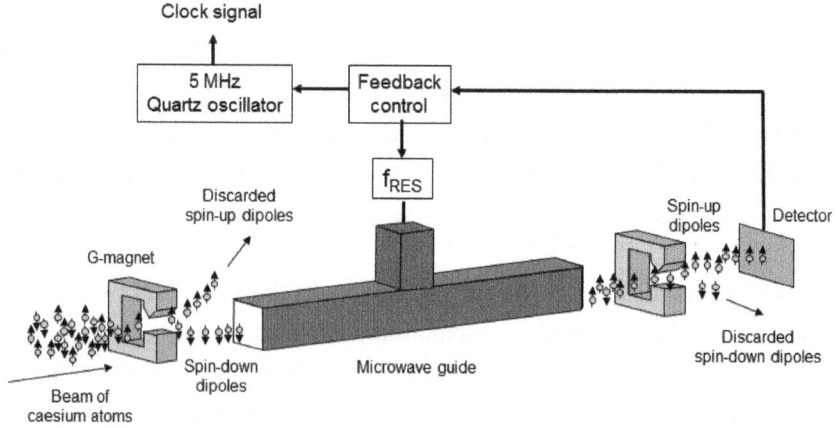

Fig. 16.4 The atomic clock.

If the atomic beam passes between the poles of a G-shaped magnet with one sharpened pole (Fig. 16.4), the beam is split into two beams (as in the Stern-Gerlach experiment), one containing magnetic dipoles with spin-up and the other with magnetic dipoles with spin-down. The beam with spin-up dipoles is discarded. The other beam passes along a microwave cavity where the frequency f_{RES} is chosen to make the caesium magnetic spin-down dipoles resonate, absorb energy and make many of them flip up to become spin-up dipoles. However, there is no change in orbital energy of the atoms. On leaving the cavity, the beam passes between the poles of another G-shaped magnet with a sharpened pole which, again, separates the atomic beam into two beams, one with spin-up dipoles and the other with spin-down dipoles. The number of spin-up dipoles is then measured by a detector. A feedback system ensures that the frequency f_{RES} doesn't wander so that the number of caesium spin-up dipoles detected remains at a maximum. The frequency f_{RES} is then divided down and used to keep check on a crystal quartz oscillator with a frequency of 5 MHz. This is the heartbeat of the atomic clock, which has an accuracy of to within a second, lasting millions of years.

Originally, the SI unit of time, the second, was based on $1/(24 \times 60 \times 60)$ of an Earth day. However, for some time, it has been known that the Earth's rotaton rate Ω_E is not perfectly steady. So, in 1967 the SI unit of time was

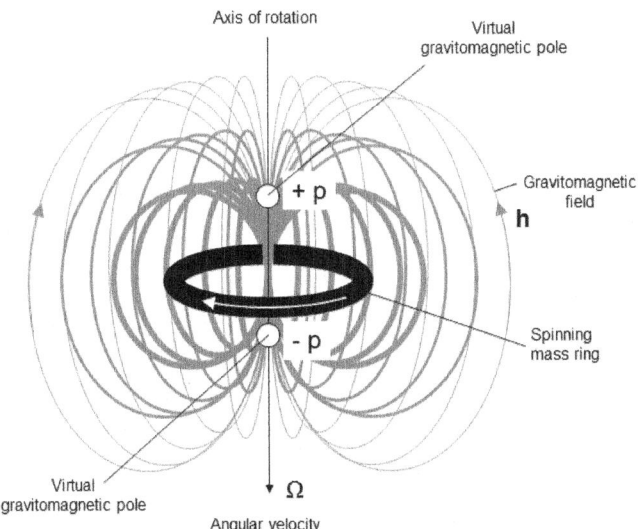

Fig. 16.5 The spinning mass ring gravitomagnetic dipole.

redefined as the second measured by a caesium atomic clock. Greenwich Mean Time will soon be replaced with Universal Time, as measured by a world system of atomic clocks. The development of our modern telecommunication systems, including mobile phones, GPS navigation, digital TV etc, would not be possible without the accuracy provided by atomic clocks.

In solid matter the random directions of any intrinsic molecular spins tends to neutralise their effect, generally leaving matter without a spin state. However, the rotation of macroscopic masses, particularly about axes of symmetry, leads to the creation of gravitomagnetic dipoles. Examples include rotating spheres, cylinders, discs and rings (Fig. 16.5). But in all cases the external effects of the gravitomagnetic fields generated are extremely weak, due to the smallness of the gravitomagnetic permeability of free space η (eqn 8.15).

In general, ordinary spinning bodies which are not in contact do not interact with one another via gravitomagnetism. Only when astronomical-sized spinning bodies are involved, such as the Sun and the Earth, is there predicted to be a tiny, but measurable, interaction between gravitomagnetic fields, called frame-dragging. The gravitomagnetic field of the rotating Earth was recently detected (see Chapter 11) in the NASA Gravity Probe-B satellite experiment.

In 1959 Paul Dirac, then the Lucasian Professor of Mathematics at Cambridge University, suggested that the neutrino might be the elusive graviton. But, given that the graviton is a boson (force carrier) while the neutrino is a fermion (particle of matter), the idea was dismissed. Nevertheless, since the neutrino and the antineutrino are both linked with the weak nuclear force, the thought that they might also be linked with gravity surely deserves further consideration. For example, might the antineutrino be an unbalanced gravitomagnetic dipole, where the positive mass content very slightly exceeds the negative mass content? This could explain the anti-neutrino's speed since, with virtually zero inertia, it would quickly accelerate to near light speed.

You may think that neutrinos are mostly confined inside neutrons within the atom but, according to scientists today, a large proportion of the energy emitted by the Sun is in the form of neutrinos generated via the weak nuclear force. Predictions indicate that about 6×10^{14} solar neutrinos strike one square metre of the Earth perpendicularly every second. But neutrinos, having near-zero rest masses and moving at nearly the speed of light mostly pass straight through the Earth without any interaction. Since the Sun rotates one wonders whether solar neutrinos, acting like gravitomagnetic dipoles, interact with the Sun's gravitomagnetic field and tend to stream out more from the Sun's polar regions forming neutrino jets. Bipolar jets have been observed to occur with more powerful stars. Equally, one wonders whether the solar neutrinos approaching the Earth interact with the Earth's gravitomagnetic field, or is the field too weak?

Contrary to what its name suggests, the weak nuclear force is also responsible for the gigantic forces arising in supernovae explosions. The Universe is awash with stellar neutrinos. Building on the pioneering work of Reines, who got the Nobel Prize in Physics in 1995, and Cowan, a number of neutrino observatories have now been built. Since neutrino detection is via the detection of scintillations in fluids and because cosmic rays (high energy protons and atomic nuclei from outer space showering the Earth) cause spurious scintillations, all new observatories are now built deep under the ground or under the sea or under the ice. This forms a barrier to cosmic rays but not to neutrinos. Using such observatories scientists have examined the stream of solar neutrinos and found that they morph into different forms of neutrinos. Since it is possible to determine the rough direction of arrival of neutrinos, sometimes the observatories are called neutrino telescopes. Neutrino-astrophysicists observed their first supernova on 23 February 1987, when a 13-second wave of neutrinos (only about ten, but more than usual) was detected emanating from a star outside the solar system, more than 11 billion times more distant than the Sun. The supernova explosion was also confirmed visually. In 2002 the US physicist Ray Davis and the Japanese physicist Masatoshi Koshiba, who separately pioneered solar neutrino detection experiments in their own countries, shared the Nobel Prize in Physics.

There are a growing number of neutrino observatories around the world. They include the Homestake Mine in the USA, the Kamiokande Neutrino Observatory (Super-K) in Japan, the Borexino neutrino detector in Italy, the Baksan Neutrino Observatory (BNO) in Russia and the Sudbury Neutrino Observatory Laboratory (SNOLAB) in Canada. I mention the latter as my uncle used to work at the nickel mine where, later, the SNOLAB was built over a mile below ground level. Modern neutrino detectors use many thousands of photomultipliers for detection. Two of the latest neutrino telescopes are NESTOR, an array built 4 km down on the seabed by Pylos, in Greece, and

ICE CUBE, an array 1.6 km under the ice at the South Pole. Even the UK MoD has got into the act with its acoustic cosmic ray neutrino experiment (ACoRNE). A small hydrophone array, off the coast of the tiny island of Rona in the north-west of Scotland, previously used to detect submarines, has been modified to detect neutrinos. Occasional neutrinos interacting with sea water create scintillations which give rise to brief heat pulses which are then detected as acoustic blips.

Nowadays, neutrino beams are being experimented with at CERN and other particle physics laboratories. It is speculated that eventually neutrino radiometers will be able to detect nuclear powered systems on land, on or under the sea and in aerospace by their neutrino signatures. If the weak nuclear force can break stars apart, then manipulation of the weak force, with the W^-, W^+ and Z boson link with mass creation, should lead to some amazing new technologies in the 21st century, with gravity control being a possibility.

17

ZERO-POINT ENERGY

We think of an atom as being in the shape of a sphere, with an outer electron shell and a central core. The atoms in a solid arrange themselves into a structure called a crystal lattice. In the lattice, the atoms form regular layers of almost touching spheres, with the distance between adjacent atom centres being about 10^{-10}m. There are many different forms of lattice. To model a crystalline solid the atoms may be represented as point masses held in place by a regular network of springs. The spring elasticity represents the flexibility of the inter-atomic bonds.

To start with, we can model a single vibrating atom as a point mass attached to a lightweight spring (Fig. 17.1) oscillating in 1-D. This classical mechanical approach is based on the empirical law that the tension T of a spring is proportional to its extension x. The relationship is

$$T = kx \qquad (17.1)$$

The constant, k, is the coefficient of elasticity of the spring. (We will take care to avoid confusing this k with the coefficient of torsional stiffness, or with Boltzmann's constant.) This famous result was discovered experimentally by Robert Hooke in about 1660. The tension and the ensuing acceleration of the mass are linked via Isaac Newton's second law of motion to give a differential equation of motion.

In 1510 Peter Henlein, of Nuremberg in Bavarian Germany, introduced the spiral spring to provide the driving power for clocks. This led to the development of small clocks and the advent of watches. During the 17th century, the clockwork revolution gained pace and London became a manufacturing centre for clocks and watches, largely explaining Hooke's interest in springs. In 1662 Hooke was appointed as the Curator of Experiments for the Royal Society, where he carried out a wide range of valuable research. Hooke was an extremely

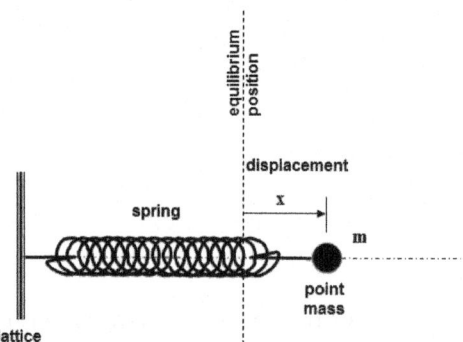

Fig. 17.1 1-D vibration of an atom.

able and clever man, but arguments with Newton over optics and gravitation made the two men bitter rivals. Hooke died in 1703, the very year that Newton became President of the Royal Society. On the positive side, Newton reinvigorated the Royal Society and placed its finances on a firm foundation. However, on the negative side, Newton took every opportunity to remove any sign of Hooke's presence during the early years of the formation of the Society, thereby dimming the memory of Hooke as a brilliant inventor and natural philosopher.

Suppose that the mass m attached to the spring is initially displaced a small distance x from its equilibrium position (Fig. 17.1). This is called a boundary condition. If released, the mass m will oscillate about the equilibrium position at a single frequency f, with amplitude equal to the initial displacement. This vibratory mode is called simple harmonic motion (SHM). The natural frequency f of oscillation is given by

$$f = \frac{1}{2\pi}\sqrt{\frac{k}{m}} \qquad (17.2)$$

The linear differential equation of motion representing the vibration can be solved and for any time t the exact position x of the point mass and its velocity v can be determined, as can be the mass's momentum p = mv. The potential energy of the mass is stored in the spring and is $\frac{1}{2}kx^2$, while its kinetic energy is $\frac{1}{2}mv^2$. Thus, for any position x the total energy E of the mass can be determined.

Modelling the atomic vibrations of a solid body in 1-D is a little more

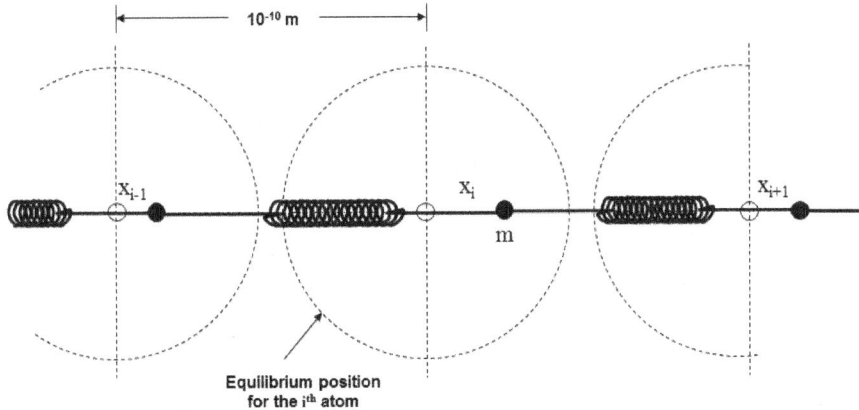

Fig. 17.2 Vibration of a line of atoms.

difficult (Fig. 17.2). This requires forming recurrence differential equations of motion for groups of three adjacent point masses along the line of atoms. Knowing the boundary conditions, the problem can then be solved. Extending the model to cover the 3-D vibrations of layers of atoms within a solid uses the same idea of interlinked springs but it becomes very complicated since various modes of vibration (different frequencies) may occur.

In developing a quantum mechanical model for the vibration of one atom we can still use the model (Fig. 17.1) of a mass attached to a spring, vibrating with SHM at a frequency f. The equation of motion still relies on Hooke's law, but the de Broglie matter wave is used to represent the mass, so that the position x of its centre is smeared out in the wave envelope along the length of the vibration.

According to Heisenberg's uncertainty principle (Chapter 12) for matter waves we cannot know the mass's position x and its momentum p simultaneously. Thus we have two unknown variables, x and p, but only one equation of motion linking them. To derive a solution, we need another equation linking x and p, so that we can eliminate one of the variables. The extra equation is provided by following the Austrian scientist Erwin Schrödinger's assumption that the probable position of the mass in the oscillation will satisfy a wave-like equation (eqn 12.4) in terms of the spring extension. This has turned out to be a very successful guess, although it is not something that can be proved theoretically. The resulting energy equation for the motion, called Schrödinger's equation (eqn 12.5), can be solved using a standard mathematical technique.

The solution shows that rather than having a continuous range of energy values E, as it does in the classical mathematical model, the quantum mechanical model of the vibrating atom has discrete energy levels E_n given by

$$E_n = (n + \tfrac{1}{2})hf \qquad \text{for } n = 0, 1, 2, 3, \ldots \ldots \qquad (17.3)$$

Ignoring, for the moment, the energy E_0 associated with the atom's ground state n = 0, we see that an atom vibrating with frequency f can emit or absorb discrete amounts of energy in integer multiples of hf, as envisaged by Planck. The model of a single atom can be extended to cater for a 1-D line of atoms. We then define the phonon as the quantum unit of energy E which can be exchanged with other atoms in a solid body, where E is given by

$$E = hf \qquad (17.4)$$

The concept of the phonon then ties in nicely with our idea of wave-particle duality, where we assume that the elastic 1-D vibrations within a solid may be replaced with particles called phonons which move along the solid at the speed of sound c_s. This is just an acoustic analogue of the electromagnetic wave-photon idea. Phonons play a part in a solid's acoustic properties and in its electrical and thermal conductivities. Heat is generally attributed to the random movement of phonons. The 1-D quantum mechanical model can be extended, with much complexity, to model the 3-D atomic vibrations of a solid body.

We must now consider the lowest energy value $E_0 = \tfrac{1}{2}hf$ for the ground state n = 0 of an atom. This is understood to occur for absolute zero temperature (T = 0 K). Since we cannot have half a phonon, E_0 is taken to be the averaged value for the zero-point energy (ZPE) of the vibration of an atom and is a direct result of the Heisenberg uncertainty principle. If the atom ceased to vibrate then the position and momentum of its centre would both be known simultaneously and this possibility is ruled out.

The result for a single vibrating atom applies to all the vibrating atoms in a solid body. Since a body in 3-D supports a number of vibratory modes then each frequency will contribute to the body's ZPE.

But mathematical models need to be tested against reality. Scientists are currently investigating what is going on inside matter when the temperature drops to near zero degrees Kelvin. By studying the wavelengths of light emitted by excited atoms, researchers have found that at very near absolute zero crystals do appear to have zero-point energy.

In the case of helium gas, at normal pressure, as it is cooled its molecular speeds reduce until, eventually, it becomes a liquid. As the temperature drops to very near absolute zero, the molecular vibrations within the liquid helium are considerably reduced, but it does not solidify. It is thought that the internal vibrations emanating from its zero-point energy keep the helium in liquid form even at absolute zero.

As Einstein pointed out in 1930, the uncertainty principle also dictates that the energy fluctuations are inversely proportional to the time resolution of the measurement period. This is expressed as

$$\Delta E \cdot \Delta t \geq \hbar. \qquad (17.5)$$

In words, this says that the error (ΔE) in measuring the energy E of a system multiplied by the error (Δt) in measuring the time interval t over which the energy measurement is made must always be greater than $\hbar$. More importantly, it means that the principle of conservation of energy can be violated for very short periods of time, providing the above inequality (eqn 17.5) is still obeyed.

So let us examine a volume of empty space at absolute zero temperature. We suppose that the volume contains no matter, no gravitational fields, no charges and no electromagnetic fields. We have a volume of space apparently containing nothing. It's a vacuum. If this volume of vacuum truly contained no energy then $\Delta E = 0$. But this would mean that $\Delta E \cdot \Delta t = 0$, which violates the inequality requirement (eqn 17.5) that $\Delta E \cdot \Delta t \geq \hbar$. So, it seems that in quantum terms $\Delta E > 0$ for the vacuum of space.

The uncertainty principle implies that a particle with energy ΔE and effective mass $\Delta E/c^2$ can spontaneously emerge out of the quantum vacuum, provided that it vanishes in a time Δt, such that the inequality (eqn 17.5) is obeyed. The lifetimes of these virtual particles are exceedingly short with very brief ghost-like appearances. Scientists now believe that the quantum vacuum is actually made up of these fluctuating virtual particles of energy in the form of photons, gravitons, gluons and so on, of every possible frequency, flitting into and out of existence. Paul Dirac, a British physicist much involved in the developments of quantum mechanics in the 1930s, called it the quantum foam. The interiors of atoms are mostly empty space. So the insides and the outsides of atoms forming a solid will be subject to a constant buffeting by the virtual particles emerging from the quantum foam, which may be the cause of atomic vibrations at very near zero degrees Kelvin.

So what are the characteristics of this quantum foam in terms of virtual photons? For a start, the foam must look the same in all directions. This means that the frequency distribution of the zero-point virtual photons popping briefly out of the quantum foam must be the same in any direction. Furthermore, the zero-point frequency distribution must be the same for any inertial frame of reference, too. In other words, no matter what constant speed observers are moving at relative to one another they must all observe the same quantum vacuum.

From Einstein's theory of special relativity we know that we can move to a new inertial reference frame and determine the new conditions there by using the Lorentz-Fitzgerald transformations. Changing from one inertial reference frame to another means a change in velocity and, as expressed in the Doppler effect, this usually means any frequencies present will change. But for our quantum vacuum the frequency distribution of the zero-point virtual photons must not be changed by a Lorentz-Fitzgerald transform. According to the US theoretical physicist Timothy Boyer (*Scientific American*, August 1988), the only frequency distribution which is not changed by a Lorentz-Fitzgerald transform is one which is dependent on the frequency cubed, since for this case the Doppler frequency shifts cancel out. With this knowledge, Boyer has derived the emissive power of the zero-point radiation from the quantum vacuum. In terms of wavelength, the emissive power $E(\lambda,0)$ over the narrow bandwidth from λ to $\lambda + d\lambda$ is given by

$$E(\lambda,0) = \frac{4\pi h c}{\lambda^5} \qquad (17.6)$$

If we consider the total radiated energy appearing out of the quantum vacuum of virtual photons, it is clear from equation 17.6 that the contribution from the high frequency virtual photons, with short wavelengths, quickly approaches infinity! Therefore the ZPE of the quantum vacuum must be infinite! But, infinities don't happen in nature, so it means that scientists have missed an important factor which limits the size of the ZPE of the quantum vacuum. This situation is reminiscent of the 'ultra-violet catastrophe', the solution of which led to the birth of quantum mechanics. Perhaps in understanding the reason why the energy of the quantum vacuum is not infinite we will discover a new realm of physics.

One solution that has been proposed is as follows. Since $½\hbar$ is the smallest amount of angular momentum that can exist, perhaps there is a smallest length L, too. By collecting the constants $\hbar$, G and c (Planck's h-bar, Newton's universal gravitational constant and the speed of light), and grouping them together, it has been suggested that the smallest length possible might be the Planck length L, given by

$$L = \sqrt{\frac{\hbar G}{c^3}} = 1.6 \times 10^{-35} \text{m}. \qquad (17.7)$$

If so, nature would impose a quantum vacuum vibration cut-off frequency f_{co} given by

$$f_{co} = \frac{c}{\lambda_{co}} = \frac{c}{L} \qquad (17.8)$$

It would mean that the vacuum of space is made of discrete lumps, with the smallest possible volume being L^3. Photons with wavelengths the size of the Planck length would have huge energies, with an energy density of the order of

$$\frac{E_0}{L^3} = \frac{½ h f_{co}}{L^3} = \frac{½ h c}{L^4} \approx 10^{114} \text{ J/m}^3 \qquad (17.9)$$

The energy density of the nucleus is of the order of 10^{34} J/m^3, but that's fairly small compared with the energy density of 10^{114} J/m^3 of the Planck wavelength virtual photons. We know how to tap nuclear energy; will we ever be able to tap quantum vacuum energy?

If we write Einstein's energy equation in the form $m = E/c^2$ it suggests that the quantum vacuum is filled with some strange form of mass with a huge density. Why aren't we aware of this? Firstly, this mass is effective mass which, if we assume a linear theory of gravity, does not have a gravitational field. And, secondly, the vacuum is uniform in all directions.

In a report written in 1996 by Dr Robert Forward for the US Air Force, he commented that the existence of electromagnetic fluctuations in the vacuum at absolute zero temperature is now accepted by most scientists, but the phenomenon is seen as a weird effect devoid of any practical applications. The idea that zero-point energy (ZPE) might be extractable from these quantum vacuum fluctuations is dismissed by many scientists as utter nonsense.

Dr Forward suggested that the situation with regard to ZPE was reminiscent of the field of nuclear energy in the early 1930s. At that time, scientists were only just beginning to understand the structure of the atom. But, in just a few decades, the esoteric, poorly understood phenomenon of the 'nuclear energy of the atom' went from being a scientific curiosity into being a major technology. "In the 1990's", wrote Dr Forward, "we are now looking at the esoteric, poorly understood phenomenon of electromagnetic fluctuation energy of the vacuum. The estimate of the Vacuum Energy Density is much higher than that associated with Nuclear Energy Density".

In Dr Forward's opinion it was essential to carry out experiments to get a much better understanding of the quantum vacuum. Studying the phenomenon in more depth, he felt, might lead to real advances in space power and propulsion technology.

18

THE QUANTUM VACUUM

We now know that we live on a minor planet rotating around a minor star in the outer arm of a swirling galaxy of stars called the Milky Way. And we know that our own galaxy is just one of an innumerable number of galaxies scattered throughout a Universe of unimaginable size. Just as the dust motes in a sunbeam eventually collect together, under the influence of Earth's gravity, to form a dust layer on the floor, so the stars in the Universe are drawn together to form galaxies around local centres of gravity. It's just a matter of scale. The vast regions of space between the stars in a galaxy and the even vaster regions of space between the galaxies are regions containing no matter and so, by definition, these regions of space encompass a gigantic vacuum.

But are these gigantic inhospitable regions of nothingness truly empty? Some regions of the vacuum may contain electromagnetic and gravitational fields which impose their own positive and negative energy distributions on space, but we will assume that these are absent. Many physicists now agree that rather than being empty the vacuum of space at zero degrees Kelvin is actually a turbulent medium in a state of agitation. The reasoning behind this is Einstein's version of the uncertainty principle for energy (eqn 17.5), which means that the energy of the vacuum cannot be zero. It is assumed that excited particles of energy suddenly burst into existence out of the vacuum and then almost immediately disappear back into it but, due to their brief existence, the averaged energy level of the vacuum is non-zero. These randomly created particles cause noise in the vacuum, the detection of which seems to confirm their brief existence. These quantum particles of positive energy appearing out of the vacuum might be photons, gravitons or gluons but, since energy is energy, out of context we can't distinguish between them. The quantum vacuum is the new name for the ether.

The idea that the vacuum is not empty is so incredible that it is natural for people to balk at the idea when they first encounter it. Surely, it's nonsense?

Even now, eighty years after the idea was first mooted, the quantum vacuum model is still tentative. It may be wrong. On the other hand, it may be that this noisy effervescence of the vacuum, due to the spontaneous creation and destruction of particles of energy throughout space, is what gives the Universe its various permeability properties.

But the quantum vacuum is even weirder. When electromagnetic and gravitational fields are present, then, under extreme conditions, space can be torn apart! In quantum terms the tearing leads to the creation of twinned pairs of energy particles, one with positive energy and the other with negative energy. That is, a particle of energy and its anti-particle twin.

The Centre for Theoretical Physics at Sussex University is one of the leading research establishments in the world studying quantum mechanical phenomena and related effects stemming from the quantum vacuum. In June 2000 Professor Ed Hinds, from the Centre, visited BAE Systems Warton to give a Project Greenglow audience the benefit of his expert knowledge on the current state-of-the-art in quantum vacuum research. Earlier in his career Professor Hinds had led a research team at Yale University which had investigated properties of the quantum vacuum. In particular they had measured an atomic-quantum vacuum effect called the Casimir-Polder force.

The thought of a vacuum teeming with virtual quantum particles seems like science fiction, so it's natural to ask what observed phenomena exist to support such a crazy notion.

One observation offered is the 'Lamb frequency shift'. In spectroscopy experiments with hydrogen gas in a strong magnetic field, the US scientist Willis Lamb noticed that the positions of the Zeeman lines in the hydrogen spectrum (associated with the energy changes which occur during electron orbit transitions resulting from heating, or cooling, the hydrogen gas) were slightly shifted in frequency from their theoretically predicted values. He suggested that within each atom the orbiting electrons are bombarded in all directions by virtual photons from the quantum vacuum (Fig. 18.1(a)) causing the electrons' actual orbital paths to fluctuate slightly about their paths predicted for a truly empty vacuum. Consequently, the electrons' energy levels also fluctuate slightly. Following an electron's transition to a new orbit, the exchange of energy is conveyed by a real photon which also fluctuates slightly in energy content from its theoretical value, resulting in a shift in the predicted frequency of the

THE QUANTUM VACUUM

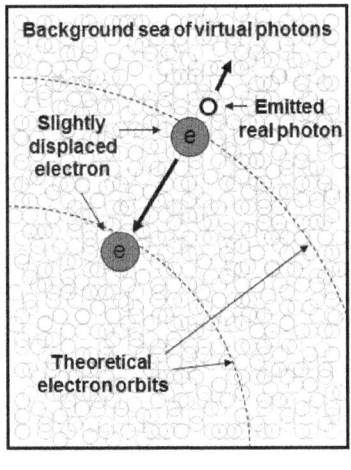 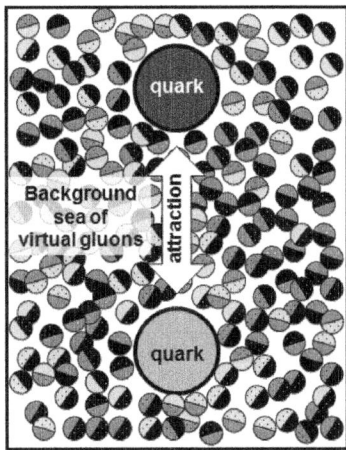

(a) The Lamb shift.   (b) Asymptotic freedom.

Fig. 18.1 Examples of the quantum vacuum.

Zeeman line. Willis Lamb won the Nobel Prize in Physics in 1955 for his observation and his quantum vacuum explanation.

Another example, where the quantum vacuum is assumed to play a part, is in the nuclear strong force (Chapter 16) between quarks (Fig. 18.1(b)).

Perhaps a more convincing observation in support of the existence of the quantum vacuum is the phenomenon of the 'Casimir force'. An uncharged conducting surface illuminated by an electromagnetic wave front, or beam of photons, experiences radiation pressure. Newton's third law of motion tells us that to every action there is an equal and opposite reaction. The radiation pressure is the response to the electromagnetic wave front being reflected from the surface. We often refer to these surfaces as mirrors, although the reflection property applies to all frequencies, not just those for light. Even in the absence of illumination such a mirror is still bombarded on all sides by virtual photons appearing out of the quantum vacuum, but because of the uniformity of the radiation pressure on both sides of the macroscopic object there is no noticeable effect.

A Casimir cavity is formed by two parallel uncharged, conducting surfaces of area A arranged a width x apart (Fig. 18.2). The Casimir force is the quantum vacuum force exerted across the cavity, arising due to the mismatch in virtual photon radiation pressure between the inside and the outside, which pushes the two surfaces together.

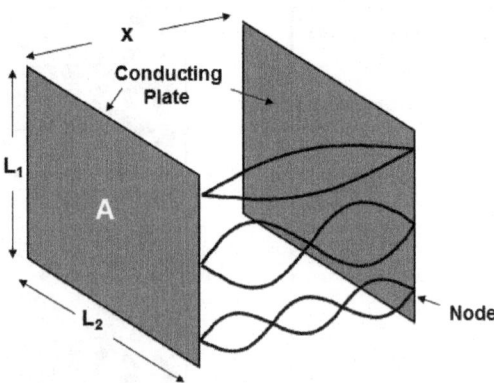

Fig. 18.2 Waveforms in Casimir cavity.

The basis for the Casimir force can be understood fairly easily, in terms of the quantum vacuum, where virtual photons of every possible frequency are assumed to be continually popping into and out of existence. But, for a virtual photon to briefly exist within the cavity it must be able to resonate within the cavity. In terms of its waveform, that means it must be able to form a standing wave with nodes on the inner conducting surfaces. So, the only virtual photons which can appear inside the cavity are those with a half-wavelength ($\lambda/2$) equal to an integer times the width x between the cavity walls. That is

$$\lambda/2 = n.x \quad \text{for } n = 1,2,3 \ldots \ldots \quad (18.1)$$

Consequently, there will be waves of some frequencies that cannot resonate in the cavity. Or, going back to discrete particles, there will be some virtual photons that can't exist in the cavity. Because some virtual photons are prohibited from making an appearance inside the cavity, there will be less radiation pressure due to the virtual photons on the walls inside the cavity than due to the radiation pressure of the unsuppressed virtual photons on the walls outside. Consequently, the conducting walls will be forced together. This prediction, based on quantum theory, was made, in 1948, by the Dutch physicist Hendrik Casimir, while working at the Philips Research Laboratory at Eindhoven, in the Netherlands.

The Casimir force F in the x-direction is predicted to be

$$F = \left(\frac{\pi^2 \hbar c}{240 x^4}\right) A \qquad (18.2)$$

The force of attraction of the walls increases as the gap width x decreases.

The predicted Casimir stress F/A, or pressure, is extremely tiny, as the following two examples show:

For width $x = 1$m, $\quad F/A = 1.3 \times 10^{-27}$ N/m^2. $\qquad (18.3)$

For width $x = 10^{-6}$m $= 1$ μm, $\quad F/A = 1.3 \times 10^{-3}$ N/m^2. $\qquad (18.4)$

Almost unbelievably, it is possible to measure the Casimir force for surfaces a distance apart of about 10^{-7}m; that's 0.1μm, or a tenth of a micron. For comparison, a human hair is about 70 μm in diameter. If the walls are any closer then the surface electrons of both surfaces begin to interfere with each other. Although the distances are tiny, many university and industrial research groups have been developing micromechanical mechanisms with these sorts of dimensions for more than thirty years. Over the last decade interest has turned to the investigation of even smaller nanotechnology devices, where dimensions are of the order 10^{-9}m – that is 0.001μm. At these dimensions the Casimir force can cause problems, so the phenomenon is receiving more and more attention.

The measuring systems needed for Casimir force experiments must be able to weigh a speck of dust on a scale pan. Also, the conducting surfaces forming the cavity must be almost perfectly flat and parallel, which is extremely difficult to achieve in practice. Finally, we must remember that the formula for the Casimir force is derived assuming that the cavity temperature is at absolute zero (0 K, or -273° C) and housed in a vacuum. If the temperature is above zero, then real (not virtual) thermal photons may penetrate the cavity housing and alter the size of the predicted force.

The Casimir force was measured for the first time in 1996 by Dr Steven Lamoreaux, at the University of Washington in the USA. Given the difficulty in forming a cavity with two perfectly flat parallel surfaces, Lamoreaux chose, instead, to use a flat surface and a spherical surface with a large radius of

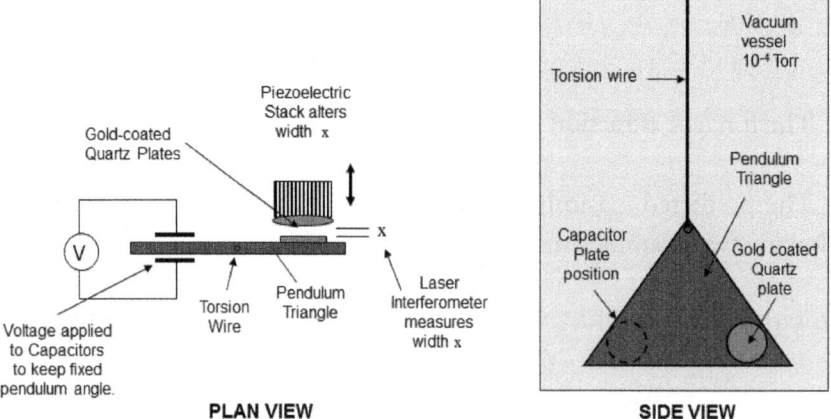

Fig. 18.3 Lamoreaux's Casimir force experiment.

curvature. Theory shows that the modified form of the Casimir force is independent of the plate area A and is

$$F = \frac{2\pi R}{3}\left(\frac{\pi^2 \hbar c}{240 x^3}\right) \quad (18.5)$$

The radius of curvature of the spherical plate is R and the minimum gap width is x. In Lamoreaux's experiment (Fig. 18.3) the gap width x could be set in the range 0.6μm < x < 11μm. The flat conducting surface was fixed to one arm of a pendulum-shaped torsion balance and the opposite, spherical-shaped, surface was fixed to a piezoelectric stack which could be expanded or contracted to alter the cavity gap width. The width x was independently measured using a laser interferometer. The other arm of the torsion balance was placed between the two parallel conducting plates of a large capacitor. This was part of a feedback system, so as the width x was altered the change of capacitor voltage needed to keep the pendulum angle fixed, and width x fixed, could be calibrated to give the Casimir force between the plates. With this apparatus, forces of order 10^{-11}N, or μdynes could be detected.

The whole apparatus was contained in a vacuum vessel and evacuated to a pressure of 10^{-4} Torr. The experiment was carried out at room temperature, not at absolute zero, so a correction factor was required. Lamoreaux's experimental results agreed to within 5% of the predicted Casimir force.

The measurement of the Casimir force between a flat surface and a spherical

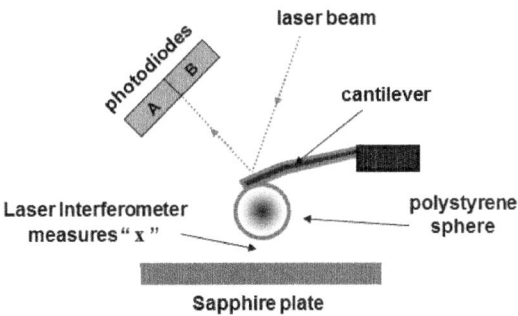

Fig. 18.4 Measuring the Casimir force with an atomic force microscope.

surface was repeated, in 1998, by Umar Mohideen and Anushree Roy, at the University of California at Riverside in the USA. In their approach they used an atomic force microscope (AFM) which could measure forces to an incredible resolution of 10^{-18}N.

An AFM uses a microscopic cantilever (about 200 μm long), which is scanned across a surface of interest. As the cantilever moves, its tip moves up and down with the undulation of the surface, like the needle on an old-fashioned gramophone record. Using a laser beam as an optical lever, to magnify the cantilever movement, the beam is shone onto a mirror fixed to the upper-side tip of the cantilever and is reflected onto a segmented photodiode which detects the deflection.

Mohideen and Roy modified their AFM by sticking a small polystyrene sphere, with a diameter of 200 μm, to the underside tip of the cantilever (Fig. 18.4). A piece of sapphire was used for the flat plate in the experiment. The sphere, the cantilever and the sapphire surface were all coated with aluminium, to a thickness of about 250 nm, to make them conducting.

The sapphire plate was moved towards the sphere in a series of discrete tiny steps. At each step, the calibration of the AFM allowed the Casimir force to be determined from the measured deflection of the cantilever. A laser interferometer measured the actual gap width x, which was in the range 0.1μm < x < 0.9μm. So the narrowest gap was on the verge of where the Casimir force conditions could legitimately be applied.

The experiment was carried out in a vacuum of 5×10^{-4} Torr and at room temperature. Their results agreed to within 1% of the predictions made with the modified Casimir force equation, with R = 100μm = 10^{-4}m.

In 1984 Dr Robert Forward was the first scientist to show that in theory it is possible to extract energy from the quantum vacuum at absolute zero temperature. This energy is known as ZPE, standing for zero-point energy. Forward envisaged using a mechanical one-shot device consisting of a collapsing charged foliated conductor, somewhat reminiscent of a voltaic pile. Although his device could not be realised in practice, it set the ball rolling for other inventors to think of devices which could be manufactured. In December 1996 the first US patent was granted to F. B. Mead and J. Nachamkin for a practical method of extracting ZPE from the quantum vacuum. Since then, further patents have been issued in connection with quantum vacuum manipulation, and start-up companies in the US are now vying for a share of the future market.

When a 2-D cavity is formed by two parallel plane walls, the Casimir force is attractive. When one wall is plane and the other is spherical the Casimir force is still attractive. However, experts in quantum vacuum theory have shown that when both walls are spherical the Casimir force is repulsive. An attractive Casimir force implies that the vacuum energy density of the cavity is less than, or negative with respect to, the vacuum energy density of the free quantum vacuum, while a repulsive Casimir force implies that the cavity contains a region of greater, or positive, vacuum energy density. Research groups around the world are already experimenting with 2-D Casimir cavities with walls of different shapes.

In 2001 NASA funded a MEMS (micro-electro-mechanical system) study to investigate the vacuum energy density balance of a box containing a row of parallel rectangular Casimir cavities. The research work is being done by Quantum Fields LLC, led by the Chief Scientist Professor Jordan Maclay, of Yale University. The cavities are gold plated, have a width of $0.1\mu m$ and share a common lid, which is supported by a tiny spring. According to theory, the position of the lid over the box dictates whether the lid is attracted, or repelled, by quantum vacuum forces. Professor Maclay hopes that by setting the lid moving, Casimir forces will take over and keep it oscillating. The question is, "If it is possible to set up an endless Casimir cavity lid oscillation, is it possible to endlessly extract energy from the motion?"

In other words, is it possible to allow the Casimir force to push the plates together and to extract energy from the work done? The problem is that the

Casimir force is a conservative force field, the strength of which depends on the gap width between the walls of the cavity. Changing the gap width changes the potential energy of the cavity. However, if the plates are returned to their starting positions, then the potential energy of the cavity returns to its original value. The question, then, is whether the Casimir cavity can be transformed in some way, during the change in energy process, to distort, or break, the conservative cycle and extract some energy, leaving the quantum vacuum to replace what has been taken from its infinite reservoir of ZPE.

We appear to extract energy from the gravitational field of the Earth in using a hydroelectric scheme. But this can't be true because the Earth's gravitational force field is also conservative. If we take a raindrop around a cycle where we change its height with respect to the Earth's surface, we know that its potential energy will change, but that when it gets back to its starting point its potential energy will be exactly the same as when it started. In fact, in the hydroelectric scheme we only use part of the gravity-height cycle. We let the water flow down, converting some of its gravitational potential energy into kinetic energy, which we make use of. We then rely on the Sun's almost infinite heat energy to lift the water back up again, to form clouds which release rain to keep the cycle going. Our hydroelectric scheme is actually extracting energy from the Sun!

One simple idea proposed to side-step the conservative field problem with regard to the Casimir force is as follows. After the walls have moved closer and the energy extracted, they can be slid sideways in opposite directions, so that there is no cavity. The gap width is then re-established before sliding the walls sideways, again, to re-form the cavity, and the whole process is repeated cyclically. However, quantum friction comes into play and spoils this idea. Yet another idea is to change the property of the cavity walls cyclically, from conducting to dielectric, in an attempt to break the conservative hold. In 1999 Dr Fabrizio Pinto, of the Jet Propulsion Laboratory at the California Institute of Technology, explained how by cyclically manipulating the dimensions of a Casimir cavity it might be possible to extract vacuum energy in an endless process. Dr Pinto has since formed the InterStellar Technologies Corporation and is now in the process of trying to commercialise his idea.

In 2001 a research team at the Bell Laboratory (Lucent Technologies), in the USA, began exploring the Casimir force. The transistor and the laser were invented at the Bell Lab. The team, led by Frederico Capasso, showed that the

Casimir force could be used to control a MEMS device. Of particular interest to the Bell scientists was the possibility that the Casimir force might be used to introduce non-linear effects into the vibration of a MEMS device, as this might provide a way of breaking the conservative cycle and enable ZPE to be extracted from the quantum vacuum.

During 2002 Dr Clive Speake, Head of Experimental Gravity and Space Research at Birmingham University, together with Dr Giles Hammond, began an experimental study of the Casimir force at near zero degrees Kelvin, using a 2-D cavity formed between a spherical surface and a plane wall attached to a superconducting torsion balance. This study was part-funded by BAE Systems as part of Project Greenglow. The longer-term aim of the Birmingham study is to explore the Casimir force with 2-D cavities of different shapes.

Crystals are known to contain their own ZPE. There are rumours from Russia and the USA that researchers have discovered that by treating certain crystals, either by contaminating them with radioactive substances or exposing them to a plasma beam, they can locally alter the ZPE of the quantum vacuum. Perhaps such crystals form ZPE diodes which may be used in a scheme to extract vacuum energy? It may sound unlikely, but all ideas need to be considered.

This is a period of research where great imagination and ingenuity is required by quantum vacuum scientists and engineers to think of new ideas and to devise experiments to test them. In his book *Tapping the Zero Point Energy*, Dr Moray King has described a number of ways in which he believes it may be possible to extract energy from the quantum vacuum. These ideas need to be examined by quantum physics experts to check on their worth.

A variation of the theoretical Davies-Unruh effect (eqn 20.18) is that a uniformly accelerating mirror will cause a distribution of photons to appear out of the quantum vacuum and be driven in front of the mirror surface. In effect, the acceleration disturbs the uniformity of the virtual photons bombarding both sides of the mirror, so that some virtual photons on the upstream side are overtaken by the mirror before they can disappear back into the quantum vacuum and so they become real photons. Thus, in theory, a vibrating mirror should create a pulse-train of photons. However, for this to be a noticeable effect the vibration of the mirror surface has to be enormous, with oscillations approaching the speed of light, and this is not thought to be mechanically feasible. In May 2011 an international team of scientists working in the Department of Microtechnology and Nanoscience at Chalmers University of Technology in Gothenburg, Sweden, reported that they had devised a way to simulate the rapid vibration of a mirror. In their experiment, conducted at a temperature of about 4 K, they used a SQUID attached to the end of a transmission line. The SQUID, or superconducting quantum interference device, has a tiny loop which allows quantum units of magnetic flux to pass through its centre. A 10 GHz signal caused the magnetic field through the SQUID loop to vibrate, simulating the vibration of a mirror at about 5% of the speed of light. But rather than the surface moving, it's where the boundary conditions of the imaginary mirror surface must be satisfied that move. Professor Chris Wilson and colleagues, at Chalmers University, claim that they have detected photons being emitted from the quantum vacuum. However, they have not tapped the ZPE and obtained some free energy. It is the vibrating energy of the imaginary mirror which converts the virtual photons from the quantum vacuum into real photons. This is a very sophisticated experiment of which only scanty details have been given above. We must now wait and see what the rest of the world's quantum physics experts make of it.

An extension of the vibrating mirror is the dynamic Casimir cavity with vibrating walls, an idea being explored by Professor Jordan Maclay. In the extreme case of a collapsing Casimir cavity it is thought that a stream of virtual photons might be squeezed out as the opposing walls approach each other and some virtual photons suddenly find they can no longer fit in between. As a ship's screw turns, on the rear side of the advancing blades low pressure regions arise and tiny vapour bubbles form as the water boils. This is the phenomenon of cavitation (Fig. 3.3). The collapsing bubbles do a great deal of damage to the blades which become pitted. Under certain conditions it has been noticed that collapsing vapour bubbles emit a short-lived pulse of light. This phenomenon is called sonoluminescence. Dr Claudia Eberlein, at the Centre for Theoretical Physics at Sussex University, has suggested that the collapsing walls of the tiny vapour bubbles might mimic a collapsing Casimir cavity, in which case sonoluminescence is a direct indication of the existence of the quantum vacuum.

In July 2006 the EU Research Programme (Framework 6), covering new and emerging science and technology, awarded €800,000 (£536,000) to a three-nation consortium to investigate ZPE and the Casimir forces of cavities with various geometries. The research programme is led by the University of Leicester in the UK, with the Universities of Pierre et Marie Curie in France, Linköping in Sweden and Birmingham in the UK. In 2012 there were forty research groups based in eleven European countries investigating many aspects of the Casimir force and holding regular workshops and conferences to speed progress.

In October 2009 the US Department of Defense (DoD) Advanced Research Projects Agency (DARPA) launched a $10 million research programme to investigate ways of manipulating the Casimir force. The five institutions involved are Yale University, Harvard, the University of California at Riverside and two US national laboratories at Argonne and Los Alamos. Researchers at Yale have expressed an interest in manufacturing new materials based on nano-scale structures containing positive and negative energy density micro-cells which may lead to a material that levitates.

A great deal of research effort and funding is now being invested in learning more about the strange properties of the quantum vacuum. The dream that we may, one day, be able to use the quantum vacuum as a source of unlimited clean energy is one reason that draws scientists on, another reason being good old-fashioned curiosity. Learning how to manipulate the quantum vacuum also opens up the possibility of a new means of propulsion. This is a mind-boggling new scientific adventure taking us into the future. But it comes with a word of warning from a few scientists who are still not convinced that the quantum vacuum exists.

19

ELEMENTS OF QUANTUM GRAVITY

If, at the lowest quantum level, space is granular with discrete dimensions based on the Planck length L (eqn 17.7), then does time also pass in discrete amounts? And, if space is inextricably linked with time, what about the granularity of space-time? And where is the discrete quantum mechanical model of gravity at the microscopic level that joins smoothly with Einstein's continuous model of gravity at the macroscopic level, in the way that discrete H_2O molecules become flowing water? For nearly a century, the theories of quantum mechanics and general relativity have developed independently. During much of this time scientists searched for a model to link the two theories together, but without much success. Part of the problem lies with the non-linearity of Einstein's model of gravity, but the search for unification was made even more difficult with the later discovery of the strong and the weak nuclear forces. Nevertheless, there are some simple links between quantum entities and gravity which are worth exploring.

Isaac Newton, in the 17th century, suspected that light was affected by gravity, suggesting that light possessed some form of mass attribute. However, he had no support for his suspicion and it wasn't until the early 20th century that experiments confirmed that he was right. In about 1660, Newton proposed the idea that light was transmitted in the form of particles, which he called corpuscles. It was an Ancient Greek idea that particles of light striking the eye caused the sensation of sight. Newton developed his particle theory to model the properties of light reflection and refraction. However, in 1680, the Dutch scientist Christiaan Huygens proposed his wave theory of light. Thus began the controversy over which theory was right. In 1801 Thomas Young at the Royal Institution in London showed that light interference patterns could be explained using Huygens' wave theory. In 1850 experiments by Léon Foucault showed that Newton's corpuscular theory of refraction was wrong. Accordingly, the particle theory of light was rejected. James Clerk Maxwell's view, expressed in 1864, that light was an electromagnetic wave, coupled with Heinrich Hertz's discovery in 1888 of other electromagnetic waves, seemed to confirm the fact that light was transmitted in waves. But, in 1900, Max Planck discovered that radiated electromagnetic energy was quantised. Based on this knowledge, in 1905, Albert Einstein introduced his own particle of light, now called the photon, to develop a successful theory to explain the photoelectric effect. Thus the argument over whether the particle, or the wave, theory of light was right was reignited. The controversy was finally resolved in the 1920s with the concept of wave-particle duality, where it was accepted that circumstances dictated whether a particle or a wave approach was the most appropriate way to model a particular phenomenon.

The particle nature of light has now been extended to cover the whole of the electromagnetic spectrum, covering all frequencies, not just the narrow band for light, with Newton's corpuscle being replaced with the photon. So, an electromagnetic wavelength, with particular frequency, unit amplitude and phase angle start, can be replaced with a photon with the same frequency. To accommodate the amplitude a distribution of photons is required. The wave and the photons all move with light speed c.

Albert Einstein tried to imagine what it would be like to ride on the crest of a wave of light. In our understanding of a light wave, as an electromagnetic wave (Fig. 7.4), this means riding on the crest of an E-wave, or an H-wave. In the frame of reference in which the light wave is stationary, the electric E-waves and the magnetic H-waves make no progress longitudinally. Neither can the E- or H-waves cycle in the transverse direction, because this would result in the wave shape changing longitudinally. No oscillations, or zero frequency f, in the rest frame of the light wave imply that everything is frozen in time. This is in agreement with the Lorentz transform for time, which predicts that events take infinitely long to occur in an inertial frame moving at light speed.

Instead of being carried along on the crest of a light wave we might think of sitting on a photon. According to Max Planck, the energy of a photon is

$$E = hf = \hbar\omega \tag{19.1}$$

Here, h is Planck's constant and f is the wave frequency. $\hbar = h/2\pi$ is called hbar and $\omega = 2\pi f$ is the angular wave frequency. The photon's energy E takes either form.

A photon has no real mass, but if we treat it as a particle (fermion) with mass m we can separate its energy into two parts, namely

kinetic energy = ½mc² (19.2)

and

spin energy = ½ ℏω (19.3)

Thus, the total energy of a photon (eqn 8.8) due to its motion is

$$E = hf = \hbar\omega = mc^2 \qquad (19.4)$$

From equations 19.1 and 19.4 we see that Newton's suspicion that light possessed the attribute of mass is confirmed because a photon has effective mass m given by

$$m = \frac{hf}{c^2} \qquad (19.5)$$

This is true for all inertial frames of reference, although in the frame of reference in which the photon is stationary the photon's effective rest mass is zero, since it has no energy.

Now the potential energy of a mass in a gravitational field **g** changes with its height in the field. For a change in height Δs the change in potential energy ΔE is given by

$$\Delta E = mg\Delta s \qquad (19.6)$$

Therefore, as a photon moves up a distance Δs through a gravitational field **g** its change in potential energy is

$$\Delta E = \left(\frac{hf}{c^2}\right) g\Delta s \qquad (19.7)$$

In quantum terms the change in a photon's energy is

$$\Delta E = h\Delta f \qquad (19.8)$$

Combining equations 19.7 and 19.8 gives

$$\Delta f = \left(\frac{g}{c^2}\right)\Delta s\ f \qquad (19.9)$$

Where Δf is the decrease in frequency f of a photon as it travels a vertical distance Δs upwards through a gravitational field **g**. For a light beam, the decrease in frequency means that the beam's frequency shifts towards the red end of the spectrum as it travels upwards through a gravitational field. The *gravitational red shift* formula (eqn 19.9) was derived by Einstein in 1907.

In 1960, five years after the death of Einstein, the first successful experiment to measure the Earth's gravitational red shift was made. The experiment, done by the US scientists Robert Pound and Glen Rebka, was carried out in the 22.56 m-high lift shaft of the Jefferson Tower in the Physics Building of Harvard University. A source of gamma rays (frequency f_0 of order 10^{22} Hz) was used to form a beam that was shone either up or down the lift shaft. The narrow band receiver was a gamma ray absorber. Since the red, or blue, shift Δf of the beam at the top, or bottom, of the tower was of order 10^7 Hz, this would have put the signal outside the bandwidth of the gamma ray absorber.

To overcome this, the source was mounted on a mobile platform which moved towards, or away from, the receiver, causing a Doppler shift in the incident beam. The Doppler red shift is

$$\Delta f = \left(\sqrt{\frac{1-\frac{v}{c}}{1+\frac{v}{c}}}\right)f_0 - f_0 \qquad (19.10)$$

The velocity v is negative for the blue shift. The platform velocity was chosen so that the blue, or red, Doppler shifts exactly cancelled the red, or blue, gravitational shifts respectively, enabling the receiver to detect the gamma ray signal.

From equation 19.9 and from a first order approximation of equation 19.10, the velocity v of the platform needed to cancel the frequency shifts is

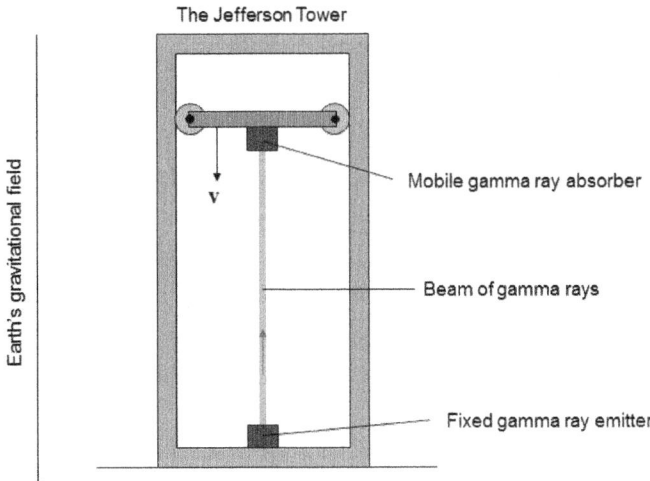

Fig. 19.1 The Pound-Rebka gravitational red-shift experiment.

$$v = \left(\frac{g}{c}\right)\Delta s \quad \text{m/s} \tag{19.11}$$

This shows that the required velocity was just over 2 mm/hour. With incredible precision, Pound and Rebka were able to adjust the velocity v so that the gamma ray absorber received the transmitted signal. Measurement of the platform's velocity v confirmed Einstein's prediction, to within 10%.

Satellite navigation mostly relies on the US global positioning system (GPS), while the Russians have GLONASS and the EU's Galileo system will be operational by 2020. The GPS user wishing to locate their position broadcasts a microwave signal which is received by at least three visible satellites, all of which transmit a signal in response. The orbiting GPS satellites are in known positions, so their exact distances from the user can be evaluated from the time delays in receiving their signals. Thus, the user is on the surfaces of three intersecting giant spheres with the satellites at their centres. With this information the position of the user in 3-D space above the Earth's surface is calculated electronically by the user's GPS device. This process requires very accurate synchronised atomic clocks, and the effect of the gravitational blue shift on timing circuits must be taken into account.

Given two identical atomic clocks, if one is placed in a GPS satellite and the other is placed in a ground station then, due to the gravitational blue shift, the ground clock has a shorter period than the clock in the satellite. So, for the same number of oscillations, the satellite clocks run faster than the ground clock and this has to be accounted for in the signal processing. Based on a caesium atomic clock with a frequency of about 9.2 GHz and with satellites in circular Kepler orbits at an altitude of 20,200 km, if the blue shift is not accounted for then errors in distance of about 1 km can accrue over one hour.

From equivalence and from equation 19.9 we can predict that light emitted from an accelerating source will increase in frequency, relative to the source, although its velocity, c, will remain unchanged.

$$\Delta f = \left(\frac{a}{c^2}\right)\Delta s\ f \qquad (19.12)$$

This is not a Doppler effect as the source and the detector are not in relative motion. It is a photon's response to linear acceleration. As a photon travels a distance Δs in the direction of the acceleration **a** it changes its frequency by Δf. The resulting increase in photonic energy (a blue shift) exactly cancels the decrease in energy of the photon with respect to the accelerating light source. The photon does not accelerate but continues to move at the constant speed of light c.

If a person in free fall shines a light in the direction of the gravitational field, equations 19.9 and 19.12 ensure that the light beam does not change frequency. In fact the beam can be shone in any direction – it makes no difference. The falling light source and the falling person are both in a force-free environment. From Newton's first law of motion, a body will continue to move in a straight line with uniform speed unless it is acted upon by a force. Since the photonic masses within the light beam move with constant speed c as they travel upwards, or downwards, in the gravitational field they also move in a force-free environment. The mystery behind why this is so is explained in the next chapter.

This chameleon property of the photon, changing colour and effective mass in response to gravitation and acceleration, is very important in physics. Perhaps in the future, the change in frequency given by equation 19.9 might be used in a stationary sensor to detect a change in gravitational field strength, while the change in frequency given by equation 19.12 could provide the basis for a solid-state accelerometer.

From statistical mechanics, vibration energy is shown to be proportional to kT, where k is Boltzmann's constant and T is the temperature. Frequency f is therefore proportional to the temperature T. From a different perspective, equation 19.12 implies that when a body accelerates towards the fixed stars the stellar light photons incident on the body will increase in frequency. So as an observer accelerates towards the fixed stars they will appear to get hotter. The same is true for the virtual photons forming the quantum vacuum. In accelerating, an observer sees an increase in the temperature of the approaching background of space. Reciprocity suggests that an observer in the rest frame will see the accelerating body get hotter and emit radiation. From either view it is as though an accelerating body experiences frictional heating as it passes through the quantum vacuum.

For a mass m to escape from the surface of a planetary body of mass M and gravitational field g (eqn 2.4) the upward force on the mass must exceed its weight. In other words, from Newton's second law of motion, the upward acceleration a of the mass must be greater than the downward acceleration g imposed by gravity. That is a > g. In energy terms, this means that the magnitude of the kinetic energy of the mass must be greater than the magnitude of its potential energy. Thus, to rise in the gravitational field requires

$$\tfrac{1}{2}mv^2 > mgr \qquad (19.13)$$

Where v is the vertical upward velocity of the mass m and r is its radial distance from the centre of the planet. Replacing g from equation 2.4, the inequality can be written as

$$\tfrac{1}{2}mv^2 > m\left(\frac{GM}{r}\right) \qquad (19.14)$$

So a mass will escape from the gravitational field of M providing its vertical velocity v satisfies

$$v > \sqrt{\left(\frac{2GM}{r}\right)} = \sqrt{2gr} \qquad (19.15)$$

On Earth's surface $g = 9.81$ m/s² and $r = R_E = 6370 \times 10^3$ m, giving an escape speed $v \approx 11180$ m/s.

For photons, moving with velocity c, to escape from the surface of a mass M, where the radius is r, requires

$$c > \sqrt{\left(\frac{2GM}{r}\right)} \qquad (19.16)$$

Therefore, providing $r > 2GM/c^2$ photons can escape from the surface, but when $r < 2GM/c^2$ they cannot. The spherical potential surface, of radius r_s, externally surrounding a mass M, given by

$$r_s = \frac{2GM}{c^2} = \frac{c^2}{2g} \qquad (19.17)$$

is called the event horizon. Note that the expression for r_s is independent of frequency and, so, applies to all photons. Photons on, or below, the event horizon cannot escape from it. To an observer outside the event horizon, the region inside would appear to be a black hole in space. The radius r_s of the event horizon surrounding a black hole is called the Schwarzschild radius, named after the German scientist, Karl Schwarzschild, who first drew attention to it in 1916.

According to astrophysicists the formation of a large black hole occurs during the final stage in the lifetime of some huge stars. When such a star is young and has plenty of hydrogen fuel to burn, the outward pressure associated with the star's radiance counters the inward pressure associated with the star's gravitational field. As the hydrogen fuel is used up and the starlight fades, the huge force of gravity begins to dominate, eventually crushing the body of the star to a small region of space and a black hole is formed. In the crushing process the atoms of the body are crushed, too, with orbiting electrons, neutrons and protons all forced into direct contact. Theoretically, the mass M of the dead star is assumed to be concentrated at the centre of the black hole, at a point in space called the singularity.

To derive the Schwarzschild radius r_s we have treated a photon as an effective mass. Suppose a photon is emitted radially outward from the singularity. In the rest frame of the singularity, after the photon has travelled a distance r_s through the gravitational field of the black hole it has undergone a frequency red shift of

$$\Delta f = \left(\frac{g}{c^2}\right)\left(\frac{c^2}{2g}\right) f = \tfrac{1}{2} f \qquad (19.18)$$

ELEMENTS OF QUANTUM GRAVITY

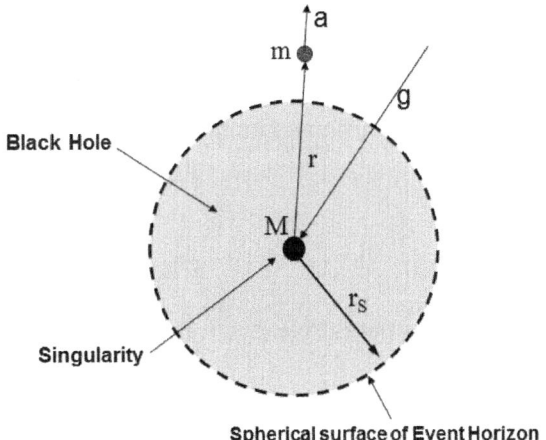

Fig. 19.2 Black hole.

Thus, on reaching the event horizon, the photon's kinetic energy of $\tfrac{1}{2}mc^2 = h(\tfrac{1}{2}f)$ is reduced to zero. Likewise, the photon's spin energy of $\tfrac{1}{2}hf = h(\tfrac{1}{2}f)$ is also reduced to zero. The photon continues to move at speed c up until the moment it reaches the event horizon when, with zero frequency, it vanishes. Conditions on the event horizon are like those envisaged by Einstein sitting on a light wave, where time dilation becomes infinite. In effect, the event horizon forms a one-way barrier allowing photons travelling towards the black hole to pass through unhindered while acting like flypaper and capturing photons projected upwards, away from the black hole. In this way, the singularity remains hidden from view.

Because the radiation stemming from a black hole is, supposedly, confined within its event horizon, many scientists assumed that the temperature of the hidden mass singularity must be indeterminate since no information about its condition was available.

However, given that a black hole is in equilibrium with its surroundings there is a way to find out its temperature. When a body of mass m falls a vertical distance Δs in a gravitational field **g** it loses potential energy $\Delta E = mg\Delta s$ (eqn 19.6) and gains kinetic energy $\tfrac{1}{2}mv^2$. If the body is brought to rest the loss of kinetic energy is converted into heat energy given by the formula

$$\Delta E = \tfrac{1}{2}k\Delta T \qquad (19.19)$$

ΔT is the rise in temperature and k is Boltzmann's constant. If the rise in temperature of a body is due to electromagnetic radiation we have to modify equation 19.19. A photon has two equal forms of energy, namely kinetic and spin energy. Conversion of the photon's kinetic energy alone into heat energy is given by

$$\Delta E = \tfrac{1}{4} k \Delta T \qquad (19.20)$$

For this case the rise in temperature is given by

$$\Delta T = \frac{4 m g \Delta s}{k} \qquad (19.21)$$

In 1970 Jacob Bekenstein at Princeton University in the USA pointed out that there are similarities between black hole physics and thermodynamics, the study of heat. Now the surface area of a black hole's event horizon is $A = 4\pi r_s^2$, where r_s is the Schwarzschild radius. Bekenstein assumed that the smallest quantum of area was $\Delta A = L^2 = (hG)/(2\pi c^3)$, where L is the Planck length (eqn 17.7).

Suppose we disperse all the mass M at the singularity of a black hole to its event horizon, so that the surface mass density is

$$\frac{M}{4\pi r_s^2} \qquad (19.22)$$

Since there is no radiation from the event horizon the mass there is at 0 K. Outside the event horizon the gravitational field remains unchanged, although inside the shell the gravitational field is zero. The elemental mass m occupying each Planck area ΔA is

$$m = \left(\frac{M}{4\pi r_s^2}\right) . L^2 \qquad (19.23)$$

A black hole is formed when the mass on the event horizon collapses onto the singularity under the intense external force of gravity. In doing so the rise in temperature T of each element of mass m will be (eqns 19.21 and 19.23)

$$T = \left(\frac{M}{4\pi r_s^2}\right)\left(\frac{hG}{2\pi c^3}\right)\left(\frac{g}{k}\right) r_s \qquad (19.24)$$

From equation 19.17 this can be written as

$$T = \frac{hg}{4\pi^2 kc} \qquad (19.25)$$

Since this applies to each elemental mass falling from the event horizon onto the singularity, equation 19.25 also gives the temperature of the black hole.

For a black hole with a mass ten times that of the Sun,

$$M = 1.99 \times 10^{31} \text{ kg} \qquad (19.26 \text{ a})$$

$$r_s = 0.85 \text{ km} \qquad (19.26 \text{ b})$$

$$g = 53.14 \times 10^{12} \text{ m/s}^2 \qquad (19.26 \text{ c})$$

$$T = 2.14 \times 10^{-7} \text{ K} \qquad (19.26 \text{ d})$$

So, the temperature of a typical black hole is not much different from absolute zero and cannot be detected in the microwave background (Fig. 15.2).

Professor Stephen Hawking considered the possibility that at the event horizon some information about the conditions below might leak out. His solution linked classical gravitational field theory with the quantum vacuum.

Hawking showed that photons could escape from the black hole if one assumed that anti-particle pairs of virtual photons were created at the event horizon, with one of the photons being absorbed by the black hole and the other photon radiating away into space (Fig. 19.3). In effect it meant that the quantum nature of the vacuum allowed the fabric of space to be torn apart across the event horizon interface. A different viewpoint is that quantum tunnelling allows one photon of the virtual photon pair to cross the event horizon barrier and escape.

Each escaping virtual photon brought with it information about conditions below the event horizon. Furthermore, as it could not annihilate with its anti-particle pair, in time Δt, it became a real photon with effective positive mass. This energy, radiated away from a black hole of temperature T, is now called Hawking radiation.

The photon swallowed by the black hole must, consequently, be an anti-

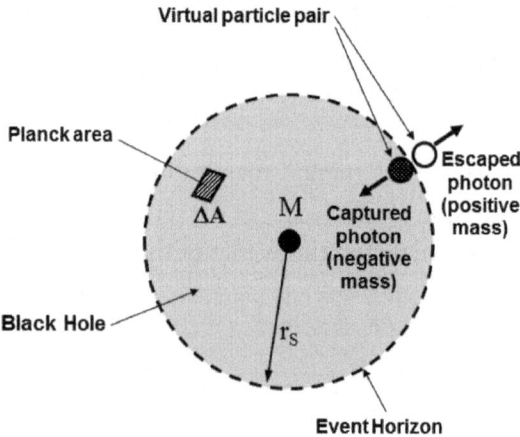

Fig. 19.3 Hawking radiation.

photon with an effective negative mass. So, in absorbing the photon the black hole loses a certain amount of mass and the negative energy stored in its gravitational field is reduced. This phenomenon is termed black hole evaporation, but the rate of mass loss is virtually negligible in terms of the Universe's calculated age of 13.7 billion years.

Astrophysicists are now convinced that a gigantic black hole lies at the centre of every galaxy. However, whether a scattering of black holes formed the seeds around which the galaxies formed or whether the galaxies themselves provided the conditions needed for black holes to form is a question currently being addressed.

These gigantic black holes can also be electrically charged and rotating providing the ultimate twinned field phenomenon of a combined magnetic-gravitomagnetic dipole. And, amazingly, to understand the physics of black holes, with their intensely large gravitational fields, we have used quantum mechanics, which deals with the physics of the very small!

20

SPECULATIVE GRAVITOMAGNETISM

It shouldn't come as too much of a surprise to learn that there is a gravitational analogue of Ohm's law. Suppose we pour a stream of particles (e.g. lead shot), all of roughly uniform size, into a column of viscous liquid (e.g. glycerine) contained in a long cylinder jar. As the particles fall under the influence of gravity, very quickly the upward force on each particle, due to the up-thrust and the viscous drag, becomes equal in magnitude to the downward force on each particle, due to its weight, so that the resultant force on each particle is zero. Thereafter, in accordance with Newton's first law, the particles have a constant falling speed called the terminal velocity. Raindrops falling through the air reach a terminal velocity.

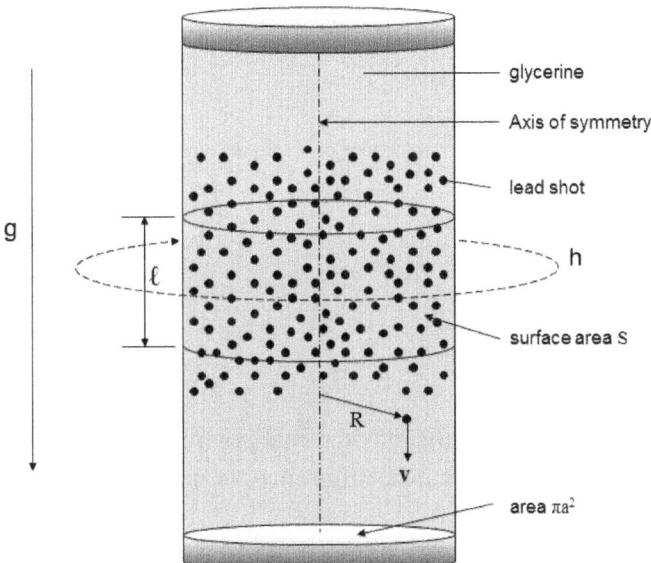

Fig. 20.1 Gravitational analogue of Ohm's law.

The terminal velocity of the mass current of particles is the analogue of the drift velocity for the current of electrons in a wire. The analogue with the electromagnetic case is not perfect, though, since an electron is assumed to be a point particle whereas a particle of mass occupies a volume, albeit assumed small. Also, conduction electrons repel each other whereas mass particles gravitationally attract each other. However, we will ignore these differences as they do not play a part in the empirical law discovered by Ohm.

To simplify the mathematical model we will assume an infinitely long column containing mass particles which are evenly spread along its length. If each particle has a mass m and the mean number of particles per unit volume is n, then the density ρ of the particle distribution is

$$\rho = nm \tag{20.1}$$

Consider a short length ℓ of the column of liquid, containing a mass M of moving particles, where $A = \pi a^2$ is the cross-sectional area and the volume $V = A\ell$. If the particles have reached a terminal velocity v, then in time t suppose each particle falls a distance $\ell = vt$. The mass current I can then take the following forms:

$$I = \frac{dM}{dt} = \frac{\rho V}{t} = \frac{\rho A \ell}{t} = \rho v A = \frac{M v}{\ell} \tag{20.2}$$

From equation 2.12 we have

$$\mathbf{g} = -\nabla \phi = -\frac{\Delta \phi}{\ell} \tag{20.3}$$

We assume that the terminal velocity v is a function of the gravitational intensity **g**. Therefore

$$v = \mu g \tag{20.4}$$

Thus, the magnitude of the gravitational potential difference $\Delta \phi$, being the analogue of the electrical potential difference, or voltage, is

$$\Delta \phi = g \ell = \left(\frac{v}{\mu}\right) \ell = \left(\frac{I}{\rho A}\right)\left(\frac{\ell}{\mu}\right) = IR \tag{20.5}$$

SPECULATIVE GRAVITOMAGNETISM

which is the gravitational form of Ohm's law (eqn 6.6). We define $\rho\mu = \sigma$ as the gravitational conductivity of the mass particles flowing in the viscous liquid so that the resistance R to the flow is (eqn 6.5)

$$R = \frac{\ell}{A\sigma} \tag{20.6}$$

The gravitational form of Ohm's law may be of minor interest, but is it of any further use? Well, the model is worth exploring further because of the new perspective that it offers. In the electromagnetic case the flow of electrons in a wire leads to the generation of a magnetic field both inside and outside the wire. Let us suppose that the flow of mass particles in a viscous liquid leads to the generation of a gravitomagnetic field.

Using our definition for **h** given in equation 8.12, modified by ½ to allow for double contributions, outside the infinitely long column the gravitomagnetic intensity at a distance r > a from the axis is

$$h = \tfrac{1}{2}\left(\frac{Mrv}{\pi r^2 \ell}\right) = \frac{1}{2\pi r}\left(\frac{Mv}{\ell}\right) = \frac{I}{2\pi r} \tag{20.7}$$

This agrees with the electromagnetic analogue derived by Messrs Biot and Savart (eqn 6.3). However, outside the column of the viscous liquid the induced gravitomagnetic field **b** will be undetectable since the gravitomagnetic permeability η_o is so small (eqn 8.15).

Inside the column, where r < a, we must ensure that only part of the mass contributes to the gravitomagnetic intensity h, so we have

$$h = \tfrac{1}{2}\left(\frac{M\left(\frac{r^2}{a^2}\right)rv}{\pi r^2 \ell}\right) = \frac{Ir}{2\pi a^2} \tag{20.8}$$

Checking with equation 6.4 shows that the result is in agreement with the electromagnetic analogue. Inside the column of liquid there is reason to suppose that the gravitomagnetic permeability η_i of the mixture is much greater in magnitude. Based on getting the kinetic energy right (eqns 20.10 and 20.19) we speculate that

$$\eta_i = \eta_o - \frac{4\pi\ell}{M} \approx -\frac{4\pi\ell}{M} \qquad (20.9)$$

This result can be compared with η derived for a point mass in equation 9.6.

The kinetic energy of the particles is contained in the gravitomagnetic **h**-field. The energy density of the gravitomagnetic field (eqn 8.13) is $-(\eta/2)h^2$. Therefore, the gravitomagnetic energy contained in the cylinder, of radius a and volume V, with approximate internal inertial permeability η_i (eqn 20.9), is

$$E = \left(-\frac{\eta}{2}h^2\right)V = \frac{1}{2}\left(\frac{4\pi\ell}{M}\right)\left(\frac{I}{2\pi a}\right)^2(\pi a^2 \ell) = \frac{1}{2}Mv^2 \qquad (20.10)$$

As shown in Figure 20.1, we are observing the particles of mass M contained in a cylinder of radius a, length ℓ and surface area S which falls under gravity through the stationary viscous liquid. The external surface area S of the cylinder is given by

$$S = 2\pi a\, \ell \qquad (20.11)$$

The gravitational analogue of Poynting's power equation (eqn 7.7) for electromagnetism is

$$P = \frac{dE}{dt} = (\mathbf{g} \times \mathbf{h}).\mathbf{S} \quad \text{watts} \qquad (20.12)$$

P is the radiated power, or rate of flow of energy dE/dt, which passes perpendicularly across the wall S of the cylinder.

In our model the particles initially undergo acceleration. During this phase the **h**-field changes and an induced gravitational field $\mathbf{g}_i$ (eqn 8.17) is created within the cylinder. Accordingly, the power equation (eqn 20.12) splits into two parts: outwards and inwards. To keep track we make use of the unit radius vector and introduce the idea of directed area. So an outward facing surface **S** is represented by $S\hat{\mathbf{R}}$, while an inward facing surface **S** is represented by $S(-\hat{\mathbf{R}})$.

The outward component due to the gravitational field **g** is

$$P_{OUT} = (\mathbf{g} \times \mathbf{h}).S\,\hat{\mathbf{R}} \qquad (20.13)$$

This is the outward flux of energy crossing the surface S.

The inward component due to the induced gravitational field $\mathbf{g_i}$ is

$$P_{IN} = (\mathbf{g_i} \times \mathbf{h}).S(-\hat{\mathbf{R}}) \qquad (20.14)$$

This is the inward flux of energy crossing the surface S.

Thus, the total rate of flux of energy crossing the surface S of the cylinder containing the particles of mass is

$$P_{TOTAL} = ([\mathbf{g} - \mathbf{g_i}] \times \mathbf{h}).S\hat{\mathbf{R}} \qquad (20.15)$$

For free fall (no viscous liquid), then $\mathbf{g_i} = \mathbf{g}$, so that there is no overall transfer of energy in either direction and, therefore, no gravitational radiation. However, the gravitomagnetic $\mathbf{h}$-field continues to change, as does the kinetic energy of the particles.

For the mass of particles falling in a gravitational field $\mathbf{g}$ through a resisting medium, during the initial acceleration phase the induced gravitational field $\mathbf{g_i} \leq \mathbf{g}$. So, there is an outward flow of energy, or gravitational radiation.

Once a terminal velocity is reached then $\mathbf{g_i} = 0$. Now, as the particles of mass fall lower in the gravitational field, there is a constant outward flux of energy, or gravitational radiation. Since the velocity of the particles is constant, the surrounding gravitomagnetic $\mathbf{h}$-field is also constant.

Another way of looking at the outward flow of gravitational radiation is as follows. As the particles fall lower in the gravitational field the energy density in the cylinder becomes more negative, displacing positive gravitational energy outwards.

Equation 20.11 gives the cylinder wall area $S = 2\pi a \ell$. When the radius $r = a$, equation 20.7, or equation 20.8, gives the gravitomagnetic field on the wall as $h = I/(2\pi a)$. Substituting for h and S in the power equation (eqn 20.15) and substituting for the mass current I (eqn 20.2) gives the outward power

$$P_{TOTAL} = (g - g_i)\left(\frac{I}{2\pi a}\right) 2\pi a \ell = M(g - g_i)v \qquad (20.16)$$

Once the terminal velocity v has been reached ($g_i = 0$), the loss of gravitational energy by the falling particles appears as outward gravitational radiation.

$$P = Mgv \tag{20.17}$$

We now need to check that our new view fits in with the conventional Newtonian view of a collective body of particles falling in a resisting medium. The mass M of particles having fallen through a distance ℓ has lost potential energy E given by

$$E = Mg\ell \tag{20.18}$$

Initially the potential energy lost by the mass M of particles leads to a gain in their kinetic energy, until finally a terminal velocity v is reached and the kinetic energy remains constant.

$$E = \tfrac{1}{2}Mv^2 \tag{20.19}$$

The power, or rate of energy, flowing out of the cylinder of mass is

$$P = \frac{dE}{dt} = \frac{Mg\ell}{t} = Mgv \tag{20.20}$$

With further reformulation we can obtain the gravitational transfer of power P in the same form as that for electromagnetism.

$$P = I^2 R \tag{20.21}$$

Straightforward Newtonian mechanics thus leads to a simpler model for the falling masses, but the important point to note is that introducing the gravitomagnetic field model has made no difference to the end results, but it does tell us more about what is happening.

In the above we have assumed that internally the viscous liquid does not absorb any of the outward radial flux of energy but, in fact, it does absorb some, causing the liquid to heat up. The same effect occurs in the electromagnetic case where the current of free electrons in a wire, falling with constant velocity in an electric field, results in the wire getting hot. Does our new model give us any clue to the heat generating mechanism?

From equations 8.20, 20.8 and 20.9 we can predict that the induced rotations within the liquid in the cylinder are given by

$$\Omega = \frac{b}{2} = \frac{vr}{2a^2} \qquad (20.22)$$

If the particles are brought to rest, the conventional view is that the kinetic energy of the particles is converted into heat energy which is absorbed by the liquid. With our new view, the kinetic energy resides in the gravitomagnetic field, which collapses when the particles are brought to rest. The large-scale induced rotational effect is rapidly broken up by viscosity, or fluid friction, into a myriad of random small-scale rotations within the liquid which we sense as heat.

Friction is an important effect in our everyday lives, because without it we wouldn't be able to walk about. Also, frictional heating has played a key part in human history, because once humans had learnt how to create fire using friction, it was the control of fire that raised them above the other animals. But does the phenomenon of friction hold another secret? If friction is linked with gravitomagnetism then gravity might be linked with heat, an idea considered by Faraday.

The two empirical laws of solid body friction are that the frictional resistance is proportional to the load and is independent of the area of the sliding surface. From the decoding of Leonardo da Vinci's notebooks, towards the end of the 19th century, we now know that da Vinci was the first person to discover the laws of friction. However, his discovery remained hidden for about 300 years and it was Guillaume Amontons, a French engineer, who first published the laws of friction in 1699. At the time the general reaction seems to have been, "So what?" Nearly a century passed before the laws were verified, in 1781, by the French scientist Charles Coulomb, who later achieved fame by showing that electrical charges obey an inverse square law. Coulomb made the added distinction between static friction (the force needed to start sliding) and kinetic friction (the force needed to maintain sliding). Demonstrating the laws of friction has now become a standard science experiment for school children.

But the laws of friction do not concern themselves with the production of heat during sliding contact between bodies. Benjamin Thompson, born in North America, was a contemporary of Benjamin Franklin and, like him, was very interested in science. During the American War of Independence, Thompson remained loyal to the crown and was, consequently, forced to flee to England after the loss of the British colonies in 1776. In 1783 he took a position working for Karl Theodor, the Elector of Bavaria, in the capital city of Munich. In 1790 the Elector expressed his gratitude to Thompson for his outstanding work, much of it to do with social reforms, by making him Count von Rumford of the Holy Roman Empire (Rumford, a town in North America now called Concord, was where the exiled Thompson owned an estate.) Count von Rumford became the Bavarian Minister of War, with responsibility for the military arsenal. While observing the boring of cannon guns in Munich, he was much taken by the colossal amount of heat generated which had to be conducted away by spraying the bore site with cold water. The accepted opinion at that time was that as the metal was cut away it released heat in the form of caloric fluid. But Rumford noted that with a blunt borer, where hardly any metal was shed, an even greater amount of heat was generated. He concluded that the heat was generated by mechanical motion, heat itself being a form of rotational or vibratory motion which had nothing to do with the release of caloric fluid. With the start of the Napoleonic War, Rumford returned to England as the Bavarian Ambassador, although the post was not recognised by the Court of St James. In 1798 he presented his idea about heat to the Royal Society in a paper entitled, 'Experimental Inquiry concerning the source of Heat excited by Friction'.

In 1799 Count von Rumford was made a fellow of the Royal Society. In the same year he was a co-founder of the Royal Institution in London, being the main person responsible for setting it up. This was where Sir Humphry Davy and Michael Faraday later made their names and many others since.

Rumford's ideas about the link between heat and friction remained dormant for nearly half a century, until they were extended by James Prescott Joule, a Manchester brewer and passionate amateur experimental scientist. During the 1840s, Joule showed that many forms of energy degenerated into heat, all of which could be linked with mechanical motion via a conversion factor J termed the mechanical equivalent of heat. In a famous experiment, Joule's apparatus consisted of a mass suspended at one end of a string, which was fed over a pulley, with the other end being wound around the vertical axle of a paddle wheel, each blade riddled with holes, in a cylindrical container of water. On releasing the mass, the paddle wheel turned and water was forced through the tiny holes creating a multitude of tiny vortex rings, decaying into turbulence, until the mass stopped moving and the water came to rest. A thermometer showed that the water had heated up. Joule argued that the increase in temperature of the water

SPECULATIVE GRAVITOMAGNETISM

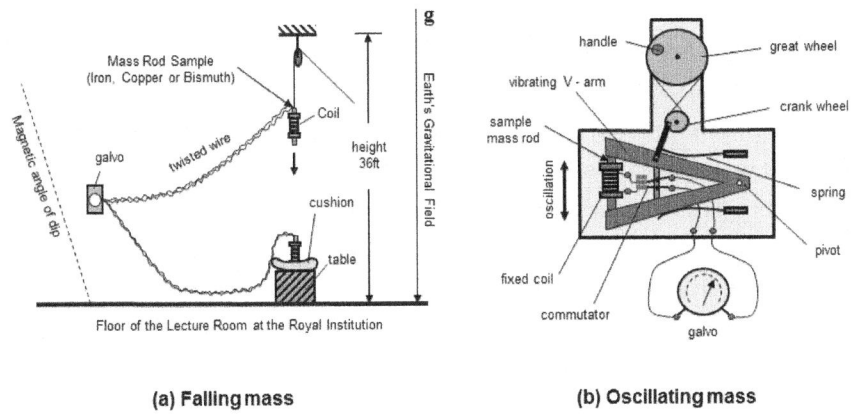

Fig. 20.2 Faraday's gravity experiments of 1851.

was caused by the frictional heat generated by stirring the water. Joule introduced his conversion factor J = 4.12 Joules/calorie, based on the assumption that the lost gravitational potential energy of the mass was equal to the increase in energy of the heated water. Even so, Joule's work did not reveal how the heat originated, other than saying it was due to friction. Nor did his work indicate any link with gravity, other than the lost gravitational potential energy in this particular case.

In our falling lead shot experiment, knowing the lost potential energy (eqn 20.18) we can use Joule's constant J to determine the rise in temperature of the liquid when the lead shot is brought to rest.

As his laboratory diary for 1849 shows, Faraday first pondered on the idea that an accelerating mass might be accompanied by a circumferential field of some kind. The idea of a gravitomagnetic field, as first envisaged by Oliver Heaviside, was forty years in the future. Faraday decided to investigate whether the circumferential field was electric and that it induced an electric current in a falling body, probably basing his idea on the analogue with electromagnetic induction. Experiments were done at the Royal Institution involving freely falling masses (Fig. 20.2(a)) contained within a coil. He assumed that any current formed around a falling mass would induce a current in the falling coil. Initially he detected just such an effect, but traced its source back to the wires connected to the galvanometer moving through the Earth's magnetic field. Twisting the wires removed this effect, but no circumferential current was detected. Faraday then investigated the possibility that a mass rapidly oscillating to and fro through a coil (Fig. 20.2(b)) might result in the induction of an oscillatory current. But, again, no effect was detected.

Ten years passed by before Faraday was willing to have another go at trying to crack the enigma surrounding gravity. He remained convinced that gravity was connected in some way with the other natural forces. Was it possible that as a mass changed its position in the Earth's gravitational field it developed an electric charge or changed its temperature? It is interesting to read Faraday's entry in his laboratory diary on 11 April 1859: "It would be strange if a body should heat as gravitation increases by nearness of distance. We conceive of heat as a positive force, and then instead of being the inverse of each other, they would seem to grow up together."

During 1859 Faraday carried out several experiments at the Shot Tower (Fig. 20.3), situated in London on the south bank of the River Thames next to Waterloo Bridge. His experiments involved raising and lowering (not free falling) lead masses within the tower and checking to see whether an electric charge appeared. It did not. He did the same with one pound of mercury, checking to see whether there was any change in temperature. But no change was detected. With hindsight we can see that with Faraday's temperature experiment he was working towards a conservation of energy experiment, already anticipated by Joule, but he hadn't fully grasped the concept. In fact, the experiment where lead shot falls freely in a column of viscous liquid can be viewed as a successful mini-version of Faraday's Shot Tower experiment. Faraday wrote up his results in a paper entitled, 'Note on the possible Relation of Gravity with Electricity or Heat', which he submitted to the Royal Society in April 1860. But, Lord Stokes, the President of the Royal Society, advised against publication and Faraday withdrew the paper.

Let us speculate. Consider a solid (non-metallic) mass cylinder along which is established a temperature gradient. The flow of heat along the cylinder can be thought of in terms of the flow of phonons (quantum units of heat energy). Since phonons have effective mass the temperature gradient constitutes an effective mass current around which a gravitomagnetic field will curl. During a period of changing temperature gradient the gravitomagnetic field will change and equation 8.17 predicts the generation of an induced gravity field opposing the change in temperature. So, if our speculation is correct, the cylinder in this condition, when placed vertically in the Earth's gravity field, ought to undergo a transient change in weight. This idea supports Faraday's intuition that gravity is linked with heat, although not quite in the way that he envisaged.

SPECULATIVE GRAVITOMAGNETISM

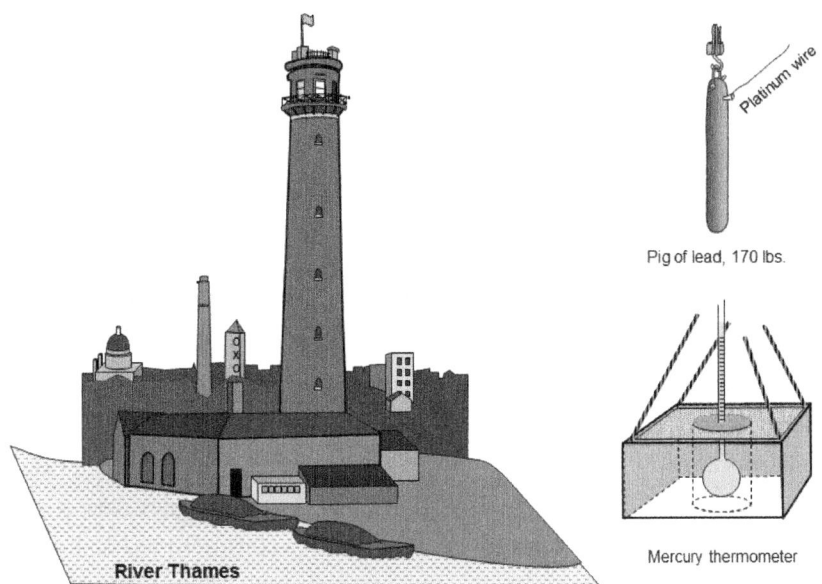

Fig. 20.3 Faraday's gravity experiments at the Shot Tower.

Other scientists have since pondered on the possibility that heat and gravity are, somehow, linked. In the 1920s the US scientist and successful electrical business entrepreneur Dr Charles Brush carried out an experimental programme involving falling bodies, a variation of Faraday's gravitational experiments, searching for a direct link between heat and gravity. Indeed, he claimed to have found an anomalous effect, but this was ignored by mainstream scientists.

John Nash, the brilliant mathematician and winner of the Nobel Prize in Economic Science in 1994, once went to see Albert Einstein (they were both at Princeton University at the time) to propose a new theory linking gravity, friction and radiation. Einstein thought that Nash's idea was absurd and he told him that he needed to go away and study more physics. But, as Einstein himself noted, absurdity is often one of the prerequisites needed for a breakthrough in understanding. Knowing too much, while searching for a breakthrough can, actually, constrain one's thinking.

With the fairly recent introduction of the atomic microscope, a better understanding of what causes friction is now being made. Contrary to what has been assumed in the past, we now learn that solid body friction has got little to do with surface roughness. Furthermore, recent studies show that friction is dependent on the true area of surface contact and, also, that the relative velocities of the sliding bodies is a factor. The underlying cause of friction is now thought to be linked to the stimulation of mutual resonance of the lattice vibrations of the two bodies in contact which arise in the sliding process.

In her *Scientific American* article (October 1996, pp. 48 – 56), 'Friction at the Atomic Scale', Jacqueline Krim noted that "Friction arising from atomic-lattice vibrations occurs when atoms close to one surface are set into motion by the sliding action of atoms in the opposing surface. (The vibrations, which are really just sound waves, are technically called phonons.) In this way, some of the mechanical energy needed to slide one surface over the other is converted to sound energy, which is eventually transformed into heat". Krim's article is almost a quantum mechanical approach to understanding friction. Consequently it makes no mention of field effects and, in particular, gravitomagnetism.

With great apprehension, remembering Einstein's comment to Nash about absurdity, what are we to make of our gravitational Ohm's law and its connection with friction?

Let us consider some known phenomena to help us understand what is going on. Suppose we have two layers of fluid of slightly different densities, say in a long glass trough. If one layer is made to move relative to the other, looking though the glass side wall we see that the surface of separation becomes unstable and breaks up into a series of line vortices (Fig. 20.4) perpendicular to the motion, rather like a series of parallel rotating cylinders. Is this a sign of induced gravitomagnetism appearing at the moving interface? In a more haphazard fashion, moving air in contact with the sea does the same thing, causing waves to form on the surface. The rotations involved continue beneath the surface, decreasing with depth.

If we could suddenly freeze the two liquids in the trough, the vortices would

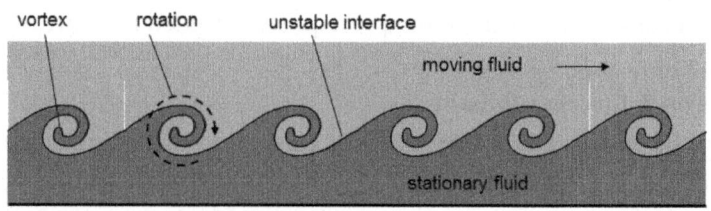

Fig. 20.4 A surface of instability.

seem to grip both layers of fluid together and provide a strong join. This is the principle behind the cold-welding of solids. When two layers of solid metal are brought into contact at relative supersonic speeds, at room temperature but at high pressure, the colossal amount of frictional heat generated at the interface between the layers causes the metals to liquefy and the molten shear layer breaks up into a series of vortices.

Suppose copper and steel layers are cold-welded together. If the interface region is exposed and the side-on face polished, then an examination of the face with an optical microscope, at a magnification of about 800, reveals the presence of the vortex pattern gripping the two layers together.

Let us now consider the friction experiment where one solid block of material slides over another solid surface. The block on top is subject to a force which causes it to move. Let us assume that the block moves with a constant sliding velocity which we can treat as a terminal velocity. From the perspective of the stationary lower surface it is speculated that the moving block induces a gravitomagnetic field **b** which at the interface is perpendicular to the direction of motion. Since the induced gravitomagnetic field is associated with rotation, this would seem to fit in with our image of the line vortices which appear at the interface between two liquid surfaces in contact, but moving at different speeds. So, with our gravitational analogue of the Poynting vector (eqn 20.12), we can envisage energy passing into the stationary block below and into the air above. However, due to the atomic structure within the sliding block we expect any induced rotations to be somewhat suppressed, so that the sliding block heats up, rather than passing the energy out. This is the frictional heat that you can feel when you push down on a copper coin and slide it across the desk top. So, in our model we see that friction is linked with gravitomagnetism.

Our gravitational analogue also means that as one body slides over the other there is force of adhesion between the two bodies, which may explain the difference between static and sliding friction. The adhesive force is the analogue of two parallel current carrying wires pulling together.

Have we really been oblivious to the existence of gravitomagnetism in the classroom friction experiment all this time? Let us think how we might modify the experiment in order to detect the presence of the gravitomagnetic field. Instead of a block of wood sliding over a desktop, we could use two thick glass blocks – those normally used to demonstrate refraction. If our model is right then as one block slides over the other a gravitomagnetic field should arise. To

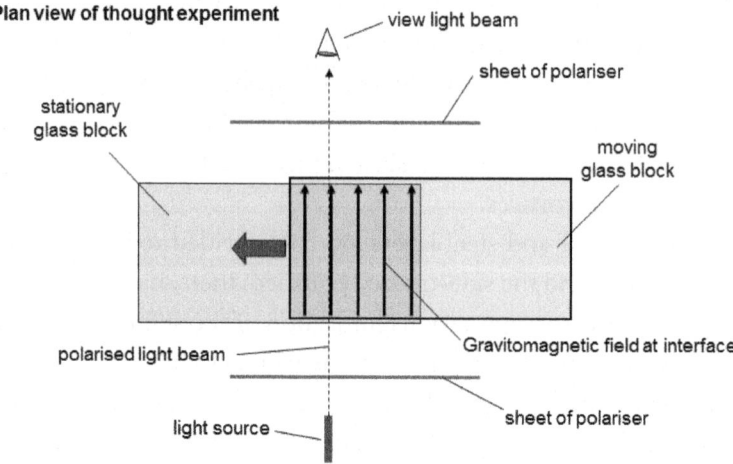

Fig. 20.5 A thought experiment for the detection of gravitomagnetism.

detect the gravitomagnetic field at the interface we might use the Faraday rotation effect. When a linearly polarised light beam runs parallel to a gravitomagnetic field we expect the angle of the plane of polarisation to be rotated (eqn 13.20). Since it doesn't matter in which direction the gravitomagnetic field points, if the beam is reflected back over the same path the polarisation angle of rotation is doubled. Suppose a beam of light is shone through a sheet of polariser (Fig. 20.5) perpendicular to the direction of motion of the top block of glass, so that it passes through the glass just above (in the moving block) or just below (in the stationary block) the interface. We can then view the polarised beam on the other side, or reflect it back and view it on the same side, to see whether any change in polarisation angle occurs. Just a change in light intensity observed on the sheet of polariser could indicate the existence of the gravitomagnetic field. (Note: Perspex blocks are not suitable as they are noticeably bi-refringent under stress. This makes them ideal for photo-elasticity experiments where revealing stress patterns is of interest, but it obscures our view of any linear polarisation rotational effects.)

Can this thought experiment be turned into a real experiment? Unfortunately, it is not a steady state experiment. As the temperature of the block increases, the internal atomic vibrations are likely to be chaotic, spoiling any chance of detecting the presence of the gravitomagnetic field. But who knows? Having raised the awareness of the possible existence of the

SPECULATIVE GRAVITOMAGNETISM

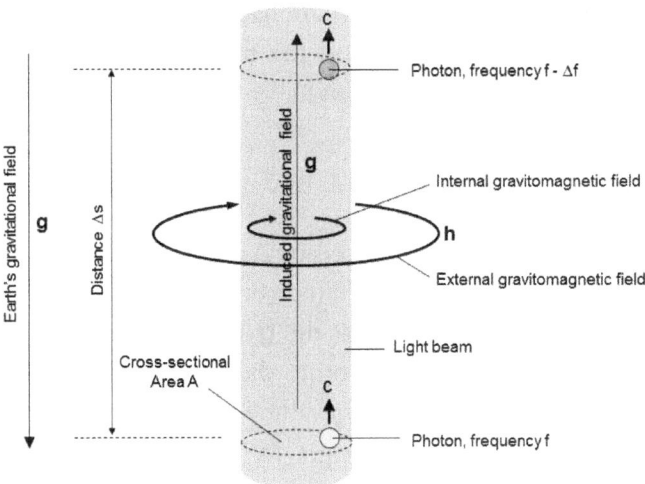

Fig. 20.6 *The constant speed of light in a gravitational field.*

gravitomagnetic field in the phenomenon of friction, it is left to others to see whether they can devise an experiment to detect it.

If gravitomagnetism and, in consequence, gravity are found to abound in such a common occurrence as friction, then Faraday's comment (diary entry no. 15808) that gravity "may prove to be one of the most changeable powers instead of one of the most unchanged" will be seen as being most perceptive.

We can adapt the model for the particles in a viscous liquid to look at the gravitational red-shift phenomenon. An infinitely long beam of photons forms an effective mass current I (for theoretical purposes the beam is assumed to be perfect) given by (eqns 20.1 and 20.2)

$$I = nmcA, \qquad (20.23)$$

Where n is the mean number of photons per unit volume, m is a photon's effective mass, c is the speed of light and A is the cross-sectional area of the beam. A circular gravitomagnetic field **h** forms around the photonic mass current (Fig. 20.6).

As the photons move up through the gravitational field there is a reduction in their frequency with distance (eqn 19.9) and, consequently, a reduction in their effective mass and a change in energy. The change of effective mass of the photon current with distance results in the change of the surrounding **h**-field with distance, which induces a gravitational field in a direction opposite to the Earth's gravitational field. Based on the modified gravitational analogue of Maxwell's equation we can write this as

$$\nabla \times \mathbf{g} = \eta\left(\frac{\partial \mathbf{h}}{\partial t}\right) = \eta\left(\frac{\partial s}{\partial t}\right)\left(\frac{\partial \mathbf{h}}{\partial s}\right) = \eta c\left(\frac{\partial \mathbf{h}}{\partial s}\right) \approx \eta c \frac{\Delta \mathbf{h}}{\Delta s} \qquad (20.24)$$

The distance ℓ has been replaced by Δs. Outside of the beam, due to the smallness of the free space value of η (eqn 8.15), the induced $\mathbf{g}$-field is virtually zero. It's only inside the beam that the $\mathbf{g}$-field is substantial, being equal in magnitude and opposite in direction to the Earth's gravity field, indicating a different value for η inside from its value outside (eqn 20.9). The mystery is thus explained. Cancellation of the gravity fields within the beam means that the photons move freely with constant speed. By changing frequency, light has created gravity!

Finally, let us consider the motion of a linearly accelerating mass observed from the following two frames of reference:

(a) Moving with the accelerating mass.   (non-inertial frame)
(b) At rest, or moving with uniform velocity.   (inertial frame)

Since gravity and acceleration are equivalent, one is tempted to replace the gravitational term, $\mathbf{g}$, in the formula for the temperature of a black hole (eqn 19.25), with the term, a, representing an acceleration. This suggests that the temperature change caused by accelerating through the vacuum of space is given by

$$T = \frac{h a}{4\pi^2 k c} \approx a(0.4 \times 10^{-20}) \text{ K} \qquad (20.25)$$

According to mathematical modelling done independently by Paul Davies and William Unruh, the result is valid and the rise in temperature due to acceleration is now referred to as the Davies-Unruh effect.

So an observer, on stepping from an inertial frame of reference to a non-inertial frame of reference, would see the background of space in the direction of the acceleration increase in temperature. Since the frequency spectrum for the temperature of the background is that for a black body (Fig. 15.1), any temperature change implies a different frequency distribution of the background radiation.

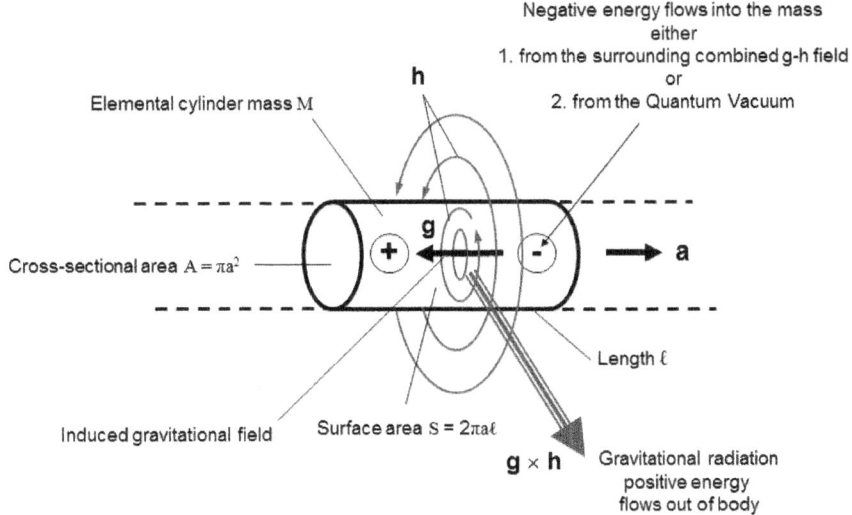

Fig. 20.7 Model of gravitational radiation by an accelerating mass.

However, for almost all accelerations the temperature rise given by equation 20.25 is negligible and would not be detectable. Nevertheless, it is interesting to speculate on the underlying mechanism behind the temperature rise. An increase in the temperature of the background means that the positive energy content of approaching space must have increased. Considering space in quantum vacuum terms, an increase in positive energy implies that some negative energy virtual photons must have disappeared while some positive energy virtual photons must have become real. Does an event horizon form ahead of an accelerating body, like a shock-wave, across which space is torn apart? And where have the negative energy virtual photons gone to?

Suppose we now observe an accelerating body from a rest frame. In 1994 Hal Puthoff and his colleagues, Bernhard Haisch and Alfonso Rueda, investigated the idea that inertia is a drag-force experienced by a mass as it accelerates through the zero-point field of the quantum vacuum. In a similar manner, we will assume that a body's inertial drag stems from the interaction of its accelerating mass with the particle-antiparticle fluctuations of the quantum vacuum. However, we will look at the idea that a body's inertia is due to the induction of a gravitational field within its mass in a direction opposing the acceleration, brought about by Hawking-type radiation. This simple model (Fig. 20.7) can then be poked about to see whether it has any merit.

From theory (eqn 3.15), we know that a gravitational dipole accelerates. So, based on Newton's third law of action and reaction, we hypothesise that when a mass is made to accelerate, internally it creates an opposing gravitational dipole by extracting virtual negative energy particles, $-\Delta E$, from the quantum vacuum as the negative mass component of the dipole. The freed virtual positive energy particles of the quantum vacuum become real particles, or gravitons, and radiate away from the accelerating body, while the negative energy particles are annihilated by the body and have to be continually replenished. However, most of the outwardly directed energy is absorbed within the body, due to the inter-atomic bonds, causing the body to heat up. Gravitons, phonons and photons are all inter-related in this process. An observer in the rest frame would see the accelerating body heat up, with the frequencies of the emitted photons following a black body distribution.

With this view, the inertia of a body is associated with the bond-breaking of virtual particle-antiparticle pairs within the confines of its mass, the continuous absorption of negative energy particles, and the continuous release of positive energy particles. In effect, an accelerating body radiates away positive mass, which is converted into heat.

The changing circular gravitomagnetic field **h** around the body may be viewed as the analogue of a thick vortex ring. It creates virtual positive and negative masses, forming a virtual gravitational dipole. So, with our simple model we can link the discrete theory of quantum mechanics with the continuous theory of extended Newtonian gravitational theory. A linearly accelerating mass acts like a tap, enabling positive energy particles to flow out from the quantum vacuum. At the same time, the negative energy particles from the quantum vacuum flow into the body, being absorbed and annihilating body mass in the process.

If we could discover a way of separating the positive and negative virtual particles of the quantum vacuum, we could use the extracted negative entities in a gravitational device to propel a spacecraft across the Universe.

21

FORMS OF GRAVITATIONAL PROPULSION

The basis of the gravitational propulsion systems described in this book is primarily the extended gravitational dipole. That is, one where the positive and negative masses are closely coupled, but not coincident. To create such a dipole we need to combine negative mass with ordinary (positive) mass in a controlled manner. If we are denied access to real negative mass then we will have to devise a way of creating virtual negative mass and use that instead.

Theoretically the simple gravitational dipole is perfectly balanced in terms of its positive and negative masses. However, in attaching an extended, or distributed, gravitational dipole device to a vehicle body to provide it with acceleration the positive masses involved, including the vehicle itself, any occupants and any cargo, will upset the mass-balance. Therefore, similar to the need to balance the mass distribution of an aircraft for flight and a ship for buoyancy, by the careful distribution of the structure and contents, not forgetting the crew and any cargo, there is a need to adjust the overall mass distribution of a gravitationally propelled vehicle so that the dipole-imposed acceleration acts as intended. Consequently, there may be a need to mass-tune the gravitational dipole system, which means being able to create more negative mass or, at least, the effect of more negative mass on demand. Such mass-tuning may also be used for steering purposes. Moreover, there will be times when it is necessary to be able to switch off, or neutralise, the dipole acceleration. If the overall mass is not balanced to zero, then the vehicle, cargo and occupants will experience some inertia. For low acceleration this might not be a problem, indeed it might be a benefit for the crew on long space flights. However, for periods of high acceleration, if there is a positive mass excess, then vehicle damage is likely due to structural stress. If the crew cannot be isolated onboard in the gravitational equivalent of a Faraday cage (a spherical version of Fig. 21.13) then they will suffer, too. High accelerations are only envisaged for spaceflight and in such cases overall zero mass distribution is really required so that the spaceship and its contents, particularly the crew, are inertia free.

Broadly, there are two approaches to implementing gravitational propulsion:

(a) A small-scale stand-alone system which generates a gravitational dipole, which can be fitted (perhaps several) to a body to provide thrust.
(b) A large-scale system built into a vehicle's superstructure which is able to generate an extended gravitational dipole to provide vehicle thrust.

At the moment this is all science fiction. With our present technology we cannot build a gravitational dipole accelerator. We haven't, yet, got any negative mass, nor do we have a clear understanding of how to create its properties. However, some simple ideas for generating a gravitational dipole do, already, exist on paper. By examining them we can begin to feel our way forwards. Also, in the investigation we will see how they might be applied.

21.1 Forward's mass toroid gravity field engine

This was an idea proposed by Dr Robert Forward in 1961. It was for illustrative purposes only and was not thought to be practicable.

Consider first the electromagnetic case where a wire toroid (Fig. 21.1), or doughnut-shaped coil, is formed by bending a solenoid into a circle. By changing the current through the wire and, hence, the induced magnetic field **B** through the wire loops, we see from Faraday's law of induction (eqn 7.4) that a transient electric field **E** is generated along the toroid's centre line. The size of the **E**-field can be increased by replacing the air core within the coils with a material of high magnetic permeability. This is standard physics!

From our twinned field notion, developed in Chapter 13, we would expect the transient **E**-field to be accompanied by a transient **g**-field. Not much thought seems to have been given to this, probably because it is not clear how to deal with the **E**-field, which is not wanted, but also because it has been assumed that the transient **g**-field would be too small to detect. But, if angular velocity Ω is equivalent to induced gravitomagnetism **b**, then the inverse of equation 13.3 suggests that a small change in **B** could give rise to a large change

FORMS OF GRAVITATIONAL PROPULSION

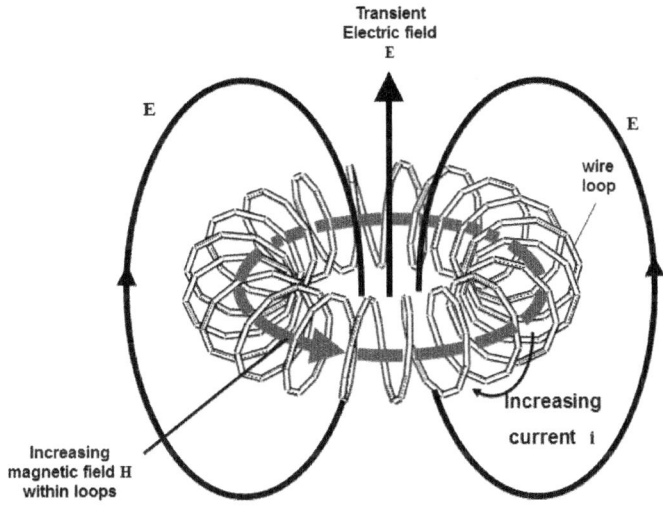

Fig. 21.1 The toroid.

in **b** and, perhaps, a detectable transient **g**-field along the toroid's centre line.

By making the toroid part of an electrical tuning circuit, so that the induced **E** and **g** signals oscillate, resonance could be used to increase the size of the effect. Due to the high inductance of the toroid, resonance can only be achieved at high frequencies, perhaps resulting in toroid vibration.

Using a balance to weigh a device undergoing a vertical oscillatory force is problematic, often resulting in a misleading non-zero averaged value being recorded. When Professor Eric Laithwaite weighed his inertial drive, a gyroscopic device, during the 145th Faraday Christmas Lecture in 1974 the balance clearly showed a change in weight. However, the effect has since been attributed to the balance's poor response to the vertical oscillations created by the device. But an interesting observation is that while a crude response balance might indicate the presence of vertical accelerations due to poor averaging, a good response balance might not do so!

In the toroid thought experiment, with electrical shielding, if an oscillatory **g**-field is present, then the averaged change in weight of a test mass immersed in the field should be zero. So any change in weight measured would be due to errors introduced by the balance.

In Dr Evgeny Podkletnov's first experiment (Fig. 13.3), three main solenoids were used to spin the YBCO superconducting annular disc. The apparatus was housed in a stainless steel cryostat which also provided electrical shielding. The three solenoids have some similarity with the toroid thought experiment, with gaps in the toroid winding. So, it's possible that an oscillatory **g**-field was created in Podkletnov's experiment, but that the balance introduced errors. If so, then Podkletnov really did observe a gravity modification effect, but misinterpreted the result.

How such an oscillating **g**-signal might be rectified to get a unidirectional **g**-field is an interesting question. Maybe in Podkletnov's experiment the three gaps between the main solenoids and the rotation of the YBCO core altered the **g**-field oscillations resulting in a small residual unidirectional **g**-force, rather than averaging to zero. But such speculations are left for others to ponder on.

In Forward's non-electric gravitational analogue, he imagined the wire toroid being replaced with a piped toroid through which an ultra-dense liquid could be pumped. To visualise this imagine copper tubing being wrapped around an anchor ring. The mass current around the toroidal tube creates a circular induced gravitomagnetic field **b** confined within the tube loops, analogous to a vortex ring. Changing the mass current through the tube results in a change in gravitomagnetic field, $\partial \mathbf{b}/\partial t$, and the analogue indicates (eqn 8.17) that a transient gravitational field **g** is induced along the toroid's centre line. In effect, this means that virtual positive and virtual negative masses are created either side of the face of the toroid, during the momentary existence of the gravitational dipole.

In passing, note that the virtual masses, which create the gravitational dipole field, are invisible so that they are a form of dark matter. Perhaps dark matter, in general, can be traced to changing gravitomagnetic fields.

Dr Forward was fully aware of the problems with his gravity field propulsion engine:

(a) Only a transient gravitational pulse is created.
(b) The gravitomagnetic permeability η of free space is extremely weak so that the induced gravitational field is, to all intents and purposes, negligible.
(c) The toroid and dense circulating liquid are too heavy to lift.

Although impractical, the Forward gravity field propulsion engine was used by

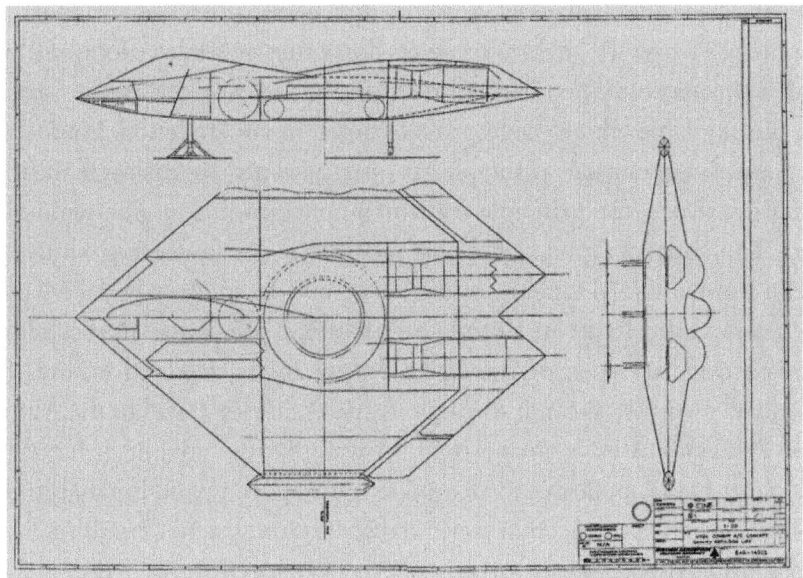

Courtesy of BAE Systems

Fig. 21.2 Blueprint of VTOL combat aircraft concept.

designers in the Advanced Projects Department of British Aerospace's Military Aircraft Division as a 'What-if' engine. This became part of Project Greenglow, a programme looking at futuristic aerospace vehicles. Fictitious engine data was prepared and, with it, several vehicle designs were produced. One aim of the study was to thrash out some ideas about significant changes in vehicle design should a gravitational propulsion engine ever be available for installation. Another aim was to see how the flight (for want of a better word) envelope might be expanded, particularly across interface-boundaries such as water-air and air-space.

The most detailed design, for which a blueprint was prepared (Fig. 21.2), was that of a vertical take-off and landing (VTOL) stealthy combat aircraft. The aircraft was a hybrid design with a Forward gravity-lift unit for VTOL but with conventional engines providing the forward thrust. The vehicle had a wing, as the study showed that for flight through the air a wing and control surfaces were still beneficial.

During the 1990s Professor John E. Allen, of Kingston University, was engaged by the Future Concepts Department, at British Aerospace Warton, as the technical consultant for an anti-gravity propulsion study.

Professor Allen dispensed with the need for conventional engine technology. In his view, hybrid concepts don't survive for very long and are best avoided. In his study he assumed that a fully enclosed engine, or propulsive unit, was available which could provide thrust in any direction. He referred to the engine's operating principle as *massdynamics* to distinguish it from thermodynamics, the principle on which conventional engine technology is based. The concepts looked at in his parametric 'Imagineering' study ranged widely: from a 'City-Hopper', or air-borne bus, to an 'Anti-podal Megaliner', which rose vertically to an altitude of 100 miles, above the atmosphere, then followed the curvature of the Earth round to its destination and, finally, descended vertically to land; through to spacecraft for travel to the Moon and Mars. Professor Allen's strong opinion was that the advent of *massdynamics* technology, being applicable to all modes of transport, would lead to the greatest revolution ever experienced in world transportation systems, resulting in a huge increase in aerospace business.

21.2   Mike's anatomy of a flying saucer

John Mike, an ex-MIT particle physicist, published his interesting book, *The Anatomy of a Flying Saucer*, in 2011. From Mike's analysis of many of the reports in the UFOCAT catalogue (see Chapter 22), he has concluded that flying saucers use a gravitational dipole for propulsion, since most effects observed during flying saucer encounters can then be explained. Whether you are a believer in the existence of flying saucers, or not, I believe that it is worth examining the theory proposed by Mike to explain his view of how a gravitational propulsion system works.

In July 1988 Robert Forward published details of a simple gravitational dipole engine (AIAA 88-3618), based on the assumption that negative matter was available. He imagined a unit containing an outer structure of positive matter within which was housed (suspended in some way) a lump of negative matter, where the surrounding positive mass exactly balanced the negative mass, cancelling out any gravitational dipole effect. Thrust in any direction by the unit was then obtained by controlled displacement of the negative mass from its equilibrium position in the direction opposite to that of the required thrust.

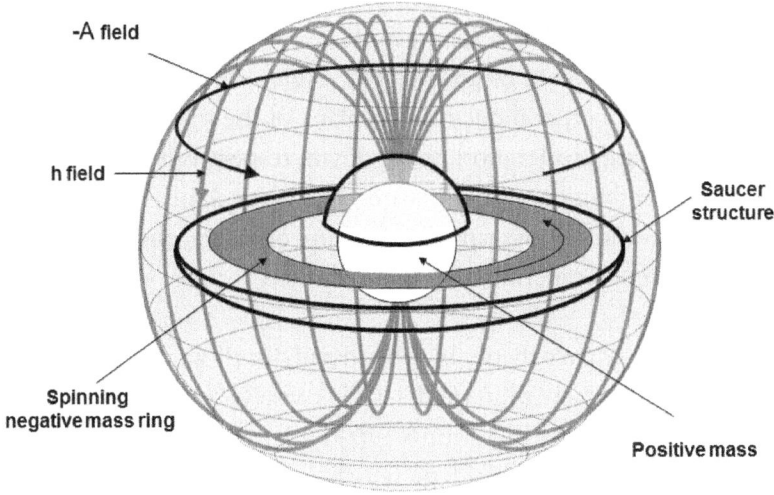

Fig. 21.3 Mike's flying saucer.

John Mike has adapted Forward's idea and has come up with a differently configured gravitational dipole which, he claims, is inertia-less. Furthermore, he believes that his version of the gravitational dipole engine is the means of propulsion used by flying saucers. In developing a theoretical model, Mike assumes that the linearised theory of gravity applies, along with the assumption that negative mass exists. Read-across from the gravitational analogue of electromagnetism can then be employed. Finally, Mike relies on a theoretical property of space referred to as Mach's principle. This is described below, but it is the way that Mike interprets the principle that is of particular interest.

Mike's gravitational dipole arrangement in a flying saucer is shown in Fig. 21.3 (see, also, Fig. 16.4). It consists of a positive mass (say a sphere) surrounded by a ring of negative mass. While the ring is centralised, so that the sphere centre is at the centre of all ring diameters, then no movement occurs. However, when the ring is slightly displaced, part of the inner edge moves closer to the sphere and a gravitational dipole effect comes into play. The combined masses will then accelerate in the direction of the ring diameter passing through the point of closest ring approach to the sphere, with the sphere in the lead.

According to the gravitational theory in this book, Mike's flying saucer cannot float above the surface of the Earth since the theory predicts (eqns

3.14(a) and 3.14(b)) that both positive and negative masses will accelerate downwards, towards the Earth's surface. However, if the positive mass at the centre of the negative mass ring of Mike's engine is allowed to move up, or down, in a direction perpendicular to the plane of the ring then a gravitational dipole forms creating acceleration up, or down, respectively. Or, if the positive mass is fixed in the flying saucer structure, then vertical thrust may be obtained by allowing the negative mass ring to move up, or down, slightly. So there are ways round the disagreement with Mike's idea to get the saucer to float.

We now look at Mike's theory in more detail, for those readers who want to see the underlying mathematical model. Otherwise you should move on to the next section. From equations 8.17 and 8.19(a) we have seen that inertia arises due to the induced gravitational field $\mathbf{g}_i$ which accompanies the change in the gravitomagnetic field of an accelerating mass m. We have also introduced the gravitomagnetic vector potential $\mathbf{A}$ (eqn 13.12). And we have seen that the induced gravity field $\mathbf{g}_i$ is equal to the rate of change of the gravitomagnetic vector potential $\partial \mathbf{A}/\partial t$. This is a general result, not just applicable to superconductors. However, mass responds to gravity, not to $\partial \mathbf{A}/\partial t$, so the force $\mathbf{F}$ of inertia is

$$\mathbf{F} = m\, \mathbf{g}_i \qquad (21.1)$$

From symmetry we might assume that a moving negative mass will generate a negative vector potential $-\mathbf{A}$ and that an accelerating negative mass will generate a $\partial(-\mathbf{A})/\partial t$ field, leading to the creation of $-\mathbf{g}_i$. So the force of inertia experienced by a negative mass (eqn 8.19(c)) is

$$\mathbf{F} = (-m)(-\mathbf{g}_i) \qquad (21.2)$$

For the elementary point gravitational dipole (Fig. 8.3(e)) the positive induced gravity, $\mathbf{g}_i$, cancels with the negative induced gravity, $-\mathbf{g}_i$, so that the accelerating dipole is inertia-less. This assumes that the positive and negative point masses are coincident.

We return to Newton's query mentioned in Chapter 11: "When a mass accelerates (in particular radial acceleration due to rotating mass), what does it accelerate relative to?" In Newton's view it was relative to absolute space, or the ether. Nowadays we might think of the ether in terms of the quantum vacuum of space. But just over a century ago, before the advent of quantum theory, the Austrian physicist Ernst Mach took a more classical view. He wondered whether acceleration might be relative to the gravitational cobweb throughout space created by the fixed stars (these days we would say the fixed galaxies) at the edge of the Universe.

As a rough model, suppose we imagine the Universe as being contained in a vast empty shell, of infinite radius, where all the galaxies are spread evenly over the shell. Although only the shell is important, based on equation 2.12, we can define the scalar gravitational vector potential of the Universe Φ_0 as

$$\Phi_0 = - \iiint_V \frac{\rho}{\gamma 4\pi r} dV \qquad (21.3)$$

Here, ρ is the mass density and V extends over the spherical volume of the Universe.

In fact, the cobweb of gravity fields inside the shell cancel (this is the gravitational analogue of Faraday's charged ice-pail experiment) so that the gravitational potential Φ_0 is uniform throughout the Universe. Being uniform and non-interactive we are not aware of the presence of Φ_0. Note that the absence of gradients is consistent with there being perfect cancellation of all gravitational fields present. Equating the energy (eqn 2.7) due to a distribution of mass in the shell with the energy given by Einstein's mass-energy formula (eqn 8.8) we conclude that

$$\Phi_0 = -c^2 \qquad (21.4)$$

For our rough model, we assume that when a small amount of matter (compared with the mass of all the galaxies in the Universe) is introduced into the ether the background gravitational scalar potential Φ_0 remains unchanged. When positive matter moves through the ether it induces a gravitomagnetic field **b** around it. And, from our definition of the gravitomagnetic vector potential, an **A** field will also be formed.

Following Mike's analogue approach, read-across from a textbook on electromagnetism allows us to define **A** as

$$\mathbf{A} = -\eta \iiint_V \frac{\rho \mathbf{v}}{4\pi r} dV \qquad (21.5)$$

Suppose we have a point mass moving with uniform velocity v, then from equations 21.5, 8.14 and 21.4,

$$\mathbf{A} = -\gamma \eta \mathbf{v} \iiint \frac{\rho}{\gamma 4\pi r} dV = \left(\frac{1}{c^2}\right) \mathbf{v} \Phi_0 = -\mathbf{v} \qquad (21.6)$$

This result is consistent with equation 13.15.

For uniform conditions, **b** and **A** cause no disturbance in the ether. However, for mass acceleration $\partial \mathbf{v}/\partial t$, from equations 8.18 and 13.12 we get

$$\nabla \times \mathbf{g}_i = \frac{\partial \mathbf{b}}{\partial t} = \frac{\partial}{\partial t}(\nabla \times \mathbf{A}) \qquad (21.7)$$

This indicates an equivalence between $\mathbf{g}_i$ and $\partial \mathbf{A}/\partial t$, but they are not the same phenomenon.

$$\mathbf{g}_i \equiv \frac{\partial \mathbf{A}}{\partial t} \qquad (21.8)$$

$\partial \mathbf{A}/\partial t$ is the reaction by the ether to the mass acceleration $\partial \mathbf{v}/\partial t$ and the mass experiences an inertial force (eqn 21.1) given by

$$\mathbf{F} = m\mathbf{g}_i \qquad (21.9)$$

Einstein liked this idea and was the first to refer to it as Mach's principle. Initially the experts in general relativity thought that Mach's principle was not incorporated into Einstein's theory, but there is now a view that it is. It is the disturbance of the classical, or quantum, ether by an accelerating mass (both inside and outside the body) that gives rise to inertia. So, here is another region of gravitational theory where classical and quantum ideas overlap (the other being the event horizon of a black hole).

In Mike's engine described above only parts of the positive and negative masses form the gravitational dipole, so Mike's flying saucer will not be inertialess at this stage. Mike accepts the idea that inertia is associated with the interaction of an accelerating positive mass with the scalar gravitational potential Φ_0 of the Universe, as suggested by Mach. But the crux of Mike's model is that he assumes that "there is no Mach's Principle for negative mass". So, in Mike's view, the properties of the ether are not symmetrical. It means that although $\partial(-\mathbf{A})/\partial t$ may exist (it may only be a mathematical construction) there is no reaction with the ether to create $-\mathbf{g}_j$, so that $\mathbf{F} = 0$ in equation 21.2. Furthermore, if Mike's view is correct, then equations 8.19(c) and (e) are not applicable.

Making the negative mass ring spin makes no obvious difference to using ring displacement to cause acceleration. From the analogue with electromagnetism we see that the spinning ring develops a gravitomagnetic dipole, where the -$\mathbf{h}$-field forms a doughnut shape, or toroid, around the ring (Fig. 21.3). The accompanying -$\mathbf{A}$ field forms great circles, or lines of latitude on the doughnut. When the flying saucer accelerates, the positive spherical mass develops a $\partial \mathbf{b}/\partial t$ field, and its accompanying $\partial \mathbf{A}/\partial t$ field, while the negative mass ring develops a $\partial(-\mathbf{b})/\partial t$ field, and its accompanying $\partial(-\mathbf{A})/\partial t$ field.

Since force fields have vector properties of magnitude and direction, the positive and negative fields will vie for dominance, with the stronger field displacing the weaker field.

A large ring spin creates a large -$\mathbf{A}$ field, which becomes a large $\partial(-\mathbf{A})/\partial t$ field during acceleration. Mike says that by making the negative $\partial(-\mathbf{A})/\partial t$ field of the ring much larger than the positive $\partial \mathbf{A}/\partial t$ field of the sphere a bubble can be created in the ether enclosing the ring and sphere, from which the positive $\partial \mathbf{A}/\partial t$ field is excluded. It is Mike's contention, therefore, based on his interpretation of Mach's principle, that the saucer cannot have any inertia. Mike assumes that for this condition the spinning ring of negative mass does not suffer from centrifugal force so a large rotation rate can be achieved, without the ring disintegrating, which, in turn, is responsible for creating the inertia-free bubble.

Mike explains in his book how his view of the flying saucer propulsion system causes the effects which UFO observers have experienced, particularly slight levitation effects and electromagnetic interference resulting in electrical systems malfunctioning.

I am not convinced with Mike's idea that Mach's principle doesn't apply to negative mass. In my view, the $\mathbf{h}$-field curled around a current of positive matter flowing to the right cannot be differentiated from the $\mathbf{h}$-field curled around a current of negative mass flowing to the left. So, why should the Mach's principle differentiate between the two? But, until we get some negative mass to test, no one can be certain of its properties.

21.3 Biefeld-Brown electro-gravitics

Legend has it that in 1928, while experimenting with an X-ray tube, US laboratory technician Thomas Townsend Brown made a curious discovery. Although a directed beam of X-rays imparted no noticeable force to objects, when the tube was switched on it experienced an impulse, or thrust, of its own. Brown traced the impulse to the very high voltage developed across the electrodes. Experimenting further, he built a capacitive device which developed a thrust towards its positive electrode when it was highly charged (Fig. 21.4). Brown thought that he had discovered a link between gravity and electricity and he discussed the effect with Paul Biefeld, a professor of physics and one-time classmate of Albert Einstein. Over the ensuing years the curious phenomenon has become known as the Biefeld-Brown effect.

The effect is real and, since its chance discovery, the phenomenon has been of continuing fringe interest to scientists and engineers who are not quite sure what to make of it. It is thought that the thrust effect is due to the combination of an ionic wind streaming from the negative electrode and to the polarisation of the dielectric between the capacitor electrodes. Tests done on capacitors, in a vacuum chamber where no ionic wind is present, show that they undergo a transient thrust, or impulse, while charging. More generally, the Biefeld-Brown effect is associated with the thrust developed by asymmetrically shaped capacitors and out of the study of these grew the study of electro-gravitics.

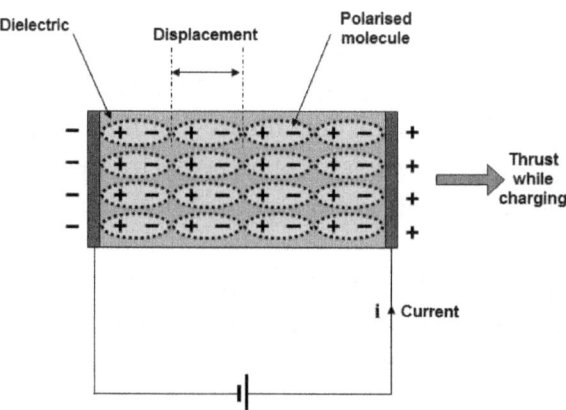

Fig. 21.4 The Biefeld-Brown effect.

In January 1996 I was contacted by Dr David King of the Electrical Engineering Division of the School of Engineering at Manchester University. One of Dr King's external PhD research students, Stavros Dimitriou, from the Technological Institute in Athens, was investigating the Biefeld-Brown phenomenon. In his experiment, Dimitriou subjected an asymmetric capacitor to ramped electrical input signals and claimed that he had detected a small thrust. I visited the university in February 1996 to watch a demonstration. Later, I arranged for Stavros Dimitriou to give a lecture at Lancaster University to allow him to describe his research work to an audience of academics and engineers. Although it was accepted that he may have detected a small effect experimentally, the audience were not persuaded with his proposed theoretical explanation for it.

The next time that I saw Stavros Dimitriou was at the first Field Propulsion Conference, held at Sussex University in January 2001. During the proceedings, Dimitriou gave a practical demonstration of his work, which he said was an alternating current (ac) version of the classic Biefeld-Brown effect. Two large parallel plate capacitors were mounted at the opposite ends of a horizontal rod so that their plates were vertical. The rod, about half a metre in length, was suspended by a string attached to its centre. The capacitors were then fed asymmetrically with a timed radio-frequency (rf) ramped wave form and the rod appeared to rotate very slightly, although not everyone was convinced.

In a 1994 essay entitled, 'The US Antigravity Squadron', Dr Paul LaViolette re-examined the idea of electro-gravitic propulsion and went on to explain why he believed that it was used on the Northrop B-2 Stealth Bomber. The claim was that the wing leading edge of the B2 was positively charged, while the jet exhausts were negatively charged, thereby creating an asymmetric capacitor in flight with a longitudinal electric field in the direction of flight. It was suggested that during the formation of the electric field along the length of the aircraft an induced gravitational field was generated, too, which provided the B2 with a Biefeld-Brown-type surge of additional thrust.

During the summer of 1996, British Aerospace, with support from AEA Technology, carried out an experimental programme, called Elmo, to investigate certain aspects of LaViolette's claim concerning the B2. A highly charged wire (~ 100 kV) was strung parallel to, and just upstream of, the leading edge of a 2-D wing section covered in aluminium foil. The wing section was then placed in the exhaust from a blow-down wind tunnel and tests done to see whether it made any difference

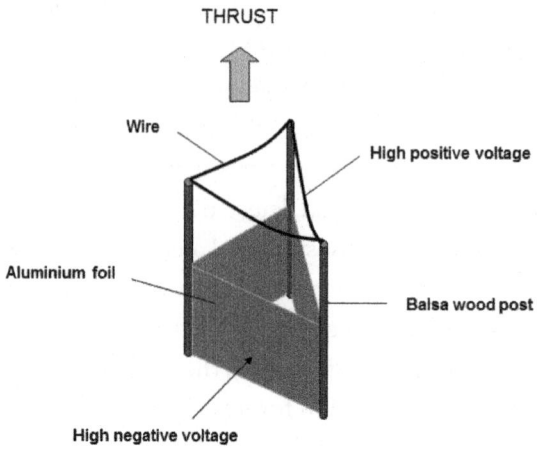

Fig. 21.5 The TDT Lifter.

to the thrust on the wing. However, no change in the aerodynamic drag on the wing was measured. So, for this configuration no Biefeld-Brown effect was noticed.

In July 2001 Trans-Dimensional Technologies (TDT), a research company based in Alabama, USA, amazed many people with their demonstration of a levitating device which they called the Lifter. This 3-D triangular shaped device was the brainchild of Jeff Cameron, the Chief Scientist at TDT. The basic lifting cell consisted of three vertical thin balsa wood poles supporting a wire strung from their tips, with a strip of aluminium foil wrapped around their lower part (Fig. 21.5). When high positive and negative direct current (dc) voltages were applied to the wire and the foil, respectively, the Lifter levitated. To obtain a greater lifting effect several basic lifting cells were ganged together. Typically, a lifter weighing 250g, including a 50g payload, would levitate when a voltage difference of 30kV was applied. The Lifter is another form of Biefeld-Brown device. Clearly the Lifter develops thrust which is directly linked to acceleration, which through equivalence is linked with gravity, but no obvious gravity field is generated.

Soon after TDT had revealed construction details of the Lifter, Jean-Louis Naudin, a prodigious French experimental scientist, replicated the work and posted details on the internet showing how to build the device. In the UK, researchers at BAE Systems and at the MoD built their own lifters, as did many other interested groups around the world.

At the moment it has not been proven that any of the electro-gravitic devices described above have any direct links with gravity. However, based on the twin field notion (Chapter 13), that a gravitomagnetic field is present within a magnetic field, it's possible that during a change in the magnetic field both electric and gravitational fields are generated. In the electromagnetic toroid case (Fig. 21.1) it was envisaged that both electric and gravitational fields are generated along the centre line of the toroid. The problem is in separating the electric field, which is not wanted and may be dangerous, from the gravitational field.

21.4(a) Microwave open cavity thrust

Illuminating a metal surface with an electromagnetic wave gives rise to a reflection, as with light reflected from a mirror. From Newton's third law of action and reaction, the metal surface gets a slight thrust during the wave absorption and reflection process. This thrust is called radiation pressure. The thrust can be increased by using an open conducting cavity, rather than merely a surface. At resonance, standing waves form within the interior of the cavity which absorbs energy from the illuminating wave. The difference in radiation pressure between the lit and unlit surface of the cavity gives rise to thrust. For the size of cavity involved, this means using microwaves, with wavelengths of the order of several centimetres.

Scientists in the USSR were among the first to investigate microwave cavity thrust. More recently, during 2001, the concept was explored further by Professor Sergei Vinogradov, from the Institute of Radiophysics and Electronics at the University of Kharkov in the Ukraine, while visiting the UK and working with Professor Paul Smith, at Dundee University. Funding support for the study came from BAE Systems, under the auspices of Project Greenglow.

Based on electromagnetic wave theory, Professors Vinogradov and Smith predicted that under stationary conditions a spherical conducting cavity, of radius 2cm, with a circular hole, of diameter 1.5cm, when illuminated by a 10GHz microwave beam should experience a thrust of about 5mN = 5×10^{-3}N. This small force would be enough to levitate a spherical cavity with a mass of 0.5g in the microwave beam.

The energy-absorbing property of the cavity is characterised by a quality, or Q, factor. At resonance, very high Q values arise and this is the effect which is

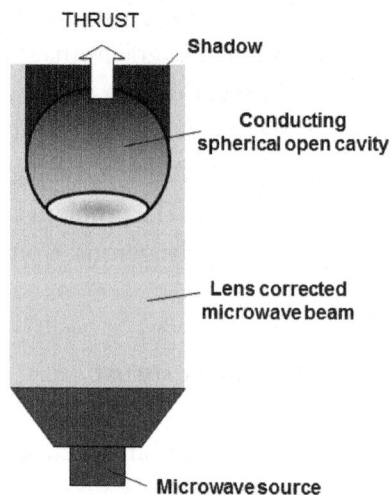

Fig. 21.6 Open cavity microwave thrust.

utilised for microwave thrust. However, the predicted frequency band over which the cavity has a high Q-factor is extremely narrow and, in practice, has to be found experimentally. During 2002 an experiment aiming to demonstrate microwave thrust was carried out by Professor Alan Phelps at Strathclyde University. It was with some excitement that an effect was initially observed but, after careful scrutiny, the scientists involved eventually concluded that the effect was more likely to be due to cavity wall heating and the subsequent formation of hot air currents. Although standard electromagnetic theory has confirmed the principle of microwave thrust, a better means of exploiting it is required.

At first sight, it might appear that microwave cavity thrust has nothing to do with gravitational propulsion. But if we think of the concept as a rocket, where the exhaust propellant streaming from the cavity consists of photons, which have effective mass, we can see that there is a link.

21.4(b) Shawyer's EM-Drive

At the time that the Dundee open cavity microwave thrust programme was coming to an end, a small UK company, called Satellite Propulsion Research (SPR) Ltd, was formed to develop a closed-cavity microwave thrust device

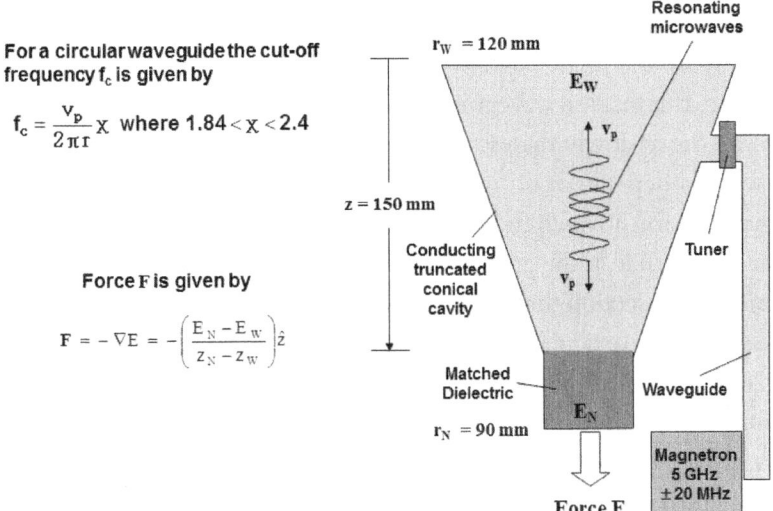

Fig. 21.7 Shawyer's EM-Drive.

called the EM-Drive. Being a closed cavity means that there is no obvious propellant used to produce thrust. The existence of the EM-Drive was first made known to the public in the December 2002 issue of *Eureka* magazine, a trade journal aimed at design and future concept engineers. The Director of SPR is Dr Roger Shawyer, formerly a programme project manager at Astrium, the European Space Satellite Company. Astrium was once owned by British Aerospace but is now part of the European Aerospace Defence Systems conglomerate, known by many as EADS.

The cavity for the EM-Drive is a truncated cone (Fig. 21.7) with the narrower region loaded with dielectric and the wider region left empty. A microwave signal with a centre frequency 5GHz and a band width ± 20MHz is fed into the cavity and is tuned for resonance, giving a quality factor Q approaching 6000. Tests, funded by the UK's Department of Trade and Industry (DTI), showed that the EM-Drive developed a thrust of 20mN in the direction from the wider to the narrower end of the cavity. The results were the same for the conical cavity arranged vertically up or down.

According to Dr Tom Shelley, the Editor of *Eureka* magazine, the amazing claim for the EM-Drive of "thrust without any expelled propellant", is due to a change in the group velocity v_g of the waves in the closed, partially dielectric loaded, conical cavity which results in a difference in radiation pressure between the two ends of the cavity, the mismatch giving rise to a thrust.

In free space the phase velocity v_p, or speed, of all electromagnetic waves is that of light c. But inside a waveguide, or cavity, higher frequency electromagnetic waves travel more slowly than lower frequency waves. When the phase velocity v_p of a wave is dependent on its frequency the regime is called dispersive and wave envelopes form which move with a group velocity v_g. Energy is transported along the wave guide at the group velocity. Depending on the dimensions of the wave-guide cross-section there is a certain frequency, called cut-off, below which the passage of that particular electromagnetic wave is prohibited.

In 2004 SPR Ltd issued a report containing a mathematical model explaining how the EM-Drive developed its thrust. The model, based on special relativity theory, was viewed with scepticism by some scientists who argued that it violated the law of conservation of momentum. When the *NewScientist* magazine published an article on the EM-Drive (8 September 2006), entitled 'Relativity Drive: The end of wings and wheels?', the editor received some caustic correspondence from a few scientists who were adamant that the EM-Drive couldn't possibly work. However, questioning the proposed theory at this stage is of secondary importance. The EM-Drive has produced thrusts of about 20mN, stimulating the need for further experimentation. As Faraday noted in his laboratory diary (entry no. 10040, 19 March 1849), "*All this is a dream*. Still, examine it by a few experiments. Nothing is too wonderful to be true, if it be consistent with the laws of nature, and in such things as these, experiment is the best test of such consistency". Details of the Shawyer experiment need to be very carefully analysed to make sure that some other unwanted effect is not responsible for the tiny thrusts measured.

The EM-Drive has been the subject of international attention by countries with major space programmes, with particular interest demonstrated by the USA and China. Chinese scientists claim to have replicated the EM-Drive and shown that it does generate a tiny thrust. In a validation test (September 2014) of a device very similar to the EM-Drive, NASA engineers confirmed that they, too, had measured a tiny thrust. So, although thought to be highly improbable by many scientists and engineers, the EM-Drive does work. Douglas Adams would have been delighted as he foresaw the invention of an Improbability Drive in his book, *The Hitchhiker's Guide to the Galaxy*.

Dr Shawyer believes that by using superconducting techniques, to reduce internal electrical resistance losses, it should be possible to increase the thrust of the EM-Drive from 20mN up to 30kN. That's enough to raise a family car against the force of gravity! If this can be achieved experimentally, then any other non-microwave sources of thrust are likely to be tiny by comparison. In a paper in 2009, Shawyer revealed that SPR were already in the process of designing a hybrid space-plane using superconducting EM-Drive thrusters for vertical lift, with forward thrust being provided by hydrogen-fuelled jet and rocket engines.

21.4(c) Thrust due to a microwave-induced gravitational dipole?

From our twinned field musings (Chapter 13) we have already concluded that gravity fields arise when electromagnetic fields change but that, in general, these induced gravity fields are tiny. One wonders, though, whether there might be certain conditions where the link between the two fields is substantial.

Consider the model (Fig. 20.6) used to explain the constancy of the speed of light in a vertical beam in a gravitational field. By replacing the light beam with a microwave beam, we could equally well argue that a gravitational field is induced within the beam, thereby neutralising the background gravitational field there. Since no overall force acts on the beam, or the photonic masses, it propagates with a constant speed c. Indeed, this principle, if true, must apply to all electromagnetic waves.

So let us speculate. Suppose we confined the microwave beam to a 1-D conducting cavity, closed at the ends. The waves input at one end would reflect up and down the cavity and with the right choice of cavity length and microwave frequency, the system could be made to resonate. If the cavity was stood vertically in a gravitational field then, by extension of our previous argument, it seems reasonable to suppose that an extended gravitational dipole would form within the cavity. Assuming that it does, what would happen if the background gravitational field was replaced with an equivalent acceleration? Would a gravitational dipole (Fig. 21.8) still be generated within the cavity? If so, providing that the microwave beam was maintained when the initial thrust mechanism was removed, would the cavity still accelerate due to the continued presence of

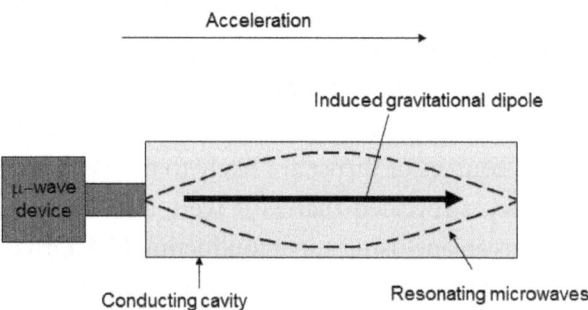

Fig. 21.8 Microwave – gravity dipole propulsion

the gravitational dipole? Clearly, as gravitational energy would be radiated away from the accelerating cavity the loss of energy would have to be replaced, or the device would cease accelerating. So the device could not go on accelerating forever, and a time would eventually be reached when the continued supply of energy would not be enough to compensate for that lost.

Standard electromagnetic theory cannot explain why the EM-Drive works. Dr. Shawyer's model relies on special relativity, but his theory is disputed. Based on the speculation above we can formulate an alternative model to explain why the EM-Drive works. Rather than employing a single frequency, the EM-Drive employs a microwave signal with a spread of frequencies $\pm \Delta f$ about a centre frequency f. The waves move at different speeds v_p in the truncated conical cavity (Fig. 21.7), but briefly group together to form and then to re-form wave envelopes which move at the group velocity v_g. We assume that the dimensions of the conical cavity are such that wave cut-off occurs for some frequencies. As the radius r of the cavity decreases, so higher frequency waves are prevented from propagating towards the narrower end of the cavity, but are reflected back towards the wider end.

In particle terms we can view the microwave signal as bunches of microwave photons with slightly different frequencies and speeds v_p which briefly come together to form groups moving with speed v_g, before dispersing to form new groups. Let us consider just one frequency. As these photons approach the cut-off region they will undergo rapid red-shifting in frequency until at the cut-off plane their frequency is zero. These photons have no effective mass, no momentum and no energy. Beyond this plane the combined microwave signal has lost some of its frequency components, which results in less energy being transported towards the narrower end of the cavity and a change in the group velocity v_g at which the remaining energy moves.

So, in the case of a spread frequency beam we must expect that at resonance the energy E_W residing in the region of the wider end of the cavity will be greater than the energy E_N residing in the region of the narrower end. Since the cone changes radius r linearly we expect the energy content to change linearly with distance along the centre line. From equation 2.11 we see that the energy imbalance gives rise to a force pointing in the direction of the narrow end.

Now let us think of the oscillating, or resonating, photonic mass current. Due to frequency cut-off the effective mass current towards the narrow end is decreasing as some photons are being extinguished, while the effective mass current towards the wide end is increasing as some photons are being resurrected. Surrounding the effective mass current is a changing gravitomagnetic field **h** which gives rise (eqn. 8.17) to an induced gravity field **g** (or gravitational dipole) pointing in the direction of the narrow end of the cone. This **g**-field provides the thrust experienced by the mass of the conical cavity. Furthermore, there will be radial gravitational radiation, so that momentum is lost from the cavity carried away by gravitons. Since the cone is metallic, forming a Faraday cage, no electromagnetic energy can escape from the cavity. If Shawyer's EM-Drive is energised near to a gravitational wave detector it might trigger a gravitational response. Perhaps changing photonic mass currents within a conducting cavity is the route to a gravitational communication system sought by Professor Raymond Chaio.

In his book, *Secrets of antigravity propulsion*, Dr. Paul LaViolette states that microwave cavity experiments were conducted by the US defence industry, starting in the late 1950s and ending in the early 1970s. Whether this research led to anything is unknown.

What other methods might be used to change the frequency content of an electromagnetic signal in order to create a gravitational dipole? Only recently (New Scientist, 15 October 2013), Ulf Peschel and his team of researchers at Erlangen-Nuremberg University, in Germany, provided experimental evidence for the existence of accelerating gravitational dipoles based on positive and negative effective mass pairs of laser-light photons. Their technique involved splitting a laser beam and sending the two signals around two different diameter fibre optic loops. Due to different radial accelerations, the photons in the two loops underwent different frequency red-shifts and, hence, developed different effective masses, leading to the formation of a stream of photonic gravitational dipoles when the two signals were recombined

As a closing thought on this section one wonders whether there is an acoustic analogue of the red shift. Do the ideas discussed above for a microwave drive read across to an acoustic version based on ultrasonics? If they do, then, since the speed of sound is much much less than the speed of light, an acoustically induced gravitational dipole would be much much stronger than its microwave counterpart. We need acoustic experts to consider the ideas and, perhaps, to try out some experiments.

21.5   Alzofon's anti-gravity screen

Science fiction often leads science fact. In 1901 H. G. Wells published his novel, *The First Men on the Moon.* In the novel, the main character is Dr Cavor, who invents cavorite, a material which can screen off gravity fields. By opening and closing cavorite panels on his spacecraft, Dr Cavor was able to journey to the Moon merely by screening off the Earth's gravitational field and using the Moon's gravitational attraction.

In July 1981 Frederick Alzofon, a specialist engineer working on infra-red and optical sensors for the Boeing Aerospace Company, published a technical paper (AIAA 81-1608) with the provocative title 'Anti-Gravity with Present Technology: Implementation and Theoretical Foundation'. The paper described how the zero point photon fluctuations of the quantum vacuum might be used to create anti-gravity screens.

Alzofon pointed out that at the microscopic level zero-point virtual photon buffeting of matter was an established fact, as observed with the Lamb shift and the Casimir force (see Chapter 18). Alzofon suggested that it might be possible to create a macroscopic effect by controlling the zero-point virtual photon fluctuations within matter. His idea was that by actively suppressing the natural level of the zero-point virtual photon fluctuations within certain specially conditioned materials, the zero-point energy (ZPE) density within such materials would be reduced below the ambient ZPE density of the external quantum vacuum. If this was possible, then such matter would have a negative energy density relative to the quantum vacuum. By using these conditioned materials as screens around a central body, any asymmetry in the zero-point vacuum energy density around the whole body would result in it experiencing a force.

At the time, in 1981, Alzofon's idea of creating negative energy density matter screens was just a dream. But, in 1985, Dick Slusher and Bernard Yorke of the AT&T Bell Laboratories generated squeezed laser light. That is, a monochromatic coherent light wave that contains regular pockets of negative energy density.

For ordinary laser light the energy density within the wave is always above or equal to the ambient energy density of the background noise attributed to the quantum vacuum (Fig. 21.9). Or, in particle terms, the photons in a laser wave are subjected to buffeting by the virtual photons popping out of the quantum vacuum, causing the wave amplitude A to suffer minor fluctuations (Fig. 21.9). Photons must obey Einstein's form of the uncertainty principle in terms of energy and time (eqn 17.5). Rewriting this condition in terms of the amplitude A and the phase angle ϕ we find that the incremental fluctuations of both, when multiplied together, must satisfy the relationship

$$\Delta A . \Delta \phi \geq (\text{constant}) \hbar \qquad (21.10)$$

Squeezed light is a quantum mechanical phenomenon achieved by redistributing the photons in a light wave. Providing the uncertainty principle is not violated we can reduce, or squeeze, the amplitude A at the expense of increasing the phase ϕ (amplitude-squeezing), or vice-versa (phase-squeezing). In either case, the effect is to make part of the wave cycle less noisy, but another part more noisy. The result is that the energy density in part of the wave can be reduced below that of the quantum vacuum noise level, while in another part the energy density is increased (Fig. 21.9). Overall, the mean energy density E/V of the wave stays the same.

Over the last thirty years, photonic experts have devised several methods to create squeezed light, but they all boil down to the relocation of the photons within a coherent laser wave. In one method laser light is shone through a cylindrically shaped crystal of lithium niobate and complicated optics is used to obtain the squeezed light. In another method laser light is shone into a resonant cavity containing a controlled gas cell which alters the apparent length

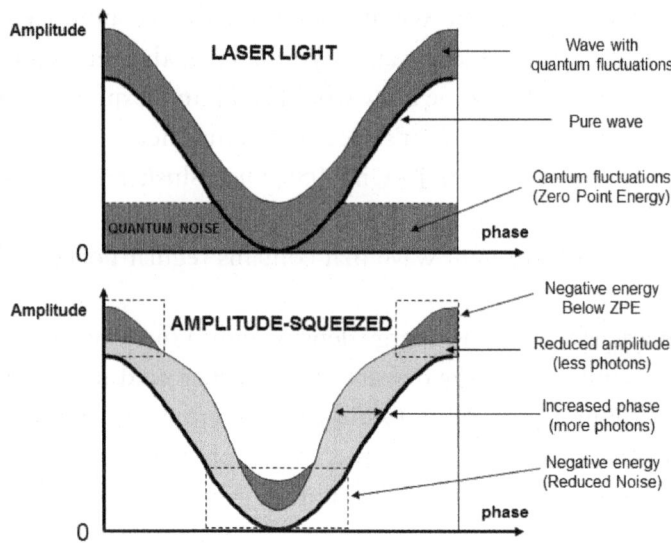

Fig. 21.9 Amplitude-squeezed light.

of the cavity in a pumping mode and squeezed light is obtained in a manner similar to holography.

One particular application for squeezed laser light is in gravitational wave interferometers (Fig. 8.9). By selecting the reduced noise portions of an amplitude-squeezed laser beam the improved sensitivity enables the interferometer greater detection of any movement of the test masses.

Squeezed laser light techniques have moved on and Professor Oskar Painter and his colleagues at Caltech in the USA have now built a solid-state device to generate squeezed light in a miniaturised silicon microchip.

The purpose of this digression has been to point out that many research groups throughout the world now experiment with squeezed laser light, which is an active method for suppressing the zero-point virtual photon fluctuations of the quantum vacuum.

As far as I know, no one has yet come up with a method for harvesting negative energy density regions. This was what Alzofon was interested in. As a non-laser expert I wonder whether it is possible to create standing waves using squeezed laser beams and, if so, can stationary pockets of negative energy regions be formed? Might these be used for Alzofon's anti-gravity screen? Also, with Alzofon in mind, has any research group created squeezed microwaves?

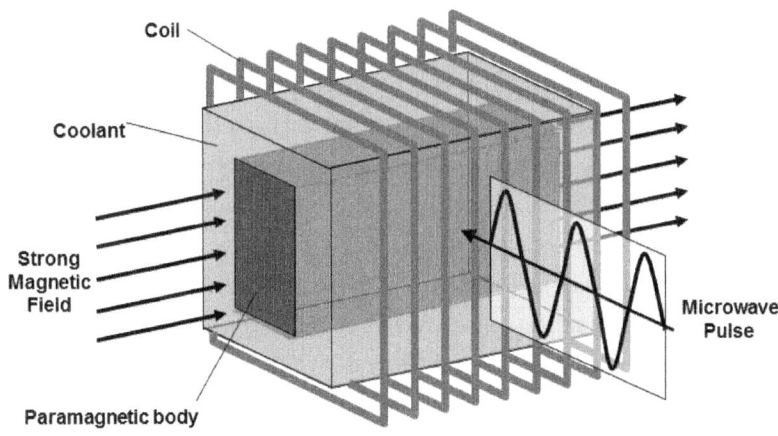

Fig. 21.10 The Alzofon anti-gravity screen.

Alzofon's proposed method for creating an active anti-gravity screen was based on a technique used to achieve really low temperatures. Standard methods of cooling like those used in fridges (evaporative cooling) can reduce the temperature of a body to fairly near absolute zero. But to achieve even lower temperatures nuclear cooling is needed. This involves using coolants made of special materials which have molecules exhibiting spin. The coolant is placed in contact with the body to be cooled and a strong external magnetic field is introduced (Fig. 21.10). The molecular dipoles in the coolant precess around the magnetic field direction, like a compass needle aligns with the Earth's magnetic field. In the aligning (forced turning) process the molecules gain a certain amount of potential energy (like an elastic band gaining energy when it is stretched).

A high-frequency (microwave) signal, at the resonant Larmor frequency, is then applied orthogonally to the external magnetic field, in a pumping action, which increases the alignment, increasing the potential energy of the molecular dipoles. The same technique is used in modern-day body scanners which have replaced many X-ray machines. The analogue is a child's swing oscillating in the Earth's gravity field. If the hanging chair is pushed horizontally, at right angles to the gravitational field, then, for the correct frequency of applying pushes (resonance), large amplitude oscillations can be achieved. This frequency corresponds to the Larmor frequency.

The temperature of the coolant is kept constant (adiabatic conditions) during the above process. When the high frequency signal is removed, the molecular dipoles revert to their original orientation in the external magnetic field, losing potential energy in the process. Under adiabatic conditions, this loss of coolant potential energy is extracted from the kinetic energy of the internal vibrations of the body to be cooled. Since the internal vibrations are associated with the thermal energy of the body, any reduction in vibrational energy results in a temperature drop.

In 1996 researchers at Lancaster University used a similar nuclear dipole cooling technique to achieve the lowest recorded temperature in the Universe, by cooling liquid helium to 90 μK (90×10^{-6} K), very near to absolute zero.

In Alzofon's scheme, he proposed that a paramagnetic body (one in which the atomic dipoles tend to line up with, and support, an external magnetic field) be held in a constant external magnetic field and be subjected to microwave pulses orthogonal to the field. Imposing adiabatic conditions he argued that, following a pulse, if the thermal changes in the body could be considerably slowed down (special materials with extremely low thermal conductivity were needed), then the loss of magnetic dipole potential energy would be made up from the kinetic energy of the background quantum fluctuations instead. If this were true, then the mean energy level of the fluctuating virtual particles would drop below the ambient level, creating a negative energy density region within the body. Such bodies would form the anti-gravity screens for Alzofon's vehicle.

21.6 Alcubierre's space-time warp-drive

Based on Einstein's theory of general relativity, the first attempt to investigate how space-time warping might be employed as a means of spacecraft propulsion was published by the Mexican physicist Miguel Alcubierre, in 1994. The theory assumes that on board the spacecraft devices can be activated which contract the region of space-time ahead of the spacecraft and expand the region of space-time behind the spacecraft.

The compressed region of space-time is a region of increased energy density, above that of the ambient conditions of empty space-time, so in relative terms it acts like a positive mass. The expanded region of space-time is a region of

FORMS OF GRAVITATIONAL PROPULSION

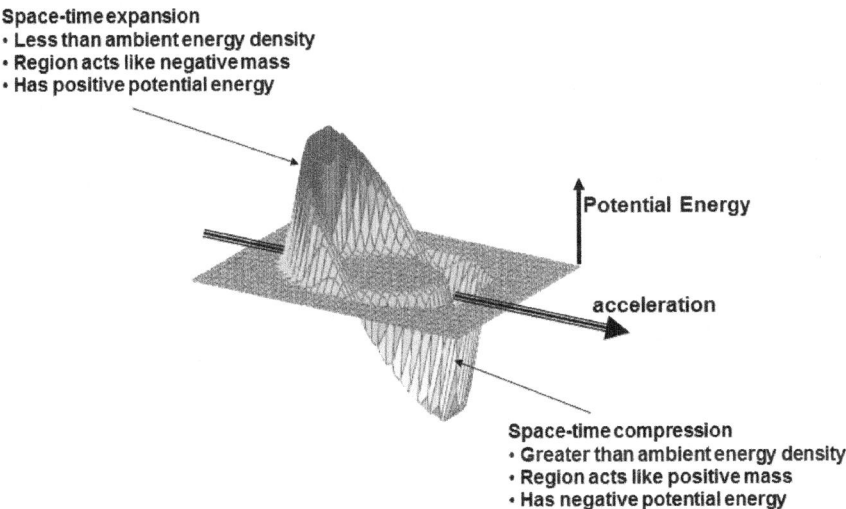

Fig. 21.11 Alcubierre's space-time warp-drive.

reduced energy density, so it acts like a negative mass. Thus, the two regions of warped space-time join together, to form an extended gravitational dipole. The spacecraft accelerates through space-time in the direction of the compression. A simple graphical form of the space-time warp, devised by E. Halerewicz, is shown in Fig. 21.11.

An external observer in the stationary frame of reference of the space-time warp would see the rest of space-time, that unaffected by the warping process, accelerate by. The spacecraft is cocooned inside a space-time bubble, separated from the rest of space-time. The pattern is similar, or analogous, to that of a point source and a nearby point sink, both held stationary in a uniform stream (Fig. 3.2). It might help to think of ordinary mass as an acceleration sink and negative mass as an acceleration source.

Now it's important that a spaceship's crew don't suffer from high g-forces during a space warp. Figure 3.5 shows a gravitational dipole consisting of a positive point mass M and a negative point mass -M, separated a distance d apart. The dipole has an acceleration (eqn 3.11) of $a = GM/d^2$. The gravitational field strength at the centre point between the positive and negative masses is

$$g = \frac{8GM}{d^2} \qquad (21.11)$$

A small mass (which doesn't disturb the overall gravitational field pattern) fixed at the centre point of the dipole, and being carried along with it at an acceleration a, would be subject to a strong gravitational field.

So, for crew comfort, the positive and negative energy density regions created by the space-time warping process must be spread over an extended length so that the spaceship's acceleration coincides with the strength of the g-field passing through the spaceship. In this way, the spaceship and crew will be in free fall and high (uniform) accelerations are not a problem.

When in free fall, the spacecraft is not subject to any inertial effects and neither are the crew. If the crew look out of the observation portholes at the background of space there is no increase in temperature because there is no motion of the spaceship relative to the space-time bubble. Also, the onboard clocks register the same time as those in the stationary reference frame.

Questions about the size and distribution of the positive and negative energy regions needed to warp space-time around a spacecraft and how such regions might be created need to be resolved. Perhaps an Alzofon screen might be used, but the technology to manufacture such a device is still way beyond us. Nevertheless, clearly there are ways of altering the ZPE condition locally. But is this akin to expanding and compressing the space-time continuum? In the event that it is, for how long can these regions be sustained? Other concerns are about what happens when light speed is approached, or exceeded? The answers to all these questions must be left to the experts in general relativity theory. In the meantime, engineers and experimentalists need to keep trying out ideas to probe the secrets of the quantum vacuum.

21.7   The worm hole

Although the concept of a worm hole through space-time is way beyond the scope of this book, it does have some points of interest relating to the gravitational dipole. And it is a form of transportation, rather like taking the lift and avoiding the long flight of stairs.

In truth, though, scientists have little idea about the technology needed to make a worm hole, connecting two regions of space far apart. Interest in the subject was aroused in recent times by the late Carl Sagan, Professor of Astronomy and Space Science at Cornell University, who was the presenter of *Cosmos*, the

TV series on astronomy. Sagan also wrote science fiction stories. In gathering material for his story, *Contact*, about an extra-terrestrial source that provided the design specifications for building a worm hole, he approached Kip Thorne, the Professor of Theoretical Physics at Caltech, to query whether the worm hole phenomenon really had any validity at all. Thorne's subsequent investigation showed that, theoretically, a worm hole might be generated if negative energy was available. This stimulated interest in the subject of worm holes and negative energy research. Sagan's story was later turned into a film by Warner Brothers.

The first worm hole to be investigated in general relativity theory was discovered by Ludwig Flamm, an Austrian physicist, in 1917. This was only two years after Albert Einstein had published his work on general relativity, in 1915, and just a few months after Karl Schwarzschild had published his paper on the event horizon surrounding a black hole.

The idea of the worm hole seems to have gained interest with the quantum view of the structure of space, down at the scale (10^{-35}m) of the Planck length L. At this scale, space is assumed to be discrete and joined up with infinitesimal bridges and percolated with infinitesimal (worm) holes, all of which are constantly changing shape, connections and positions. To form a picture in your mind of what the quantum vacuum looks like at any instant, think of gruyère cheese, massively reduced in scale. Note that a bridge is the surface covering of a worm hole. The virtual photon fluctuations which we now think make up the quantum vacuum occur in the bridges, not in the worm holes.

We are really only interested in the possible existence of large-scale bridges and worm holes through space which humans (or other intelligent beings) can use. The idea of taking a shortcut through space-time began with the concept of the black hole. Although the black hole is a 3-D gravitational phenomenon, it is usual to think of the severe distortion that it causes in the fabric of space-time as a 2-D surface funnelling down to the singularity. In this context, the surface is part of a bridge.

Suppose the mass singularity of a black hole lies on the surface of a mirror, then there will be a mirror image showing the funnel extending back, behind the mirror, and opening out in the distance. Back in 1935, Albert Einstein and Nathan Rosen used the theory of images to investigate the case of two coincident black holes, joining two separate parts of space-time together. They called their space-time structure the Einstein-Rosen bridge. The problem with the Einstein-

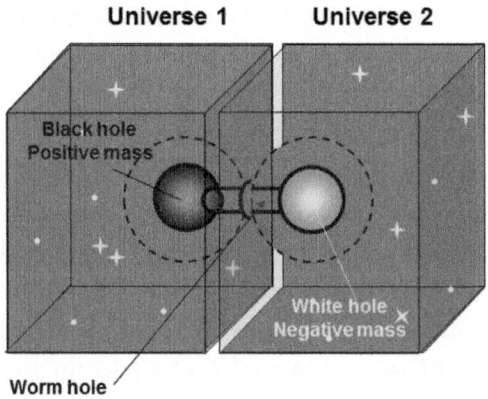

Fig. 21.12 The worm hole – extended gravitational dipole.

Rosen bridge, with the funnel, or worm hole, acting as a passageway through space-time, is that it's blocked by the mass of the coincident black holes.

Another way of thinking about the worm hole is as the connection between a black hole and a white hole, forming a special gravitational dipole (Fig. 21.12). Theoretically, a gravitational dipole forms a propulsion system but we suppose that our special version is fixed in space, somehow. The white hole is a very dense negative mass. Some physicists have ruled out white holes, arguing that they don't satisfy the second law of thermodynamics. However, we will assume that white holes can exist and leave the experts to argue about it.

In very simplistic terms, people and goods arrive at the black hole, pass through the connecting tunnel, or worm hole, and leave from the white hole.

To travel through a natural worm hole you need to take your own space with you as, theoretically, inside the hole is not part of space. Consequently, there are no quantum vacuum fluctuations inside a natural worm hole and the zero-point energy density inside is zero. The outside surface of the worm hole is part of space and the zero-point energy density there is colossal. So a cross-section acts like a Casimir cavity. The energy gradient across the wall of the worm hole is enormous, creating a huge force on the wall.

To keep a worm hole open, against the ZPE radiation pressure gradient, the walls may have to be lined with negative mass (Fig. 21.13). The negative mass lining and the positive mass wall form a gravitational dipole layer which exerts an outward force. If a section of the dipole layer were to break off it would accelerate away. Thus, the dipole layer reduces the energy gradient across the wall.

FORMS OF GRAVITATIONAL PROPULSION

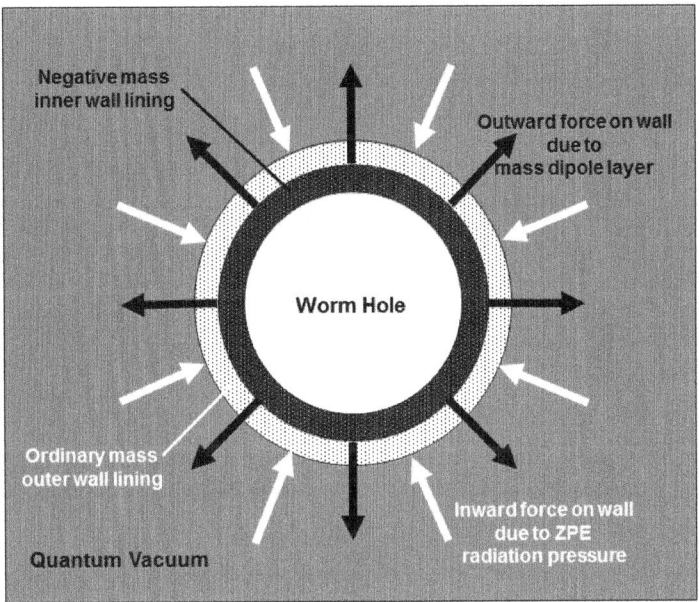

Fig. 21.13 Cross-section of a worm hole showing the dipole layer wall.

Because of the inverse square law nature of positive and negative mass of the dipole layer it means that the radial gravitational field inside the worm hole is zero.

The obvious problem with the idea of the worm hole described above is the black hole causes a blockage to any traveller wanting to pass through the worm hole and the white hole annihilates any traveller, with positive mass, that does manage to get through. What we need is our special gravitational dipole to be formed with virtual black and white holes, which don't interfere with the passage of a body.

In fluid dynamics we considered the fluid dipole with real positive and negative sources. We also looked at the vortex ring (Fig. 3.4) and realised that it could be viewed as a fluid dipole with virtual positive and negative fluid sources. A ping-pong ball cannot pass through a real fluid dipole, but it can pass through a virtual one!

This suggests the idea of a gravitomagnetic vortex-type ring acting as a virtual gravitational dipole. But remember, to create a gravitational field, we must change the gravitomagnetic field ($\nabla \times \mathbf{g} = \partial \mathbf{b}/\partial t$). Perhaps, in the future, we will be able to build a machine which is capable of generating closely

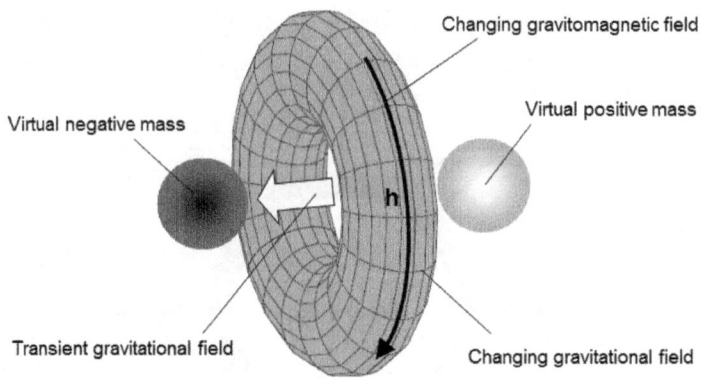

Fig. 21.14 Gravitational mass transport through core.

separated virtual positive and negative masses (or, even mini black and white holes) as its gravitomagnetic field strength **h** changes (Fig. 21.14).

Like the fluid dipole, the machine would have to be held in place, otherwise it would zoom off. Since the device would only work in a transient mode, it means that there would be a limited period when the transport gravitational field, or worm hole, was open for transportation in one direction, or the other.

This form of worm hole is more like the one described by Dean Devlin and Roland Emmerich in their book *Stargate*, which appeared in the MGM film, released in 1994.

In our search for a breakthrough in understanding of gravitational physics that will enable us to control gravitational fields, we must think up and test ideas which lie outside current wisdom. Clearly the answer that we are searching for is not obvious, or it would have been discovered long ago. Quite likely the breakthrough will come from a quirky result in an experiment in an area not originally thought to be linked with gravity. When we finally discover nature's gravitational secret, as eventually we will (probably by stumbling over it, if history is any guide!), the answer will probably seem very strange, at first, until we realise that it follows the known patterns of the other force fields. Whether any of the concepts described above hold a germ of the truth that we seek remains to be seen.

22

THE SEARCH FOR GRAVITY CONTROL

It seems to me that for the last half century gravity research activity in the UK (and possibly worldwide) has focused mostly on the detection of gravitational waves. There has also been some interest in the development of gradiometers for surveying and navigational purposes. But little thought has been given to exploring the more fundamental aspects of gravity which might lead to the discovery of how to control gravity fields. Over many years this area of experimental science has stultified, consequently receiving little funding support, further stifling academic interest.

On Friday 26 January 2001 I attended the inaugural meeting of the Institute of Physics' (IoP) Gravitational Physics Group, held at their HQ in London. I had been invited by the Chairman, Professor Mike Cruise of Birmingham University, to talk about Project Greenglow. The meeting was well attended with representatives from many UK universities.

I listened carefully to all the presentations. On the theoretical side, most of the interest was focused on studying gravitational waves, using computer-aided numerical models to investigate disturbances in space-time associated with the interaction of massive bodies. On the experimental side, interest was mostly focused on detecting gravitational waves on Earth or in near space and in space experiments to test the validity of general relativity.

I was the last speaker. As the only industrial representative and not being a gravitational physicist it was a slightly daunting task to talk about Project Greenglow. I explained that we, at BAE Systems, were particularly interested in stimulating research which might lead to a way of manipulating gravitation, rather like the way electromagnetic effects are used today. This had implications for gravitational propulsion and for the transmission and reception of gravitational waves and umpteen other uses. We realised that the research was extremely speculative, but were keen to get things moving. I described the research work that was already underway at several universities in the UK in support of the

Greenglow programme. What the audience made of my talk wasn't clear, since at the end of my presentation there was a deathly hush. However, the committee of the Gravitational Physics Group demonstrated their support for the BAE initiative by co-opting IoP member Dr Walter Johnston, an engineer from BAE Systems Warton, onto their committee as their industrial liaison officer.

When, prior to the inaugural meeting at the IoP, BAE Systems first released details of Project Greenglow, with its aim as the search for gravity control, it received a mixed reception by professional scientists and engineers, the media and the public. Some people were nonplussed, others enthusiastic. Unfortunately, gravity control is associated with anti-gravity, a topic which many in the media love to ridicule. For example, during the year 2000 the editorial page of a monthly UK engineering magazine, which really ought to have known better, commented, "The military wing of BAE Systems has confirmed it has launched an anti-gravity research programme called Project Greenglow. The scientific establishment is bemused. 'One can only conclude that at a high level of these organisations there are people who don't have a very sound grounding in fundamental physics'." In fact, the studies undertaken for Project Greenglow were being done by physicists, engineers and mathematicians at several universities in the UK, mostly assigned with the highest 5★ government rating for research. A similar article appeared in the business column of a popular daily newspaper in 2002, commenting on Project Greenglow and its possible effect on BAE Systems' share price. It included the following, "He (*the Chief Executive*) is investing in an anti-gravity machine. Of course, any fool knows such machines cannot exist: they break nature's laws…" That sort of comment could have affected the share price and resulted in BAE Systems management killing off Project Greenglow. Fortunately, it didn't. I wondered whether the journalist who wrote the article had ever flown in an aeroplane, a machine designed and built by engineers to overcome gravity?

The suggestion that it might be technically possible to counteract the force of gravity is seen as bizarre by many people, resulting in the nervous response by some of poking fun at the idea. One amusing quip being, "I read a book on anti-gravity. I couldn't put it down." Ridiculing gravity research is not new, it was around when Newton made his breakthrough in understanding. The English language hasn't helped the situation, either. Although gravity implies something serious, levity, another name for anti-gravity, implies something frivolous. Scientists and engineers, conscious of the need to take care of their professional reputations (important for promotion and research funding reasons), have, quite naturally, tended to avoid a subject linked with frivolity. The fact that the immutability of gravity has baffled the minds of some brilliant scientists in the past is probably another reason why many researchers have avoided the subject. So, it's no wonder that the subject of gravity control, a legitimate area of experimental study, has been shunned by many academics.

Progress in aeronautics during the 19th century suffered in the same way, being treated as unbelievable and a huge joke by many people. Fortunately, in the early years a few inventive engineers and sportsmen were so interested in the art of flying that they were willing to risk being ridiculed, as well as injured, as they sought to make advances. Once the breakthrough in powered flying machines had been made, two things led to further rapid developments. The first was realising that flying machines could be used as a decisive weapon of war, so that antagonistic governments invested heavily in the technology. The second was realising that surplus military flying machines and pilots could be exploited for commercial use. Although it's true that there were mishaps on the way, that is the nature of technical development. But, just look at the mighty aerospace industry today!

Equally, the investigation of the idea of extracting energy from the zero-point field of the quantum vacuum has stretched many people's minds to the limit. Is this incredible idea a joke, too, or can you really get energy from apparently nothing? Many scientists now accept the existence of the quantum vacuum and some scientists are interested in the possibility of being able to extract energy from the quantum vacuum's zero-point field. One of the Greenglow research studies was involved in the investigation of the zero-point field, a phenomenon which may also have a bearing on gravitational propulsion. Around the world, research teams are investigating various techniques to extract energy from the quantum vacuum. The supporting technology, in the form of micro-electro-mechanical systems (MEMS) and nanotechnology, is already well established and some patents for vacuum energy devices have already been granted. If the researchers are successful, then I expect that the development of vacuum energy (so called *free energy*) devices will follow the rapid development pattern of motor cars, personal computers and mobile phones, being driven by a public desire to own one for one's own use.

To be fair to the UK media, there were some constructive and supportive articles about Project Greenglow. However, what really amazed me was the general public's interest. I was completely overwhelmed by requests from people around the world for more information about our programme. I always felt that the value of the publicity and the goodwill that Greenglow engendered for British Aerospace/BAE Systems far outweighed the small cost of the research programme.

TV programmes like *Star Trek* have spurred some scientists on to investigate the physics behind the USS Enterprise's warp drive. Disney's film, *The Black Hole*, made the public more aware of the phenomenon, while films like *Contact* and *Stargate* even sparked academic interest in the exotic study of worm hole physics. So the media clearly has an important role to play in persuading scientists and engineers to take a serious interest in the study of how to control and manipulate gravitational fields. This is not just about science fiction programmes, but also about good quality science programmes covering the subject, too. In fact, the UK has some extremely good science communicators.

The kite and the balloon paved the way for the start of aeronautics, but gave way to the powered aircraft. Similarly, the rocket has paved the way for access to space, but now we need to replace it with something better that is not thrown away after using it. We need a safer, greener and cheaper way of getting into near Earth orbit. Building a substantial space station, with some commercial promise, then becomes more realistic. From there, travel to the Moon and throughout the solar system becomes easier. Even so, for exploration of the solar system and beyond, the major problem to be overcome is speed as the distances involved are vast. Spacecraft using ion-drives and solar sails are already being developed. Even the use of nuclear powered rockets has been considered. Gravitational propulsion technology could provide the means of getting to near Earth orbit under low acceleration, more like a comfortable train ride than a dangerous fairground ride. And it could provide the means for accelerating interplanetary spaceships to huge speeds, assuming some form of energy supply can be drawn from the quantum vacuum. Also, the crew could enjoy Earth-like gravitational conditions onboard during the journey. In fact, without gravitational propulsion technology and some means of extracting energy from the quantum vacuum it seems most unlikely that humans will ever be able to explore deep space. But, although gravitational propulsion is likely to be a front

THE SEARCH FOR GRAVITY CONTROL

runner for future space vehicles, as far as we know the breakthrough in understanding how to control gravity has not yet been made. Or has it?

Towards the end of the Second World War and the start of the Cold War there were a growing number of reports of unidentified flying objects (UFOs), including flying saucers. Many of the reports were made by the public in the USA, and in March 1952 the United States Air Force (USAF) began Project Blue Book, aimed at collecting details of UFO sightings. Dr J. Allen Hynek, an academic astrophysicist, was employed by USAF as a consultant for the project. Initially, Dr Hynek was convinced that most sightings could be explained but, gradually, he became less sure and more puzzled by the UFO phenomenon. He was also dissatisfied with the way that the UFO data was handled by USAF and this led him, in 1973, to form the Center for UFO Studies (CUFOS) with access to the public. In 1967 USAF sponsored the Condon Committee to investigate UFO sightings. The Co-Principal Investigator was Dr David R. Saunders, a professor of psychology at Colorado University. He began a catalogue of UFO sightings known as the UFOCAT, which was given to CUFOS in 1976. The NICAP (National Investigations Committee on Aerial Phenomena) study, which started in 1956, collected a considerable amount of UFO data. This was given to CUFOS in the early 1980s, following the demise of NICAP. Today, the UFOCAT contains more than 209,000 worldwide eye-witness accounts of UFO shapes and sizes, including saucers and deltas, amazing UFO flight characteristics tracked by radar and strange effects caused by the presence of UFOs, such as the loss of traction by nearby road vehicles, due to their slight levitation, the interference of vehicles' electrical circuitry and even the paralysis of some observers. Is this all a figment of the imagination, or is there something real going on?

On 28 July 1952 Winston Churchill, the British Prime Minister, wrote a minute to the Secretary of State for Air, asking "What does all this stuff about flying saucers amount to? What can it mean? What is the truth? Let me have a report at your convenience." It seems that the answer Churchill received was somewhat misleading, claiming that all UFO sightings could be explained, which was untrue. Sixty years later are we, the public, any the wiser?

During the 1950s interest in the idea of gravity control intensified, partly due to the possibility that flying saucers might be using gravity propulsion. In the USA more than fifty separate research programmes were initiated, with two-thirds being classified military funded programmes. Most US aircraft companies had some involvement, including Bell, Boeing, Convair, General Electric, Grumman, Hughes and Lockheed. However, by the mid-1970s it seems that the US aerospace firms had all ended their gravity research programmes with no hint of any breakthrough having been made, although some have suggested that the programmes weren't terminated but had gone black. Today, rumours abound that the US military has not only developed gravitational control technology but that they have already built a fleet of interplanetary spaceships, each the size of an aircraft carrier. But there is no hard evidence to support this speculation.

In 1989 Bob Lazar sensationally claimed that the strong nuclear force was the basis of a gravitational propulsion system for flying saucers that were being flight-tested at the secret Groom Lake test facility (Area 51) in the Nevada desert. Lazar claimed to be a US scientist who had worked on the programme and said that the technology was developed from back-engineered alien technology (see 15, Reference Books). This startling revelation, coupled with Lazar's apparent lack of scientific credentials, resulted in mainstream scientists dismissing Lazar's claim as a fantastical hoax. Naturally, scientists do not want to appear gullible in front of their peers so, to avoid any embarrassment, Lazar's claim was never raised at any Project Greenglow meetings. But in ignoring Lazar's outlandish claim of alien technology have we been diverted away from considering the possibility that the strong nuclear force is linked with gravitational propulsion?

On the unclassified front, details from the Commission on the Future of the US Aerospace Industry, issued in November 2002 by the US Congress, highlighted the US government's continued interest in making a breakthrough in the development of a new means of spacecraft propulsion for exploration of the solar system and in making a breakthrough in new energy sources (zero-point energy). However, since 2009, fiscal problems in the west have meant the inevitable temporary scale-back in funding of space programmes. In the US, President Obama has cancelled the Shuttle Program, the Return to the Moon Program and the Manned Mars Mission. In Obama's view, voiced in July 2011, "NASA needs new technology breakthroughs to revitalize its mission to explore the universe." Of course, there are those that believe that the huge cut-backs are a covert signal that the US has already made a breakthrough in gravitational control technology and that a new space programme is now being implemented in secret.

The Soviet statistician Nikolai Dmitrievich Kondratieff, in studies carried out in the 1920s, was the first person to identify the economic-industrial 55/65-year cycles of boom and bust – the up-waves and the down-waves which characterise the capitalistic system. Analysis of the last 300 years seems to suggest that each new up-wave is characterised by the introduction of a new form of power/transportation system which leads on, in turn, to a new form of society. Based on the Kondratieff cycle, some futurologists have predicted that a breakthrough leading on to a new form of powered propulsion system is likely to occur sometime around 2015, marking the beginning of the next up-wave. Even if a breakthrough in controlling gravity occurs then, experience from the aerospace industry suggests that the timescale from designing and testing a vehicle with a new propulsion system through to first flight would take thirty years! But, the impact on aircraft and space technology would be profound, no matter what the timescale. Other terrestrial applications could burgeon as well, even if, with our current level of ignorance, we cannot envisage them at the moment.

Obviously, to have any chance of mastering the control of gravity requires venturesome fundamental research in the attempt to make a breakthrough in scientific understanding. An important first step, therefore, is to encourage the scientific and technical communities' interest in the idea of controlling gravity and getting them to formulate and carry out the fundamental research. It might be better to think in terms of a race, as the excitement of the competition may spur research groups on. Although some scientists will be attracted by the fundamental nature of the research and the glittering prize, funding is a major inducement. NASA's Breakthrough Propulsion Physics (BPP) Program and BAE Systems' Project Greenglow were both attempts to provide a widely publicised programme of research on gravitational control, backed with small amounts of funding. But these two initiatives have now ended. Today, there is the privately funded Tau Zero Foundation in the USA which provides a centre for researchers interested in inter-stellar flight. The foundation was started in 2008 and is led by Marc Millis, who was formerly the Head of the NASA BPP Program. Also in the USA, there is the 100-Year Starship Program, begun in 2011, which is a small funded initiative by DARPA (US Defense Advanced

Research Project Agency), to get academics to start thinking about the technologies needed for a starship, which might be built at the beginning of the next century. Disappointingly, there are no competing programmes in the UK, Europe, or elsewhere.

During the European Renaissance, governments of the continent's seafaring countries supported oceanic voyages of discovery and exploration. The success of these daring voyages led to the advent of the merchant adventurers driven on by the stories of exotic lands and the desire to acquire vast wealth. On the other side of the world, the Chinese authorities, who had previously supported similar oceanic voyages of discovery, officially prohibited such adventures and so China began a period of stagnation, followed by gradual decline. The great voyages of discovery by the Europeans led to the development of new technologies, particularly in navigational aids and time-keeping. In a like manner, the exploration of near-space around the Earth during the US and Soviet Russian space programmes fostered new technologies, including electronic miniaturisation, the development of advanced computers, navigational satellites and robotics. Obvious commercial benefit has largely been restricted to the business of communication and Earth observation satellites, but the technologies spawned have changed the way we live. We can reasonably expect that the next phase of deep-space exploration will lead to another range of new technologies.

Given their new-found wealth from energy exportation, we are likely to see the Russians re-establish their interest in advanced space research. The Chinese, rapidly emerging as a huge economic power, have committed large amounts of funding to their conventional space programme. After the USA and Russia, the Chinese were the third nation to land a vehicle on the Moon in December 2013. The Indians are likely to follow suit fairly soon. The Europeans, the Japanese, the Brazilians and even the Iranians, although squeezed financially, are all expanding their space research programmes. Space technology is big business in military and commercial terms. Even the UK now has a Space Agency. A new Space Race is underway, with many participants! It seems quite reasonable to assume that many countries are also looking at the possibility of gravity control for spaceship propulsion in the future.

Gravitational propulsion could herald a new age of discovery, this time with the manned exploration of the surfaces of the planets and moons in the solar system. The initial period of government-funded planetary explorations will be followed by the advent of a new breed of space adventurers driven on by the excitement of exploring new worlds. Private funding of space transportation systems will follow, with the lure of making a fortune. In place of wooden ships bringing back gold, silver and rare plants from the New Worlds across the oceans, giant space tugs will bring back cargos of precious raw materials needed to replenish the Earth's dwindling stocks, mined from planets, moons and asteroids across the heavens. In the not-too-distant future an inter-continental trip may take less than an hour, a visit to the Moon may take less than a day while a journey to Mars may take, perhaps, a week at the most! Or you might like to break your journey and spend a day, or so, in a space hotel enjoying the view of Earth below and an uninterrupted view of the stars.

Perhaps we are closer to a breakthrough in gravity control technology than we realise. The particle physicists might provide us with a surprise breakthrough in understanding. Or, maybe, the breakthrough will arise from a spin-off from quantum vacuum research. It could result from government-funded military/space programmes or academic research sponsored by government/industry. It might stem from a moment of inspiration by one gifted individual, or result from a collaborative effort between scientists. No matter how it occurs, the breakthrough in understanding how to control gravity will eventually be made. Its discovery will lead to the next fantastic adventure in the twin fields of aerospace science and engineering!

ACKNOWLEDGEMENTS

Many people were involved with the British Aerospace gravity research programme, which evolved into the BAE Systems' Project Greenglow. My colleagues at Warton deserve a mention. The late Tom Smith, Head of the Advanced Technology Group, took an early interest. Geoff Salkeld, George Seyfang, Brian Chilton and Brian Beele helped me get things started and remained supportive throughout the venture. The late Jim Robertson was also supportive. Les Green, Mark Spore and Steve Roe were the wind tunnel engineers at Warton who tested Sandy Kidd's inertial thrust machine. John Lowe was the Warton Chief Librarian who sent my 'University-Industry Gravity' report to NASA.

Professor Brian Young, the late Director of Strategic Projects at Warton, took an early interest in the gravitational study and gave it his support. Brian proudly claimed that he was actually a nuclear physicist by training, which gave him some insight into the possibility that gravitational control was linked with quantum mechanics. Brian used some of the results of the gravity study in his inaugural lecture as the Visiting Professor of Aerospace at Salford University. Formerly the Technical Director of British Aerospace's Military Aircraft Division, Brian was responsible for leading the research and development programme which led to the multinational Typhoon programme.

Professor John E. Allen was formerly the Chief of Future Projects at British Aerospace's Kingston site, where the VTOL Harrier and the Hawk Jet Fighter were designed and built. Long before the start of the gravity study at Warton, John had already developed an interest in the possibility of gravity propulsion. In a chapter in a book published in 1970, about the future of aeronautics (see Reference Books), John drew attention to a UK government report, issued in the 1920s, listing expected breakthroughs in science and technology which included auto(anti)-gravity. As he noted at the time, all of the items on the list have come to pass, except for auto-gravity. John had a close involvement with the British Aerospace gravity research programme, publishing a peer-reviewed paper on the work and, through lectures and personal contacts, has continued to publicise its importance.

ACKNOWLEDGEMENTS

Peter Liddell, Head of Advanced Projects at Warton, was responsible, with some slight misgivings, for the 'What-if' anti-gravity vehicle study, following a request by Brian Young. Dave Cundy conjured up the fictitious anti-gravity engine data and Martin Kennedy produced a technical drawing of the VTOL anti-gravity combat aircraft. Alan Groves, of Graphic Support at Warton, painted some inspiring pictures of our anti-gravity concepts.

Professor Ian MacDiarmid, Dr Chris C. R. Jones, Geoff Salkeld and Peter Beckett helped me get funding from the Warton technologist research budget to start the university gravity research programme. Lambert Dopping-Hepenstal, the Head of Research and Development at Warton, was also supportive. This research programme formed the basis of Project Greenglow, a name suggested by Tony Miller, of the Warton Advanced Technology Department. After the British Aerospace merger with Marconi, in November 1999, to form BAE Systems, Project Greenglow was expanded to become a headquarters-backed university research programme reporting to Dr Bill Martin and assisted by Dr Marcus Naraidoo. I retained my project management role. Funding for Project Greenglow was arranged by Dr Carl Loller, the Head of the BAE Systems Sowerby Research Centre in Filton, Bristol. Dr Brian Wardrop (BAE Systems Technology Centre, Great Baddow) administered the programme, supported with great enthusiasm by Dr Vaughan Stanger (a writer of science fiction stories).

Project Elmo, our electrostatic-wing study, was carried out at Warton and led jointly by Brian Probert from the Warton Aerodynamics Office and Dr Chris Jones from the Warton Electronic Warfare Office with major technical support provided by AEA Technology. Dr Gary Proudfoot and Dr Andrew Holmes, of AEA Technology, first demonstrated a pink-glowing, free atmospheric plasma to us, at Culham, in July 1996. Dr Simon Scott, of the BAE Systems Sowerby Research Centre, at Bristol, was primarily responsible for our electrical plasma studies. Added support in aerodynamics was provided by Dr Graham Johnson of the Sowerby Research Centre, by Brian Probert of the Warton Aerodynamics Office and by Dr Terry Cain, of DERA (formerly RAE Farnborough).

I spent some of the early part of my retirement performing a 'brain-dump' and writing a detailed account of Project Greenglow, describing what had happened and what we'd learnt during the programme. Professor Robin Tucker

(Lancaster University), Dr H. Ron Harrison (retired Senior Lecturer, City University, London), Dr Andrew May (MoD/Qinetiq), George Seyfang (BAE Systems, retired), Geoff Salkeld (BAE Systems, retired) and Charles Harmer (BAE Systems, retired) all read various draft versions of my report and pointed out errors and suggested improvements.

However, the report was not really in a form which could be turned into a book and made available for others to see what we had done. To overcome this I decided to write a book about the search for gravity control, describing the background leading to the research pathways, giving some details of the latest experimental results, suggesting new ideas for experiments and filling in with details from Project Greenglow, as appropriate.

John Wright (Technology Acquisition Manager, BAE Systems), Professor Ian MacDiarmid (Executive Technical Specialist, BAE Systems) and Andy Bunce (Head of Public Relations, BAE Systems) obtained written company permission allowing me to use anything to do with Project Greenglow and related studies in this book. However, this does not necessarily mean that BAE Systems supports what I have written.

Professor Robin Tucker read an early draft version of the book and I took note of his comments. He has helped with more of my queries since, resulting in a number of changes to the book and, hopefully, to its improvements. Even so, Robin would probably not wish to be associated with some of my wilder speculations. Dr H. Ron Harrison read through a nearly completed draft version of the book and spotted several errors, which I have corrected. Ron has an interest in gravitational theory, too, and has published his own book on the subject (see Reference Books).

Others not mentioned in the book that helped along the way (many of whom have now retired) include John Whalley, Dr David Ashcroft, Dr Tony Llewellyn, Neil Kiley, Nick Colosimo, Nick Shepherd, Steve Nunn, Paul Milner, Dr Doug King, Mick Butler and Graham Roe, all based at BAE Systems (Warton). Also, helpful were Alan Levenston, John Ackroyd, Dr Andrew Wright, Dr Simon Scott, Dr Rinaldo Leo and Professor Terry Knibb at the Sowerby Research Centre at BAE Systems (Filton), in Bristol. Pat Callaway and his colleagues at BAE Systems (Brough) provided details of the Cayley Flight re-enactment. Rob Chambers and Alan Malvern, both from BASE at BAE Systems (Plymouth), helped us by providing us with details on advanced

ACKNOWLEDGEMENTS

gyroscopes. Rob Chambers also has a great interest in gravitational propulsion. Peter Laurie provided us with an early introduction to the Casimir force. Dr Eddie Williams and Dr Mike Provost, of Rolls-Royce Ltd, participated in some of the Greenglow activities. Richard Obousy, of Qinetiq, was also supportive. And there were many others, too, and I apologise for not mentioning you. During my retirement I have received continuing encouragement from Dave Waring (BAE Systems (Warton), retired) to get my book finished and published. Tony Cuthbert, an inventor with a close interest in new forms of propulsion and new sources of energy, has also maintained regular contact with me.

I had a very traumatic experience with the Windows 'Blue Screen of Death', when my hard drive crashed. I lost much of my support material, many references and the latest version of my book at the time. I learnt a very hard lesson: continually back up all work! My son, Nicholas, helped to piece some things back together for me and helped with many aspects of computing. My children, Nicholas, Claire, Richard and Emma, have all put up with my long obsession with gravity and the vortex (a feature of my PhD study) with remarkably good grace. Lastly, I acknowledge the encouragement and support of my ex-wife, Doris, when I first set out on my journey to forge a career in mathematics and engineering more than half a century ago.

Ronald Evans

Lytham St Annes
Lancashire
England, UK

December 2013

SELF PUBLISHING

Greenglow is a semi-technical book which describes the history of gravitational research up until the present time. We don't yet know how to control gravity and the book speculates on where a breakthrough in understanding might be made. Although the book contains equations and many explanatory diagrams, it is not an academic textbook, but more a scientific adventure story. I wrote to nearly 50 UK literary agents, seeking support to get the book published, but none were willing to help.

I considered the option of self-publishing. Here I had the good fortune to contact Jeremy Thompson of Troubador Publishing Ltd. After viewing a draft copy of my manuscript, he said that his company would be willing to publish the book. Since I felt that the Greenglow story was well worth telling, I took the plunge and accepted his offer. Much of the credit for producing the book must go to Naomi Green, the Production Controller at Troubador. My book was clearly a special project for her and she persevered with my text, equations and diagrams, plus numerous changes, to produce a good looking book. Other members of staff at Troubador were also helpful in the long road to publication. I'm pleased with the result and I hope that you, dear reader, are too.

POSTSCRIPT

I'm conscious of how many times the words *late* and *retired* appear in my Introduction and in my Acknowledgements. The breakthrough in understanding in gravitational physics, showing how gravity can be manipulated, did not happen during my working lifetime. Even after a confirmed breakthrough, there will still be much work needed to develop the technology for gravity control. My generation of aerospace engineers has missed the chance to be the first to design and build aerospace vehicles exploiting this futuristic means of propulsion. The search has now passed to you, the next generation of scientists and engineers. I hope that my book will inspire you and help you realise the fantastic and exciting opportunity that is waiting to be grabbed.

REFERENCE BOOKS

1.   *The Hutchinson Dictionary of Scientists*
                        Helicon Publishing Ltd, 1996
                        ISBN: 1-85986-216-0
2. Allen, J. E. (Editor) — *The Future of Aeronautics*
                        Hutchinson & Co. (Publishers) Ltd, 1970
                        ISBN: 0-09-100920-0
3. Asimov, I. — *Asimov's New Guide to Science*
                        Penguin Books, 1984
                        ISBN: 0-14-007621-2
4. Berkson, W. — *Fields of Force*
                        Routledge and Kegan Paul, 1974
                        ISBN: 0-7100-7626-6
5. Calder, N. — *The Key to the Universe*
                        British Broadcasting Corporation, 1977
                        ISBN: 0-563-17091-3
6. Clarke, A. C. — *Profiles of the Future*
                        Victor Gollancz Ltd, 1982
                        ISBN: 0-575-03210-3
7. Cohen, I. B. — *The Birth of a New Physics*
                        Heinemann Ltd, 1961
8. Cook, N. — *The Hunt for Zero Point*
                        Arrow Books, 2002
                        ISBN: 978-0-09-941498-8
9. Davidson, J. — *The Secret of the Creative Vacuum*
                        C. W. Daniel Company Ltd, 1989
                        ISBN: 0-85207-202-3
10. Davies, P. C. W. — *Space and Time in the Modern Universe*
                        Cambridge University Press, 1977
                        ISBN: 0-521-2915-8

REFERENCE BOOKS

11. Davies, P. C. W. *The Search for Gravity Waves*
    Cambridge University Press, 1980
    ISBN: 0-521-23197-2
12. Davies, P. *Superforce: The Search for a Grand Unified Theory of Nature*
    Unwin Paperbacks, 1985
    ISBN: 0-04-539006-1
13. Davies, P. *How to Build a Time Machine*
    Penguin Books, 2002
    ISBN: 0-14-100534-3
14. Forward, R. L. *Future Magic*
    Avon Books, 1988
    ISBN: 0-380-89814-4
15. Good, T. *Alien Liaison*
    Arrow Books, 1992
    ISBN: 0-09-985920-3
16. Gribbin, J. *In Search of Schrödinger's Cat*
    Black Swan, 1991
    ISBN: 0-552-12555-5
17. Gribbin, J. *Erwin Schrödinger and the Quantum Revolution*
    Bantam Press, 2012
    ISBN: 9780593067765
18. Harrison, H. R. *Gravity: Galileo to Einstein and Back*
    Universal Publishers, Boca Raton, Florida, 2006
    ISBN: 1-58112-932-7
19. Hey, A. J. G. & Walters, P. *The Quantum Vacuum*
    Cambridge University Press, 1987
    ISBN: 0-521-31845-9
20. Kidd, A. R. *Beyond 2001 – The Laws of Physics Revolutionised*
    Sidgwick & Jackson, 1990
    ISBN: 0-283-99925 X
21. King, M. B. *Tapping the Zero Point Energy*
    Paraclete Publishing, 1993
    ISBN: 0-9623356-0-6
22. Laithwaite, E. R. *Engineer Through the Looking Glass*

	British Broadcasting Corporation, 1980
	ISBN: 0-563-12979-4
23. LaViolette, P. A.	*Secrets of Antigravity Propulsion*
	Bear & Company, 2008
	ISBN: 978-1-59143-078-0
24. Lorrain, P. & Corson, D.	*Electromagnetic Fields and Waves* (2nd Edition)
	W. H. Freeman & Company, 1970
	ISBN: 0-7167-0331-9
25. Mallett, R.	*The Time Traveller*
	Corgi Books, 2007
	ISBN: 978-0-55215-575-5
26. McEvoy, J. P. & Zarate, O.	*Quantum Theory for Beginners*
	Icon Books Ltd, 1996
	ISBN: 1-874166-37-4
27. Mike, J.	*The Anatomy of a Flying Saucer*
	Lightning Source UK Ltd, 2011
	ISBN: 9781463598068
28. Szames, A.	*L'Effet Biefeld Brown*
	ASZed, 1998
	ISBN: 2-913377-01-7
29. Taylor, J. G.	*New Worlds in Physics*
	Faber and Faber, 1974
	ISBN: 0-571-10258-1
30. Wheeler, J. A.	*A Journey into Gravity and Spacetime*
	W. H. Freeman & Company, 1990
	ISBN: 0-7167-5016-3
31. Will, C. M.	*Was Einstein Right?*
	Oxford University Press, 1986
	ISBN: 0-19-282203-9

NAMES INDEX

A

Adams, Douglas, 266
Alcubierre, Miguel, 274
Alexander the Great, 11
Allen, John E., xxi, 253-254
Alzofon, Frederick, 270-274
Amontons, Guillaume, 237
Ampère, André, 63
Arago, Dominique François, 63, 72
Archimedes, 12
Aristarchus of Samos, 14
Aristotle, 1, 11
Avogadro, Amadeo, 140

B

Baker, Robert, xxiii, 161
Barnett, Samuel, 141
Becquerel, Henri, 176
Bekenstein, Jacob, 228
Bennet, Abraham, 59
Bernoulli, Daniel, 32, 54, 169
Biefeld, Paul, 260
Biot, Jean Baptiste, 64
Birdsall, Graham, xxiii
Bissell, Phil, xx
Blackett, Patrick, 141
Bohr, Niels, 180
Boltzmann, Ludwig, 168
Bonaparte, Napoléon, 62
Bond, James, 15
Bondi, Herman, 38

Borelli, Giovanni Alfonso, 17
Born, Max, 130, 133
Bose, Satyendra Nath, 137, 174
Boulton, Matthew, 60
Boyer, Timothy, 204
Boyle, Robert, 57, 169
Braginsky, Vladimir, 124
Brahe, Tycho, 16
Brown, Thomas Townsend, 260
Brush, Charles, 241
Bunsen, Robert, 62
Burton, David and Anne, 125

C

Cameron, Jeff, 262
Capasso, Frederico, 215
Carlisle, Anthony, 62
Cartan, Élie, 9
Cartmell, Matthew, xxi
Casimir, Hendrick, 210
Cavendish, Henry, 31, 62
Chadwick, James, 179
Champollion, Jean-François, 112
Chiao, Raymond, 98, 161
Childress, David Hatcher, 149
Childress, Jamie, xxiii, 149
Churchill, Winston Spenser, 285
Columbus, Christopher, 56
Compton, Arthur, 129
Cook, Nick, xx
Cooke, Steve, 149
Copernicus, Nicolaus, 14
Corson, Dale, xvi
Coulomb, Charles, 29, 237
Cowan, Clyde, 184, 197
Cruise, Mike, xxiii, 281
Csikai, Gyula, 184
Curie, Marie, 176

Curie, Paul, 176
Curie, Pierre, 176
Cuthbert, Tony, xxii, 149

D

Daimler, Gottlieb, 55
Da Vinci, Leonardo, 237
Davies, Paul Charles William, 216, 246
Davis, Ray, 197
Davisson Clinton J., 130
Davy, Humphry, 62
Dawson, Grant, xv
De Broglie, Louis-Victor, 128
De Broglie, Maurice, 128
De Coriolis, Gaspard, 51
Dee, John, 14
De Haas, Wander, 141
Dehnen, Heinz, 124
De Matos, Clovis, 150
Democritus, 176
Devlin, Dean, 280
Dimitriou, Stavros, xxii, 261
Dirac, Paul, 135, 196, 204
Doppler, Christian, 222
Du Fay, Charles, 58

E

Eberlein, Claudia, 217
Eddington, Arthur, 8
Edwards, Tony, x, xvii, xxii
Einstein, Albert, xii, 3, 7, 81, 93, 99, 128-141, 164, 173-174, 203, 220-222, 241, 258, 277
Emmerich, Roland, 280
English, Malcolm, xxii
Ennis, Graham, xxii-xxiv
Ericsson, John, 36
Essen, Louis, 194
Euler, Leonhard, 42

F

Faraday, Michael, xii, 62, 68-70, 79, 159, 183, 239-240, 245, 266
Fermi, Enrico, 180
Feynman, Richard, 132
Fitzgerald, George, 164
Flamm, Ludwig, 277
Flamsteed, John, 27
Fleming, John Ambrose, 68
Forward, Robert Lull, xix-xxiii, 40, 97, 206, 214, 250-252
Foucault, Léon, 108
Fourier, Jean Baptiste, 65
Franklin, Benjamin, 59
Friedman, Jerome, 187

G

Galilei, Galileo, 1, 15
Galvani, Luigi, 61
Gauss, Karl Friedrich, 45
Geer, Mike, xxii
Geiger, Hans, 178
Gell-Mann, Murray, 187
Gerlach, Walter, 183
Germer, Lester H., 130
Gilbert, William, 16, 56
Glashow, Sheldon, 190
Goeppert-Mayer, Marie, 182
Goudsmit, Samuel, 182
Gray, Stephen, 57
Green, George, 45
Gross, David, 189

H

Hahn, Otto, 180
Haisch, Bernhard, 247
Halerewicz, E., 275
Halley, Edmund, 28

NAMES INDEX

Hammond, Giles, 216
Hansson, Anders, xix, xxii
Harrison, H. Ron, xi
Harrison, John, 19
Hathaway, George, 152
Hauksbee, Francis, 57
Hawking, Stephen, 17, 229
Heaviside, Oliver, xviii, 83, 102
Heisenberg, Werner, 131, 135
Henlein, Peter, 199
Henry, Joseph, 73
Henson, William, 54
Hero of Alexandria, 13
Hertz, Heinrich, 77, 174, 220
Higgs, Peter, 190
Hilgenberg, Ott Christoph, 33
Hinds, Ed, 208
Hipparchus of Rhodes, 3
Hoffmann, Banesh, 41
Holt, Alan, xxii
Homer, 52
Hooke, Robert, 21, 199
Horrocks, Jeremiah, 17
Horsfield, Don, xv
Hough, Jim, 98
Hoyle, Fred, 52
Hubble, Edwin, 33, 116
Hughes, David E., 76
Hulse, Russell, 96
Humphrys, John, xxiii
Huygens, Christiaan, 17, 20, 219
Hynek, J. Allen, 285

J

Jeans, James, 170
Johnston, Walter, xxi, 161, 282
Joliot, Frédéric, 176
Joliot-Curie, Irène, 176
Joule, James Prescott, 238

K

Kant, Immanuel, 62, 115
Kendall, Henry, 187
Kepler, Johannes, 16
Ketterle, Wolfgang, 138
Kidd, Sandy, xviii
King, David, 261
King, Moray, 216
Knight, Gowin, 60
Knoll, Max, 135
Koczor, Ron, 148
Kondratieff, Nikolai Dmitrievich, 287
Koshiba, Masatoshi, 197
Krim, Jacqueline, 242
Kunsman, Charles, 130

L

Laithwaite, Eric Roberts, xvi, 85, 91, 125, 251
Lamb, Willis Eugene, 208
Lambert, Johann, 116
Lamoreaux, Steven, 211
Laplace, Pierre, 31
Larmor, Joseph, 140
LaViolette, Paul, 261, 269
Lazar, Bob, 286
Leake, Jonathan, xxii
Leibniz, Gottfried, 2
Lenard, Philip, 173
Lense, Josef, 119
Lenz, Heinrich, 71
Lewis, Alun, xx
Li, Ning, xxiii, 145, 148
London, Fritz, 122
Lorentz, Hendrik Antoon, 68, 164
Lorrain, Paul, xvi
Lucretius, Titus, 169
Lyons, Harry, 194

M

Maccone, Claudio, xxii
Mach, Ernst, 257
Maclay, Jordan, 214
Magnus, Heinrich, 54
Mallett, Ronald, 157-161
Marrison, Warren, 193
Marsden, Ernest, 178
Matthews, Robert, 144
Maxwell, James Clerk, 10, 73, 75, 80, 155, 162, 169, 219
May, Andrew, xxi
Maybach, Wilhelm, 55
McInnes, Colin, xxi
Mead, Franklin B., 214
Meek, James, xxiii
Meissner, Walther, 144
Mendeleyev, Dimitry Ivanovitch, 186
Michell, John, 31
Michelson, Albert Abraham, 163
Mike, John, 38, 254-259
Millis, Marc, xxi, 287
Mohideen, Umar, 213
Morgan, Harvey, 126
Morley, Edward, 163
Murad, Paul, xxiii, 161

N

Nachamkin, J., 214
Nagaoka, Hantaro, 178
Nash, John, 241
Naudin, Jean-Louis, 262
Newton, Isaac, xii, 2, 20, 23, 28, 118, 128, 182, 200, 219
Newton, Jeffrey, xv
Nicholson, William, 62
Nobel, Alfred, 131
Noble, Alan, 125
Noever, David, 148

O

Obama, Barack, 286
Obousy, Richard, xxii
Odysseus, 52
Oersted, Hans Christian, 62
Ohm, Georg, 64
Oldenburg, Henry, 57
Onnes, Kammerlingh Heike, 144
Owen, Gari, xxi

P

Painter, Oskar, 272
Parry, Jack, 194
Parsons, Charles, 36
Pauli, Wolfgang, 177, 184, 186
Penzias, Arno, 175
Peregrinus, Peter, 56
Peschel, Ulf, 269
Petit, Jean-Paul, xxii
Phelps, Alan, 264
Pilkington, Mark, xxiii
Pinto, Fabrizio, 215
Pixii, Hippolyte, 73
Planck, Max, 171, 220
Plato, 11
Podkletnov, Evgeny, 144, 150, 152, 251
Poincaré, Jules Henri, 27
Politzer, David, 189
Polnarev, Aleksander, 124
Pope, Nick, xxii
Pound, Robert, 222
Powell, Baden, xvii
Poynting, John Henry, 76
Priestley, Joseph, 60, 62
Ptolemy of Alexandria, 3
Puthoff, Hal, xxii, 247

NAMES INDEX

R

Rabi, Isador, 194
Ramsey, Arthur S., xvi
Rebka, Glen, 222
Reines, Frederick, 184, 197
Richardson, Owen, 140
Robertson, Glen, 148
Robertson, Norna, 98
Röntgen, Wilhelm, 177
Rosen, Nathan, 277
Rowland, Henry, 99
Roy, Anushree, 213
Rubbia, Carlo, 190
Rueda, Alfonso, 247
Ruska, Ernst, 135
Rutherford, Ernest, 176, 179

S

Sagan, Carl, 52, 276
Sagnac, Georges, 109, 112
Salam, Abdus, 190
Sample, Ian, xxii, 144
Saunders, David R., 285
Savart, Felix, 64
Schiff, Leonard, 120
Schreiber, Ulrich, 124
Schrödinger, Erwin, 131-132
Schuster, Arthur, 141
Schwarzschild, Karl, 226
Schweigger, Johann, 63
Shawyer, Roger, 265-268
Shelley, Tom, 265
Slusher, Dick, 271
Smith, Paul, 263
Solvay, Ernest, 128
Speake, Clive, xxiii, 216
Stedman, Geoffrey, 115
Stefan, Josef, 168
Stephenson, Gary, xxiii

Stephenson, George, 36
Stern, Otto, 183, 194
Stevin, Simon, 15
Stokes, George, 76, 240
Strassmann, Fritz, 180
Stringfellow, John, 54
Strutt, John William (Lord Rayleigh), 170
Sturgeon, William, 63
Szalay, Sándor, 184
Szames, Alexandre, xxii

T

Tajmar, Martin, 150
Taylor, Joseph, 96
Taylor, Richard, 187
Thirring, Hans, 119
Thompson, Benjamin (Count von Rumford), 237
Thompson, Ron, xviii
Thomson, George P., 130
Thomson, Joseph John, 99, 103, 176
Thomson, William (Lord Kelvin), 46, 163
Thorne, Kip, 124, 277
Torr, Doug, 145
Tucker, Robin, xxi, 125, 158

U

Uhlenbeck, George, 182
Unruh, William George, 216, 246

V

Van der Meer, Simon, 190
Van Musschenbroek, Pieter, 59
Vigier, Jean-Pierre, xxii
Vinogradov, Sergei, 263
Volta, Alessandro, 61
Von Fraunhofer, Joseph, 183
Von Guericke, Otto, 57, 162

Von Helmholtz, Hermann, 53
Von Kleist, Ewald, 59
Von Lauchen, Georg, 14

W

Watt, James, 60
Weber, Joseph, 96
Wegener, Alfred, 33
Weinberg, Steven, 190
Wells, Herbert George, 270
Wheatstone, Charles, 70
Wheeler, John, 31
Wien, Wilhelm, 168
Wilczek, Frank, 189
Wilson, Chris, 217
Wilson, Robert, 175
Wright, Orville and Wilbur, 55
Woods, Clive, 149-150

Y

Yorke, Bernard, 271
Young, Brian, xix
Young, Thomas, 111, 128, 219

Z

Zeeman, Pieter, 183
Ziegler, Spencer, xxi

SUBJECT INDEX

A

Acceleration Accelerometer, 224; Centrifugal force, 6, 20; Centripetal acceleration, 20; Coriolis force, 51; Equivalence with gravity, xiii, 4; Inclined plane, 15 ; Roller coaster, 6

Almagest, 3

Analogies, xii, 29

Ancient civilisations Arabic, 3, 56; Babylonian, 11, 19; Chinese, 12, 56; Egyptian, 12; Greek, 1, 11-14, 52, 56, 162, 176, 219

Atom Bohr's discrete electron orbit model, 180, 181; Electron, proton and neutron, 176, 178, 179; Fission, 180; Kelvin's knotted ether vortex ring model, 163; Magic numbers for electron shells, 180; Magic numbers for proton and neutron shells, 182; Quantum numbers, 180; Quark, 185, 187-189, 209; Rutherford's solar system model, 178; Strong nuclear force, 179, 184-185, 188; Thomson's plum pudding model, 178; Weak nuclear force, 184-185, 190, 197

Atomic devices Atomic clock, 194; Atomic force microscope (AFM), 213, 240; Atom interferometer, 136; Atom laser, 137, 174; Fission, 180; Laser, 136; Laser beam intensity, 159; Laser interferometer, 97; Nuclear power, 180; Race for atomic bomb, 180

Avogadro Atomic weight, 140; Einstein's estimate, 140; Hypothesis, 140; Number N_A, 140

B

BAe/BAE Systems sites Brough, 292; Filton, 291; Great Baddow, 291; Kingston, 290; Plymouth, 292; Warton, xv, xxiii, 290

BAe-British Aerospace Merger with Marconi, xv, xxi; Round table report on UK gravity research, x

Black hole Bekenstein temperature, 228; Davies-Unruh effect, 216, 246; Einstein-Rosen bridge, 277; Evaporation, 230; Hawking radiation, 229-230; Laplace's Invisible star, 31; Michell's Dark star, 31; Schwarzschild's Event horizon, 226; Schwarzschild radius r_S, 226; Singularity, 226; Wheeler's Black hole, 31; White hole, 278; Worm hole, 276-277

Boeing, xxiii; Project GRASP, 149

Bose-Einstein condensate (BEC), 137, 174

Boson – force carrier, 185-186; Gluon, 187; Graviton, 186; Higgs, 191-192; Photon, 82, 111, 165, 173, 186, 220; W^-, W^+, Z, 190

Breakthrough research programmes BAE Systems' Project Greenglow, xxi, xxii, 281; DARPA 100-year Starship Program, 287; NASA Breakthrough Propulsion Physics Program (BPP), xxi, 282; US Tau Zero Foundation, 287

C

Casmir force, 209-213; Atomic force microscope experiment, 213; Bell Lab MEMS experiment, 215; Birmingham University experiment, 216; Casimir-Polder force, 208; Casimir cavity, 209, 214; Cavities with positive and negative energy densities, 214; Cavitation, 36, 217; Chalmer's University Experiment, 217; DARPA

Research Program, 218; Davies-Unruh Effect, 216; Dynamic Casimir Cavity, 217; Extracting energy from the vacuum, 205, 214, 216; EU Research Programme, 217; Lamb Frequency Shift, 208-209; Lamoreaux's experiment, 211-212; Jet Propulsion Laboratory study, 215; Nanotechnology, 211; NASA MEMS study, 214; Sonoluminescence, 217

Clocks Atomic clock, 194-196; Galileo's simple pendulum, 16-17; Gravity driven clockwork, 18; Harrison's Chronometers, 19; Hooke's law, 199; Huygens's pendulum regulation, 18; Piezoelectricity, 176; Quartz crystal oscillator, 193; Spring driven clockwork, 199

Conservation laws Angular momentum, 21; Colour charge, 188; Electric charge, 103; Energy, 41; Mass, 103; Momentum, 41

Continental drift Pangea, 33; Plate tectonics, 33

D

Dimensional analysis MLT, 85-86

Dipoles Coil/Solenoid, 63; Electret, 35; Fluid Doublet, 35, 37; Gravitational dipole, 40-41, 195, 249, 254-258, 267-279; Jet engine, 37, 41; Magnet, 31; Vortex ring, 37

E

Earth's properties Chandler's wobble, 115; Gravitational acceleration **g**, 24; Gravitomagnetic field **h**, 84, 121; Mass M_E, 24; Modified gravitational acceleration, 118-119; Natural magnetic field **H**, 125; Radius R_E, 24; Rotation rate Ω_E, 19, 115

Einstein's special relativity, xvi, 99; Constant speed of light c, 76, 86; Inertial frames of reference, 99; Lorentz-Fitzgerald Transforms, 164, 204; Mass-energy formula $E = mc^2$, 82, 102, 177, 179, 191, 192, 203, 221; Mass-velocity formula, 102; Rowland's rotating disc experiment, 99; Time dilation, 165

Einstein's theory of gravity Alcubierre's space-time warp-drive, 274-276; Expansion of Universe, 33, 116, 175; General relativity, xvi, 7; Link with gravitomagnetism, 10; Mach's principle, 118, 257-259; NASA Gravity Probe-B experiment, 121-124; Predicted bending of starlight, 8; Space-time curvature causes gravity, 7, 8; Time travel, 7, 160; Torsion, 9

Einstein and quantum theory Energy form of the uncertainty principle, 203; Gravitational red shift, 221-223; Photoelectric effect, 173-174, 220;

Electrostatic charge Amber, 56; Cells and batteries, 62; Charge Q, 29; Charge density ρ, 48, 73, 75; Coulomb's Inverse square law, 30-31; Dielectrics, 75; DuFay's and Franklin's views, 58-59; Electric field **E**, 30; Electron charge e, 139; Electron mass m_e, 139; Electron spin $\pm½\hbar$, 182, 186; Frictional electricity, 57; Gold-leaf electroscope, 59; Leyden jar, 59; Permittivity ε, 29; Von Guericke's sulphur sphere, 57

Electrodynamics Biot-Savart's law, 64; Conductivity σ, 65; Current i, 63; Drift velocity, 64; Electrolysis, 62; Franklin's Lightning conductor, 59; Galvani's animal electricity, 61; Galvanometer, 63; Gray's experiments, 57-58; Lorentz force, 69; Ohm's law, 65-67; Oersted's discovery, 62-63; Resistance R, 65; Solenoid/coil, 63; Spin polarisation in gases, 193; Voltage V, 65; Voltaic pile, 61

Electro-gravitics Biefeld-Brown effect, 260; Dimitriou's asymmetric capacitor, 261; LaViolette's B2 bomber claim, 261; Naudin's

SUBJECT INDEX

Lifter, 262; Project Elmo - BAE Systems/AEA, 261-262; Trans-Dimensional Technologies (TDT) Lifter, 262

Electromagnetics Analogue with fluid dynamics, 67, 73; Discovery of electrical induction, 70-72; Displacement current, 75; Duality between **E** and **H**, 67; Electromagnet, 63; Faraday's dynamo, 72; Fleming's left-hand rule of induction, 68-69; Maxwell's equations, 74-75; Power Generation Industry, 73

Electromagnetic waves Communications industry, 78; Energy density, 80; Hertz experiment, 76-78; Maxwell's prediction, 76; Poynting Power transfer, 76, 234; Radiation pressure, 81

Energy Gradient gives force, 26; Gravitational potential energy, 25; Gravitomagnetism and kinetic energy, 101; Positive and negative energy density, 80

Equivalence Gravity and acceleration, 4-6; Induced gravitomagnetism **b** and angular velocity Ω, xiii, 85-86

Ether Greek quandary- Vacuum or plenum?, 162; Maxwell's luminiferous ether, 162; Michelson-Morley experiment, 163; Special relativity ignores ether, 164-165

European Aerospace Defence Systems (EADS) Astrium, 265

European Space Agency (ESA) Centre, xxi; HYPER mission, 138

F

Faraday Dynamo, 72; Electrolysis, 62, 69; Induction experiments, 70-72; Last experiment on magneto-optics, 183; Gravity experiments, vii-viii, 79, 239-240; Gyro, xvii; Magnetic rotation of polarised light, 159-160; Simple motor, 67-69 ; Verdet's constant, 160

Fermion – particle of matter Baryon, 185; Electron, 176; Lepton, 185; Neutrino, 177-178, 184; Neutron, 179, 187; Proton, 179; Quark, 187

Fluid dynamics Archimedean screw, 36; Archimedes and buoyancy, 13; Bernoulli Effect, 32, 54; Cavitation, 36; Continuity, 43-44, 47; Coriolis force, 51; Density ρ, 33; Dipole/Fluid doublet, 35; Euler's equation of motion, 42; Gas and steam turbine, 36; Hele-Shaw water table, 35; Incompressibility, 42; Jet engine, 37; Line Vortex, 53; LOW pressure region, 51; Magnus Effect, 54; Screw and Propeller, 36; Shock wave, 42; Source and sink, 43, 46; Streamlines, 35, 42-43; Sycamore seed, 54; Viscosity, 42; Vortex ring, 37; Vorticity vector ζ, 53; Vorton source absence, 73; Windmill, 48-49

Force-field constants Big G, 23; Density ρ, 33; Electric permittivity ε, 29; Gravitational permeability γ, 23; Gravitomagnetic permeability η, 85-86, 234; Magnetic permeability μ, 32; Thermal conductivity κ, 34

Force-field sources Electric charge Q, 29; Fluid source q, 43, 46; Gravitational mass M, 22; Gravitomagnetic monopole p, 85; Heat source q, 34; Magnetic monopole m, 32

Forward *Future Magic*, xix; Gravitational dipole propulsion, 40, 249; Gravitational wave apparatus, 97; Mass toroid gravity field engine, xx, 150, 250-253; Quantum vacuum, 206

Friction Amonton's laws, 237; Caloric fluid, 65, 238; Cold welding, 243; Frictional generation of electricity, 57; Frictional shear layer, 243; Heat due to mechanical motion, 238; Joule's idea of energy conversion, 238; Lattice vibrations of bodies in sliding contact, 241-242; Mechanical equivalent of heat J, 238-239; Speculative link with gravitomagnetism, 243-245; Static and Kinetic friction, 237

307

G

Galileo Correspondence with Kepler, 16; Dispute with church, 15, 17; Inclined plane experiments, 15; Inertia, 3; Leaning tower of Pisa, 15; Simple pendulum, 16; Telescope and moons of Jupiter, 16; View on Solar system, 14-16

Gauss's divergence theorem Divergence and convergence, 45; Flux of velocity, 44; Green's derivation, 45-46

Gravity Analogue of Maxwell's equations, 9, 80, 84; Analogue of Ohm's law, 231-232; Analogue of Poynting vector, 234; Big G, 23, 31; Centre of gravity, 13; Dark matter and dark energy, 39; Einstein's non-linear theory, 6-9; Equivalence with acceleration, 4-6; Free fall in gravity field, 5, 235; Gradar, xvi, 98; Gradiometry, 138; Gravitational dipole acceleration, 40, 89; Gravitational field **g**, 22; Gravitational permeability γ, 23; Gravitational potential, 25; Gravity gravitates, 83; Higgs boson as source of mass, 191; Inertial and gravitational mass, 104; Jerk – the rate of change of acceleration, 87; Modified by Earth's rotation, 118-119; Negative energy density of field, 80; Newton's inverse square law, 24; Newton's gravitational theory, 20-28; Podkletnov and gravity shielding, 144, 147-148; Positive (ordinary) mass as source, 22; Possible existence of negative mass, 38-40; Space curvature simulation of gravity, 6-7; Superconductivity and gravity, 145-147; Thoughts on the unification of the forces, vii, xii, 62-63; Tides due to Earth, Moon and Sun, 28; Weight, 26

Gravitomagnetism Analogue between magnetism **H** and gravitomagnetism **h**, 84; Analogue between electric current i and mass current I, 84; Analogue of Biot-Savart's law, 158, 233; Detecting gravitomagnetism, 107, 124, 138; Equivalence with angular velocity, xiii, 85-86; Gravitomagnetic permeability η, 85-86, 234; Heaviside's speculation, xviii, 84; Induced gravitomagnetic field **b**, 86; Laithwaite's unfinished experiment, xix; Lense-Thirring frame-dragging, 119-120, 123-124; Mach's principle, 118, 255, 257-258; Mallett's proposed ring laser gyro experiment, 157-160; Morgan's experiment, 126-127, 156; NASA Gravity Probe-B experiment, 121-124, 196; Newton's bucket, 117; Possibility of time travel, 160; Rotation sensing, 107; Self-energy and kinetic energy, 101; Speculative link with friction, 243-244; Twinned fields, 139; US-Soviet proposed Foucault experiment, 124; Vector gravitomagnetic potential **A**, 146

Gravitational propulsion Alcubierre's space-time warp-drive, 274-276; Allen's 'imagineering' study, 253-254; Alzofon's anti-gravity screen, 270-274; BAe 'What-if' anti-gravity study, 253; Biefeld-Brown's electro-gravitics, 260; Dundee open cavity microwave thrust, 263-264; Forward's gravitational dipole, 254; Forward's toroid, 250-252; H.G. Wells and the Cavorite panel, 270; Microwave induced gravitational dipole?, 267-269; Mike's anatomy of a flying saucer, 254-259; Shawyer's closed cavity EM-Drive, 264-267; Tube transport by worm hole, 276-280

Gravitational waves Analogue of electromagnetic wave, 93; Chiao's gravity radio, 98, 161; Dipole radiation, 95; Graviton, 95; Laser Gravitational wave interferometer, 97, 272; Podkletnov's gravity beam, 150-152; Quadrupole radiation, 95; Radiation by accelerating mass, 104-105, 247-248, 268-269; Weber's bar antenna, 96

Gyroscope Demonstration at the Royal Institution, xvi, 91-93; Fibre optic gyroscope

SUBJECT INDEX

(FOG), 116; Flywheel, 107; Forced precession, xvii, 107, 121; Foucault's experiments, 108-109; Gravity Probe-B gyroscopes, 121; iPhone, 117; Ring laser gyroscope (RLG), 113-115, 158; Rotation sensing, 107; Sagnac's experiment to detect rotation, 109, 112-113; Sagnac beat frequency, 114; Ultra Gyroscope-1 (UG-1), 115, 124; Vibrating structure gyro (VSG), 116

H

Heat and electricity experiment Faraday's experiments, 240; Leduc effect, xx

Heat radiation experiments Black body cavity, 167; Black body spectrum, 169; Boltzmann's constant k, 168; Colours of heated body, 167; Emissive power $E(\lambda,T)$, 168; Emissive power at absolute zero $E(\lambda,0)$, 204; Explain the photoelectric effect, 173-174; Planck's constant h, 172; Planck's energy quantum $E = hf$, 172; Planck's radiation formula, 171-172; Remnant of heat from the Big Bang, 175; Statistical mechanics, 170; Stefan's experimental fourth power law, 168; Temperature equilibrium, 167; Ultra-violet catastrophe, 171, 205; Wien's displacement law, 168; Wien's empirical radiation formula, 170

I

Inertial thrust machines Controversy over gyro theory, xvi; Laithwaite's Royal Institution Lecture, 91-93, 251; Testing Sandy Kidd's machine, xviii

Infinitesimal calculus Ancient Greek introduction, 13; Discrete increment Δ, 64, 221-222; Multi-variable increment ∂, 74; Newton and Leibnitz, 13; Single variable increment d, 74

Institute of Physics (IoP) Gravity Group, 281

K

Kondratieff Boom and bust cycles of business, 287; Dawn of age of electrical power, 73; Decline of age of steam power, 55, 162; Internal combustion engine and the oil age, 55; Jet engine and the age of air travel, 37, 41; Start of age of space travel, 288-289

L

Light Effective mass of light, 155, 221, 245; Optical lever, 213; Optical pumping, 193; Particle and wave duality, 219-220; Radiation pressure, 81-82; Slow-light, 160; Speed of light c, 76; Sphere of influence of point source, 16, 22 ; Starlight deflected by Sun, 8
Lockheed Martin, xv, 123
Lunar Society of Birmingham, 60

M

Magnetism Barnett effect, 141-142; Coil/Solenoid dipole, 63; Eddy currents, 72, 190; Einstein-de Haas effect, 141-142; Hypothetical monopoles, 31; Induced field **B**, 86; Inverse square law for monopoles, 32; Larmor frequency of precession, 140; Lorentz force, 69; Magnetic compass, 55; Magnetic field **H**, 31, 48, 64, 67; Magnetic trap, 138; Magnetite, 55; Natural magnetic dipole, 31; Permeability μ, 32; Vector magnetic potential **A**, 146

Massdynamics, 254

Maxwell Analogue for extended Newtonian gravity theory, 80; Equations for electromagnetism, 71-75; Experiment to detect the gyroscopic effect of a conducting coil, 155; Prediction of electromagnetic waves, 76

Micro-electro-mechanical systems (MEMS), 214, 216

Microwave thrust Dundee open cavity experiment, 263-264; *Eureka* magazine, 265; *NewScientist* magazine, 266; Shawyer's closed-cavity thrust, 264-266; Speculative gravitational dipole, 267-269

N

Neutrinos First detected, 184; Neutrino Beams, 198; Pauli's speculation, 177-178; Solar source, 196-197; Stellar bi-polar jets, 197; Supernovae source, 197; Telescopes and observatories, 197-198

Newton Calculus, 13; Disputes with Hooke, 200; Magnus effect, 54; President of the Royal Society, 200; *Principia*, 28

Newtonian dynamics Chaos and n-body problem, 27; Explaining Kepler's laws, 16; Force, 25, 26, 28, 104; Inertia, 3, 28, 247; Inverse square law of gravity, 24; Laws of motion, 28; Newton's bucket, 117; Tidal theory, 28

Newtonian optics Corpuscular theory of light, 219; Early Prism experiment, 182; Fraunhofer's dark lines, 183

Nobel Prize, 131

P

Particle physics CERN particle accelerator, 190, 193; Fermilab accelerator, 192; Particle spin state, 138, 185; Rutherford's research group, 178; Search for Higgs boson, 191; Speculation that gravity is linked to strong force, 188, 189, 286; Stanford Linear Accelerator Center (SLAC), 187; Strong Nuclear Force, 179, 184-185, 188; Weak nuclear force, 184-185, 190, 197; Unification of weak and electromagnetic forces, 190

Photon Compton effect, 129-130; Effective mass, 5, 155, 221; Phonon, 181, 202, 240; Photoelectric effect, 220; Photon spin energy $\hbar\omega$, 95, 221; Planck's energy quantum $E = hf$, 172; Positive and negative energy particles, 208, 229-230, 247-248; Red shift, 222

Piezoelectric effect, 176

Podkletnov Boeing GRASP study, 149; Gravity impulse generator, 150-152; Gravity shielding claim, 144-145; Hathaway Consulting Services, 152-153; NASA Delta-G experiment, 148-149; Sheffield University experiment, 149-150; Tajmar and de Matos experiment, 150

Powered flight *Aerial Steamer*, 55; *Flyer 1*, 55

Q

Quantum mechanics Continuous energy levels, 200; Discrete energy levels, 202; Einstein's uncertainty principle for energy, 203; Electron spin ½$\hbar$, 136, 182-184, 221; Heisenberg's uncertainty principle, 135; Heligoland, 131; Matrix mechanics, 133, 135; Mendeleyev's Periodic Table, 186; Pauli's exclusion principle, 186, 193; Quantum mechanical motion, 20, 202; Quantum numbers, 180, 186; Quantum tunnelling, 133; Schrödinger's equation, 132; Simple harmonic motion (SHM), 200; Solid body vibration, 201; Wave function ψ, 132-133; Wave mechanics, 133, 135

Quantum ether Acceleration and Davies-Unruh effect, 216, 246; Can energy be extracted from the vacuum?, 205, 214, 216; Casimir force, 209-213; Dirac's quantum foam, 204; Gravitational radiation during acceleration, 104-105, 216, 247-248, 268-269; Lamb frequency shift, 208-209; Planck

SUBJECT INDEX

length L and cut-off, 205; Quantum vacuum, 203; Sonoluminescence, 217; Spontaneous virtual particles, 203; Tearing the vacuum apart, 229-230; Vacuum energy predicted to be infinite, 205; Zero-point energy (ZPE), 202

Quark Asymptotic freedom, 189; Colour charge, 187-188; Make up of proton and neutron, 187; Types, 187

R

Radioactivity Discovery by Becquerel and Curies, 176; Emitted particles labelled alpha, beta, gamma, 177; Pauli's speculation of neutrino, 177-178; Rutherford's research group, 178-180; Scintillation, 178, 184, 198

Royal Institution Foundation, vii, 238

Royal Society European Renaissance, 10, 14; Foundation, 15; Motto, 16; Philosophical Transactions, 57; The first Curator of Experiments, 21, 199

S

Satellite Navigation systems, 223

Science Fiction films Disney film *The Black Hole*, 284; MGM film *Stargate*, 280, 284; WB film *Contact*, 277, 284; *Star Trek* USS Enterprise, 284

SI International Standard Amount of matter – Mole (M), 140; Capacitance – Farad (F), 29; Charge – Coulomb (Q), 29; Current – Amp (A), 63; Energy – Joule (J); Force – Newton (N), 25, 31; Inductance – Henry (H), 32; Length L – Metre (m), 23; Mass M – Kilogram (kg), 23; Power – Watt (W), 76; Temperature T – Kelvin (K), 34; Time – Second (s), 23;

Steam locomotion *Novelty*, 55; Rainhill Trials, 55; *Rocket*, 55

Space-time Curvature of space-time, 8; Space-time in 3-D and 4-D, 7; Time lines, 7; Time travel, 7, 160

Squeezed laser light, 271-272; Lithium niobate crystal, 271; Optically pumped resonant cavity gas cell, 272

States of matter Ancient Greek view, 162; Modern view, 162

Stealth Dawson Report on Radar Camouflage, xv; F117 Nighthawk, xv; Northrop B-2 Stealth Bomber, 261; Plasma sheath, xxi

Superconductivity Discovery by Onnes, 144; Cooper pairs of electrons, 193; London moment, 122; Meissner levitation, 144; Speculative link with gravity, 144-145; SQUID magnetometer, 122, 217; Type II materials, 147; Type II magnetic vortex array, 147; Yttrium barium copper oxide (YBCO), 145, 147-148, 151, 153

Synchronicity in science, 21, 73, 178, 190

T

Thermodynamics, Relative heat sources and sinks, 34; Second law of thermodynamics, 34; Thermal conductivity κ

Time measurement Atomic clock, 193-194; Burning candle, 12; Chronometer, 19; Clepsydra/water clock, 12, 15; Gravity driven mechanical clock, 18; Pendulum regulated clock, 18; Quartz crystal oscillator, 193-194; Sand timer, 12; Spring driven clocks and watches, 18, 199; Sundial, 12

Torsion Torsional stiffness, 30; Torsion balance, 30, 31, 153

U

UFO Alien technology, 286; Area 51-Groom Lake test facility, 286; Churchill's query about flying saucers, 285; Condon

Committee, 285; Mike's *The anatomy of a flying saucer* 254-259; Project Blue Book, 285; National Investigations Committee on Aerial Phenomena (NICAP), 285; UFOCAT, 285; UFO Studies Center (CUFOS), 285

UK Department of Trade and Industry (DTI), 265; Ministry of Defence (MoD), xxi, xxii; National Physical Laboratory (NPL), 194

US Congress Commission on future of US Aerospace Industry, 286; Defense Advanced Research Project Agency (DARPA), 160, 217, 287; Department of Defense (DoD), xxiii, 217; National Bureau of Standards (NBS), 194; President Obama cancels Space programs, 286

Waves Amplitude A, 109; Beats, 129; Frequency f, 109; Group velocity v_g, 129, 160, 265, 268; Nodes and anti-nodes, 113; Phase angle φ, 109-110; Phase velocity v_p, 129, 266, 268; Sine wave, 109; Standing waves, 113; Travelling waves, 110; Wave envelope, 124, 134; Wave groups, 129, 134; Wave interference, 110-111, 136; Wavelength λ, 110; Wave speed c, 110

V

Vector and scalar Curl operator $\nabla \times$, 52-53; Divergence operator $\nabla \cdot$, 45; Gradient operator ∇, 26; Scalar, 23; Vector, 24

Vortex phenomena Atlantic hurricane, 50; Bermuda Triangle, 50; Charybdis, 52; Corryvreckan, 52; Dust eddy, 50; Indian cyclone, 50; Jupiter's Red Spot, 52; Line vortex, 53; Maelstrom, 52; Pacific typhoon, 50; Shear layer, 242; Starting vortex at airport, 53; Tornado, 50; Vortex ring, 37; Waterspout, 50; Whirlpool Galaxy M51, 52

W

Wave-particle duality Compton's photon, 129-130; De Broglie's matter waves, 129-131; Einstein's explanation of photoelectric effect, 173-174, 220; Electron beam, 130-131; Electron beam microscope, 135; Newton's corpuscle of light, 112, 219; Planck's energy quantum E = hf, 172; X-ray beam, 131, 176; Young's fringes, 111-112